T0324913

Graduate Texts in Mathematics 19

Editorial Board
J.H. Ewing   F.W. Gehring   P.R. Halmos

Paul R. Halmos

A Hilbert Space Problem Book

Second Edition, Revised and Enlarged

Springer-Verlag
New York Berlin Heidelberg London Paris
Tokyo Hong Kong Barcelona Budapest

Editorial Board

P.R. Halmos
Department of
  Mathematics
Santa Clara University
Santa Clara, CA 95053
USA

F.W. Gehring
Department of
  Mathematics
University of Michigan
Ann Arbor, MI 48104
USA

J.H. Ewing
Department of
  Mathematics
Indiana University
Bloomington, IN 47405
USA

Mathematics Subject Classifications (1991): 46-01, 00A07, 46CXX

Library of Congress Cataloging in Publication Data
Halmos, Paul R. (Paul Richard), 1916–
    A Hilbert space problem book.
    (Graduate texts in mathematics; 19)
    Bibliography: p.
    Includes index.
    1. Hilbert spaces—Problems, exercises, etc.
I. Title. II. Series.
QA322.4.H34   1982   515.7′33   82-763
                                          AACR2

© 1974, 1982 by Springer-Verlag New York Inc.
All rights reserved. No part of this book may be translated or reproduced in any form
without written permission from Springer-Verlag, 175 Fifth Avenue, New York,
New York 10010, U.S.A.

Typeset by Composition House Ltd., Salisbury, England.
Printed and bound by R. R. Donnelley & Sons, Harrisonburg, VA.
Printed in the United States of America.

9 8 7 6 5 4 3 2

ISBN 0-387-90685-1 Springer-Verlag New York Heidelberg Berlin
ISBN 3-540-90685-1 Springer-Verlag Berlin Heidelberg New York

To J. U. M.

Preface

The only way to learn mathematics is to do mathematics. That tenet is the foundation of the do-it-yourself, Socratic, or Texas method, the method in which the teacher plays the role of an omniscient but largely uncommunicative referee between the learner and the facts. Although that method is usually and perhaps necessarily oral, this book tries to use the same method to give a written exposition of certain topics in Hilbert space theory.

The right way to read mathematics is first to read the definitions of the concepts and the statements of the theorems, and then, putting the book aside, to try to discover the appropriate proofs. If the theorems are not trivial, the attempt might fail, but it is likely to be instructive just the same. To the passive reader a routine computation and a miracle of ingenuity come with equal ease, and later, when he must depend on himself, he will find that they went as easily as they came. The active reader, who has found out what does not work, is in a much better position to understand the reason for the success of the author's method, and, later, to find answers that are not in books.

This book was written for the active reader. The first part consists of problems, frequently preceded by definitions and motivation, and sometimes followed by corollaries and historical remarks. Most of the problems are statements to be proved, but some are questions (is it?, what is?), and some are challenges (construct, determine). The second part, a very short one, consists of hints. A hint is a word, or a paragraph, usually intended to help the reader find a solution. The hint itself is not necessarily a condensed solution of the problem; it may just point to what I regard as the heart of the matter. Sometimes a problem contains a trap, and the hint may serve to chide the reader for rushing in too recklessly. The third part, the longest, consists of solutions: proofs, answers, or constructions, depending on the nature of the problem.

The problems are intended to be challenges to thought, not legal technicalities. A reader who offers solutions in the strict sense only (this is what was asked, and here is how it goes) will miss a lot of the point, and he will miss a lot of fun. Do not just answer the question, but try to think of related questions, of generalizations (what if the operator is not normal?), and of special cases (what happens in the finite-dimensional case?). What makes the assertion true? What would make it false?

Problems in life, in mathematics, and even in this book, do not necessarily arise in increasing order of depth and difficulty. It can perfectly well happen that a relatively unsophisticated fact about operators is the best tool for the solution of an elementary-sounding problem about the geometry of vectors. Do not be discouraged if the solution of an early problem borrows from the future and uses the results of a later discussion. The logical error of circular reasoning must be avoided, of course. An insistently linear view of the intricate architecture of mathematics is, however, almost as bad: it tends to conceal the beauty of the subject and to delay or even to make impossible an understanding of the full truth.

If you cannot solve a problem, and the hint did not help, the best thing to do at first is to go on to another problem. If the problem was a statement, do not hesitate to use it later; its use, or possible misuse, may throw valuable light on the solution. If, on the other hand, you solved a problem, look at the hint, and then the solution, anyway. You may find modifications, generalizations, and specializations that you did not think of. The solution may introduce some standard nomenclature, discuss some of the history of the subject, and mention some pertinent references.

The topics treated range from fairly standard textbook material to the boundary of what is known. I made an attempt to exclude dull problems with routine answers; every problem in the book puzzled me once. I did not try to achieve maximal generality in all the directions that the problems have contact with. I tried to communicate ideas and techniques and to let the reader generalize for himself.

To get maximum profit from the book the reader should know the elementary techniques and results of general topology, measure theory, and real and complex analysis. I use, with no apology and no reference, such concepts as subbase for a topology, precompact metric spaces, Lindelöf spaces, connectedness, and the convergence of nets, and such results as the metrizability of compact spaces with a countable base, and the compactness of the Cartesian product of compact spaces. (Reference: [87].) From measure theory, I use concepts such as σ-fields and L^p spaces, and results such as that L^p convergent sequences have almost everywhere convergent subsequences, and the Lebesgue dominated convergence theorem. (Reference: [61].) From real analysis I need, at least, the facts about the derivatives of absolutely continuous functions, and the Weierstrass polynomial approximation theorem. (Reference: [120].) From complex analysis I need such things as Taylor and Laurent series, subuniform convergence, and the maximum modulus principle. (Reference: [26].)

This is not an introduction to Hilbert space theory. Some knowledge of that subject is a prerequisite; at the very least, a study of the elements of Hilbert space theory should proceed concurrently with the reading of this book. Ideally the reader should know something like the first two chapters of [50].

I tried to indicate where I learned the problems and the solutions and where further information about them is available, but in many cases I could find no reference. When I ascribe a result to someone without an accompanying bracketed reference number, I am referring to an oral communication or an unpublished preprint. When I make no ascription, I am not claiming originality; more than likely the result is a folk theorem.

The notation and terminology are mostly standard and used with no explanation. As far as Hilbert space is concerned, I follow [50], except in a few small details. Thus, for instance, I now use f and g for vectors, instead of x and y (the latter are too useful for points in measure spaces and such), and, in conformity with current fashion, I use "kernel" instead of "null-space". (The triple use of the word, to denote (1) null-space, (2) the continuous analogue of a matrix, and (3) the reproducing function associated with a functional Hilbert space, is regrettable but unavoidable; it does not seem to lead to any confusion.) Incidentally kernel and range are abbreviated as ker and ran, their orthogonal complements are abbreviated as $\ker^\perp$ and $\operatorname{ran}^\perp$, dimension is abbreviated as dim, and determinant and trace are abbreviated as det and tr. Real and imaginary parts are denoted, as usual, by Re and Im. The "signum" of a complex number z, i.e., $z/|z|$ or 0 according as $z \neq 0$ or $z = 0$, is denoted by sgn z.

The zero subspace of a Hilbert space is denoted by 0, instead of the correct, pedantic $\{0\}$. (The simpler notation is obviously more convenient, and it is not a whit more illogical than the simultaneous use of the symbol "0" for a number, a function, a vector, and an operator. I cannot imagine any circumstances where it could lead to serious error. To avoid even a momentary misunderstanding, however, I write $\{0\}$ for the set of complex numbers consisting of 0 alone.) The *co-dimension* of a subspace is the dimension of its orthogonal complement (or, equivalently, the dimension of the quotient space it defines). The symbols $\bigvee$ (as a prefix) and $\vee$ (as an infix) are used to denote spans, so that if M is an arbitrary set of vectors, then $\bigvee M$ is the smallest closed linear manifold that includes M; if M and N are sets of vectors, then $M \vee N$ is the smallest closed linear manifold that includes both M and N; and if $\{M_j\}$ is a family of sets of vectors, then $\bigvee_j M_j$ is the smallest closed linear manifold that includes each M_j. Subspace, by the way, means closed linear manifold, and operator means bounded linear transformation.

The arrow in a symbol such as $f_n \to f$ indicates that a sequence $\{f_n\}$ tends to the limit f; the barred arrow in $x \mapsto x^2$ denotes the function φ defined by

$\varphi(x) = x^2$. (Note that barred arrows "bind" their variables, just as integrals in calculus and quantifiers in logic bind theirs. In principle equations such as $(x \mapsto x^2)(y) = y^2$ make sense.)

Since the inner product of two vectors f and g is always denoted by (f, g), another symbol is needed for their ordered pair; I use $\langle f, g \rangle$. This leads to the systematic use of the angular bracket to enclose the coordinates of a vector, as in $\langle f_0, f_1, f_2, \ldots \rangle$. In accordance with inconsistent but widely accepted practice, I use braces to denote both sets and sequences; thus $\{x\}$ is the set whose only element is x, and $\{x_n\}$ is the sequence whose n-th term is x_n, $n = 1, 2, 3, \cdots$. This could lead to confusion, but in context it does not seem to do so. For the complex conjugate of a complex number z, I use z^*. This tends to make mathematicians nervous, but it is widely used by physicists, it is in harmony with the standard notation for the adjoints of operators, and it has typographical advantages. (The image of a set M of complex numbers under the mapping $z \mapsto z^*$ is M^*; the symbol $\overline{M}$ suggests topological closure.)

Operator theory has made much progress since the first edition of this book appeared in 1967. Some of that progress is visible in the difference between the two editions. The journal literature needs time, however, to ripen, to become understood and simplified enough for expository presentation in a book of this sort, and much of it is not yet ready for that. Even in the part that is ready, I had to choose; not everything could be fitted in. I omitted beautiful and useful facts about essential spectra, the Calkin algebra, and Toeplitz and Hankel operators, and I am sorry about that. Maybe next time.

The first edition had 199 problems; this one has $199 - 9 + 60$. I hope that the number of incorrect or awkward statements and proofs is smaller in this edition. In any event, something like ten of the problems (or their solutions) were substantially revised. (Whether the actual number is 8 or 9 or 11 or 12 depends on how a "substantial" revision is defined.) The new problems have to do with several subjects; the three most frequent ones are total sets of vectors, cyclic operators, and the weak and strong operator topologies.

Since I have been teaching Hilbert space by the problem method for many years, I owe thanks for their help to more friends among students and colleagues than I could possibly name here. I am truly grateful to them all just the same. Without them this book could not exist; it is not the sort of book that could have been written in isolation from the mathematical community. My special thanks are due to Ronald Douglas, Eric Nordgren, and Carl Pearcy for the first edition, and Donald Hadwin and David Schwab for the second. Each of them read the whole manuscript (well, almost the whole manuscript) and stopped me from making many foolish mistakes.

Santa Clara University   P.R.H.

Contents

17. UNILATERAL SHIFT

18. CYCLIC VECTORS

19. PROPERTIES OF COMPACTNESS

PROBLEMS

CHAPTER 1

Vectors

1. Limits of quadratic forms. The objects of chief interest in the study of a
Hilbert space are not the vectors in the space, but the operators on it. Most
people who say they study the theory of Hilbert spaces in fact study operator
theory. The reason is that the algebra and geometry of vectors, linear functionals, quadratic forms, subspaces, and the like are easier than operator
theory and are pretty well worked out. Some of these easy and known things
are useful and some are amusing; perhaps some are both.

Recall to begin with that a bilinear functional on a complex vector space **H**
is sometimes defined as a complex-valued function on the Cartesian product
of **H** with itself that is linear in its first argument and conjugate linear in the
second; cf. [50, p. 12]. Some mathematicians, in this context and in other
more general ones, use "semilinear" instead of "conjugate linear", and,
incidentally, "form" instead of "functional". Since "sesqui" means "one
and a half" in Latin, it has been suggested that a bilinear functional is
more accurately described as a sesquilinear form.

A quadratic form is defined in [50] as a function $\varphi^{\sim}$ associated with a
sesquilinear form φ via the equation $\varphi^{\sim}(f) = \varphi(f, f)$. (The symbol $\hat{\varphi}$ is used
there instead of $\varphi^{\sim}$.) More honestly put, a quadratic form is a function ψ for
which *there exists* a sesquilinear form φ such that $\psi(f) = \varphi(f, f)$. Such an
existential definition makes it awkward to answer even the simplest algebraic
questions, such as whether the sum of two quadratic forms is a quadratic
form (yes), and whether the product of two quadratic forms is a quadratic
form (no).

Problem 1. *Is the limit of a sequence of quadratic forms a quadratic
form?*

2   **2. Schwarz inequality.** One proof of the Schwarz inequality consists of the verification of one line, namely:

$$\|f\|^2 \cdot \|g\|^2 - |(f, g)|^2 = \frac{1}{\|g\|^2} \left\| \|g\|^2 f - (f, g)g \right\|^2.$$

It might perhaps be more elegant to multiply through by $\|g\|^2$, so that the result should hold for $g = 0$ also, but the identity seems to be more perspicuous in the form given. This one line proves also that if the inequality degenerates to an equality, then f and g are linearly dependent. The converse is trivial: if f and g are linearly dependent, then one of them is a scalar multiple of the other, say $g = \alpha f$, and then both $|(f, g)|^2$ and $(f, f) \cdot (g, g)$ are equal to $|\alpha|^2 (f, f)^2$.

This proof of the Schwarz inequality does not work for sesquilinear forms unless they are strictly positive. What are the facts? Is *strict* positiveness necessary?

> **Problem 2.** *If φ is a positive, symmetric, sesquilinear form, is it necessarily true that*
>
> $$|\varphi(f, g)|^2 \leqq \varphi(f, f) \cdot \varphi(g, g)$$
>
> *for all f and g?*

3   **3. Representation of linear functionals.** The Riesz representation theorem says that to each bounded linear functional ξ on a Hilbert space **H** there corresponds a vector g in **H** such that $\xi(f) = (f, g)$ for all f. The statement is "invariant" or "coordinate-free", and therefore, according to current mathematical ethics, it is mandatory that the proof be such. The trouble is that most coordinate-free proofs (such as the one in [50, p. 32]) are so elegant that they conceal what is really going on.

> **Problem 3.** *Find a coordinatized proof of the Riesz representation theorem.*

4   **4. Strict convexity.** In a real vector space (and hence, in particular, in a complex vector space) the *segment* joining two distinct vectors f and g is, by definition, the set of all vectors of the form $tf + (1 - t)g$, where $0 \leqq t \leqq 1$. A subset of a real vector space is *convex* if, for each pair of vectors that it contains, it contains all the vectors of the segment joining them. Convexity plays an increasingly important role in modern vector space theory. Hilbert space is so rich in other, more powerful, structure, that the role of convexity is sometimes not so clearly visible in it as in other vector spaces. An easy example of a convex set in a Hilbert space is the *unit ball*, which is, by definition, the set of all vectors f with $\|f\| \leqq 1$. Another example is the *open unit ball*, the set of all vectors f with $\|f\| < 1$. (The adjective "closed" can be used to distinguish the unit ball from its open version, but is in fact used only when unusual emphasis is necessary.) These examples are of geometric interest even in the extreme case of a (complex) Hilbert space of dimension 1; they reduce then to the closed and the open unit disc, respectively, in the complex plane.

If $h = tf + (1 - t)g$ is a point of the segment joining two distinct vectors f and g, and if $0 < t < 1$ (the emphasis is that $t \neq 0$ and $t \neq 1$), then h is called an *interior* point of that segment. If a point of a convex set does not belong to the interior of any segment in the set, then it is called an *extreme point* of the set. The extreme points of the closed unit disc in the complex plane are just the points on its perimeter (the unit circle). The open unit disc in the complex plane has no extreme points. The set of all those complex numbers z for which $|\operatorname{Re} z| + |\operatorname{Im} z| \leq 1$ is convex (it consists of the interior and boundary of the square whose vertices are 1, i, -1, and $-i$); this convex set has just four extreme points (namely 1, i, -1, and $-i$).

A closed convex set in a Hilbert space is called *strictly convex* if all its boundary points are extreme points. The expression "boundary point" is used here in its ordinary topological sense. Unlike convexity, the concept of strict convexity is not purely algebraic. It makes sense in many spaces other than Hilbert spaces, but in order for it to make sense the space must have a topology, preferably one that is properly related to the linear structure. The closed unit disc in the complex plane is strictly convex.

Problem 4. *The unit ball of every Hilbert space is strictly convex.*

The problem is stated here to call attention to a circle of ideas and to prepare the ground for some later work. No great intrinsic interest is claimed for it; it is very easy.

5. Continuous curves. An infinite-dimensional Hilbert space is even roomier than it looks; a striking way to demonstrate its spaciousness is to study continuous curves in it. A *continuous curve* in a Hilbert space **H** is a continuous function from the closed unit interval into **H**; the curve is *simple* if the function is one-to-one. The *chord* of the curve f determined by the parameter interval $[a, b]$ is the vector $f(b) - f(a)$. Two chords, determined by the intervals $[a, b]$ and $[c, d]$, are *non-overlapping* if the intervals $[a, b]$ and $[c, d]$ have at most an end-point in common. If two non-overlapping chords are orthogonal, then the curve makes a right-angle turn during the passage between their farthest end-points. If a curve could do so for every pair of non-overlapping chords, then it would seem to be making a sudden right-angle turn at each point, and hence, in particular, it could not have a tangent at any point.

Problem 5. *Construct, for every infinite-dimensional Hilbert space, a simple continuous curve with the property that every two non-overlapping chords of it are orthogonal.*

5

6   **6. Uniqueness of crinkled arcs.** It is an interesting empirical fact that the example of a "crinkled arc" (that is, a simple continuous curve with every two non-overlapping chords orthogonal—cf. Solution 5) is psychologically unique; everyone who tries it seems to come up with the same answer. Why is that? Surely there must be a reason, and, it turns out, there is a good one. The reason is the existence of a pleasant and strong uniqueness theorem, discovered by G. G. Johnson [80]; for different concrete representations, see [81] and [146].

There are three trivial senses in which crinkled arcs are not unique. (1) Translation: fix a vector f_0 and replace the arc f by $f + f_0$. Remedy: normalize so that $f(0) = 0$. (2) Scale: fix a positive number α and replace the arc f by αf. Remedy: normalize so that $\|f(1)\| = 1$. (3) Span: fix a Hilbert space H_0 and replace H (the range space of f) by $H \oplus H_0$. Remedy: normalize so that the span of the range of f is H. In what follows a crinkled arc will be called normalized in case all three of these normalizations have been applied to it.

There are two other useful ways in which one crinkled arc can be changed into another. One is reparametrization: fix an increasing homeomorphism φ of $[0, 1]$ onto itself and replace f by $f \circ \varphi$. The other is unitary equivalence: fix a unitary operator U on H and replace f by Uf. Miracle: that's all.

Problem 6. *Any two normalized crinkled arcs are unitarily equivalent to reparametrizations of one another.*

7   **7. Linear dimension.** The concept of dimension can mean two different things for a Hilbert space H. Since H is a vector space, it has a *linear* dimension; since H has, in addition, an inner product structure, it has an *orthogonal* dimension. A unified way to approach the two concepts is first to prove that all bases of H have the same cardinal number, and then to define the dimension of H as the common cardinal number of all bases; the difference between the two concepts is in the definition of basis. A *Hamel basis* for H (also called a *linear basis*) is a maximal linearly independent subset of H. (Recall that an infinite set is called linearly independent if each finite subset of it is linearly independent. It is true, but for present purposes irrelevant, that every vector is a finite linear combination of the vectors in any Hamel basis.) An *orthonormal* basis for H is a maximal orthonormal subset of H. (The analogues of the finite expansions appropriate to the linear theory are the Fourier expansions always used in Hilbert space.)

Problem 7. *Does there exist a Hilbert space whose linear dimension is* $\aleph_0$?

8   **8. Total sets.** A subset of a Hilbert space is *total* if its span is the entire space. (Generalizations to Banach spaces, and, more generally, to topological vector

6

spaces are immediate.) Can a set be so imperturbably total that the removal of any single element always leaves it total? The answer is obviously yes: any dense set is an example. This is not surprising but some of the behavior of total sets is.

Problem 8. *There exists a total set in a Hilbert space that continues to be total when any one element is omitted but ceases to be total when any two elements are omitted.*

9. Infinitely total sets. The statement of Problem 8 has a natural broad generalization: for each non-negative integer n, there exists a total set in Hilbert space that continues to be total when any n of its elements are omitted but ceases to be total when any $n + 1$ elements are omitted. The result is obvious for $n = 0$: any orthonormal basis is an example. For $n = 1$, the statement is the one in Problem 8. The generalization (unpublished) was discovered and proved by R. F. Wiser in 1974.

Can a set be such that the removal of every finite subset always leaves it total? (Note: the question is about *sets*, not sequences. It is trivial to construct an infinite sequence such that its terms form a total set and such that this remains true no matter how many terms are omitted from the beginning. Indeed: let $\{f_0, f_1, f_2, \ldots\}$ be a total set, and form the sequence $\langle f_0, f_0, f_1, f_0, f_1, f_2, f_0, f_1, f_2, f_3, \cdots \rangle$.) A sharper way to formulate what is wanted is to ask whether there exists a linearly independent total set that remains total after the omission of each finite subset.

The answer is yes; one way to see it is to construct a linearly independent dense set. To do that, consider a countable base $\{\mathbf{E}_1, \mathbf{E}_2, \cdots\}$ for the norm topology of a separable infinite-dimensional Hilbert space (e.g., the open balls with centers at a countable dense set and rational radii). To get an inductive construction started, choose a non-zero vector f_1 in $\mathbf{E}_1$. For the induction step, given f_j in $\mathbf{E}_j$, $j = 1, \cdots, n$, so that $\{f_1, \cdots, f_n\}$ is linearly independent, note that $\mathbf{E}_{n+1}$ is not included in $\bigvee\{f_1, \cdots, f_n\}$ (because, for instance, the span is nowhere dense), and choose f_{n+1} so that it is in $\mathbf{E}_{n+1}$ but not in $\bigvee\{f_1, \cdots, f_n\}$.

Another example of an "infinitely total" set, in some respects simpler, but needing more analytic machinery, is the set of all powers f_n in $L^2(0, 1)$ (i.e., $f_n(x) = x^n$, $n = 0, 1, 2, \cdots$). See Solution 11.

Problem 9. *If a set remains total after the omission of each finite subset, then it has at least one infinite subset whose omission leaves it total also.*

10. Infinite Vandermondes. The Hilbert space l^2 consists, by definition, of all infinite sequences $\langle \xi_0, \xi_1, \xi_2, \cdots \rangle$ of complex numbers such that

$$\sum_{n=0}^{\infty} |\xi_n|^2 < \infty.$$

7

The vector operations are coordinatewise and the inner product is defined by

$$(\langle \xi_0, \xi_1, \xi_2, \cdots \rangle, \langle \eta_0, \eta_1, \eta_2, \cdots \rangle) = \sum_{n=0}^{\infty} \xi_n \eta_n{}^*.$$

Problem 10. *If* $0 < |\alpha| < 1$, *and if*

$$f_k = \langle 1, \alpha^k, \alpha^{2k}, \alpha^{3k}, \cdots \rangle, \qquad k = 1, 2, 3, \cdots,$$

determine the span of the set of all f_k's *in* l^2. *Generalize (to other collections of vectors), and specialize (to finite-dimensional spaces).*

11   **11. T-total sets.**

Problem 11. *Does there exist an infinite total set such that every infinite subset of it is total?*

12   **12. Approximate bases.**

Problem 12. *If* $\{e_1, e_2, e_3, \cdots\}$ *is an orthonormal basis for a Hilbert space* **H**, *and if* $\{f_1, f_2, f_3, \cdots\}$ *is an orthonormal set in* **H** *such that*

$$\sum_{j=1}^{\infty} \|e_j - f_j\|^2 < \infty,$$

then the vectors f_j *span* **H** *(and hence form an orthonormal basis for* **H**).

This is a hard one. There are many problems of this type; the first one is apparently due to Paley and Wiener. For a related exposition, and detailed references, see [114, No. 86]. The version above is discussed in [14].

8

CHAPTER 2

Spaces

13. Vector sums. If **M** and **N** are orthogonal subspaces of a Hilbert space, then **M** + **N** is closed (and therefore **M** + **N** = **M** ∨ **N**). Orthogonality may be too strong an assumption, but it is sufficient to ensure the conclusion. It is known that something is necessary; if no additional assumptions are made, then **M** + **N** need not be closed (see [50, p. 28], and Problems 52–55 below). Here is the conclusion under another very strong but frequently usable additional assumption.

> **Problem 13.** *If* **M** *is a finite-dimensional linear manifold in a Hilbert space* **H**, *and if* **N** *is a subspace (a closed linear manifold) in* **H**, *then the vector sum* **M** + **N** *is necessarily closed (and is therefore equal to the span* **M** ∨ **N**).

The result has the corollary (which it is also easy to prove directly) that every finite-dimensional linear manifold is closed; just put **N** = 0.

14. Lattice of subspaces. The collection of all subspaces of a Hilbert space is a *lattice*. This means that the collection is partially ordered (by inclusion), and that any two elements **M** and **N** of it have a least upper bound or supremum (namely the span **M** ∨ **N**) and a greatest lower bound or infimum (namely the intersection **M** ∩ **N**). A lattice is called *distributive* if (in the notation appropriate to subspaces)

$$\mathbf{L} \cap (\mathbf{M} \vee \mathbf{N}) = (\mathbf{L} \cap \mathbf{M}) \vee (\mathbf{L} \cap \mathbf{N})$$

identically in **L**, **M**, and **N**.

There is a weakening of this distributivity condition, called modularity; a lattice is called *modular* if the distributive law, as written above, holds at least when $N \subset L$. In that case, of course, $L \cap N = N$, and the identity becomes

$$L \cap (M \vee N) = (L \cap M) \vee N$$

(with the proviso $N \subset L$ still in force).

Since a Hilbert space is geometrically indistinguishable from any other Hilbert space of the same dimension, it is clear that the modularity or distributivity of its lattice of subspaces can depend on its dimension only.

Problem 14. *For which cardinal numbers m is the lattice of subspaces of a Hilbert space of dimension m modular? distributive?*

15. Vector sums and the modular law. Two possible kinds of misbehavior for subspaces are connected with each other; if one of them is ruled out, then the other one cannot happen either.

Problem 15. *For subspaces* M *and* N *of a Hilbert space, the vector sum* $M + N$ *is closed if and only if the modular equation*

$$L \cap (M \vee N) = (L \cap M) \vee N$$

is true whenever $N \subset L$.

16. Local compactness and dimension. Many global topological questions are easy to answer for Hilbert space. The answers either are a simple yes or no, or depend on the dimension. Thus, for instance, every Hilbert space is connected, but a Hilbert space is compact if and only if it is the trivial space with dimension 0. The same sort of problem could be posed backwards: given some information about the dimension of a Hilbert space (e.g., that it is finite), find topological properties that distinguish such a space from Hilbert spaces of all other dimensions. Such problems sometimes have useful and elegant solutions.

Problem 16. *A Hilbert space is locally compact if and only if it is finite-dimensional.*

17. Separability and dimension.

Problem 17. *A Hilbert space* H *is separable if and only if* $\dim H \leq \aleph_0$.

18. Measure in Hilbert space. Infinite-dimensional Hilbert spaces are properly regarded as the most successful infinite-dimensional generalizations of finite-dimensional Euclidean spaces. Finite-dimensional Euclidean spaces have, in addition to their algebraic and topological structure, a measure; it might be useful to generalize that too to infinite dimensions. Various attempts have been made to do so (see [92] and [132]). The un-

10

sophisticated approach is to seek a countably additive set function μ defined on (at least) the collection of all Borel sets (the σ-field generated by the open sets), so that $0 \leq \mu(M) \leq \infty$ for all Borel sets M. (Warning: the parenthetical definition of Borel sets in the preceding sentence is not the same as the one in [61].) In order that μ be suitably related to the other structure of the space, it makes sense to require that every non-empty open set have positive measure and that measure be invariant under translation. (The second condition means that $\mu(f + M) = \mu(M)$ for every vector f and for every Borel set M.) If, for now, the word "measure" is used to describe a set function satisfying just these conditions, then the following problem indicates that the unsophisticated approach is doomed to fail.

Problem 18. *For each measure in an infinite-dimensional Hilbert space, the measure of every non-empty ball is infinite.*

CHAPTER 3

Weak Topology

19. Weak closure of subspaces. A Hilbert space is a metric space, and, as such, it is a topological space. The metric topology (or norm topology) of a Hilbert space is often called the *strong* topology. A base for the strong topology is the collection of open balls, i.e., sets of the form

$$\{f : \|f - f_0\| < \varepsilon\},$$

where f_0 (the center) is a vector and ε (the radius) is a positive number.

Another topology, called the *weak* topology, plays an important role in the theory of Hilbert spaces. A subbase (not a base) for the weak topology is the collection of all sets of the form

$$\{f : |(f - f_0, g_0)| < \varepsilon\}.$$

It follows that a base for the weak topology is the collection of all sets of the form

$$\{f : |(f - f_0, g_i)| < \varepsilon, i = 1, \cdots, k\},$$

where k is a positive integer, $f_0, g_1, \cdots, g_k$ are vectors, and ε is a positive number.

Facts about these topologies are described by the grammatically appropriate use of "weak" and "strong". Thus, for instance, a function may be described as weakly continuous, or a sequence as strongly convergent; the meanings of such phrases should be obvious. The use of a topological word without a modifier always refers to the strong topology; this convention has already been observed in the preceding problems.

Whenever a set is endowed with a topology, many technical questions automatically demand attention. (Which separation axioms does the space satisfy? Is it compact? Is it connected?) If a large class of sets is in sight (for example, the class of all Hilbert spaces), then classification problems arise. (Which ones are locally compact? Which ones are separable?) If the set (or sets) already had some structure, the connection between the old structure and the new topology should be investigated. (Is the closed unit ball compact? Are inner products continuous?) If, finally, more than one topology is considered, then the relations of the topologies to one another must be clarified. (Is a weakly compact set strongly closed?) Most such questions, though natural, and, in fact, unavoidable, are not likely to be inspiring; for that reason most such questions do not appear below. The questions that do appear justify their appearance by some (perhaps subjective) test, such as a surprising answer, a tricky proof, or an important application.

Problem 19. *Every weakly closed set is strongly closed, but the converse is not true. Nevertheless every subspace of a Hilbert space (i.e., every strongly closed linear manifold) is weakly closed.*

20. Weak continuity of norm and inner product. For each fixed vector g, the function $f \mapsto (f, g)$ is weakly continuous; this is practically the definition of the weak topology. (A sequence, or a net, $\{f_n\}$ is weakly convergent to f if and only if $(f_n, g) \to (f, g)$ for each g.) This, together with the (Hermitian) symmetry of the inner product, implies that, for each fixed vector f, the function $g \mapsto (f, g)$ is weakly continuous. These two assertions between them say that the mapping from ordered pairs $\langle f, g \rangle$ to their inner product (f, g) is separately weakly continuous in each of its two variables.

It is natural to ask whether the mapping is weakly continuous jointly in its two variables, but it is easy to see that the answer is no. A counterexample has already been seen, in Solution 19; it was used there for a slightly different purpose. If $\{e_1, e_2, e_3, \cdots\}$ is an orthonormal sequence, then $e_n \to 0$ (weak), but $(e_n, e_n) = 1$ for all n. This example shows at the same time that the norm is not weakly continuous. It could, in fact, be said that the possible discontinuity of the norm is the only difference between weak convergence and strong convergence: a weakly convergent sequence (or net) on which the norm behaves itself is automatically strongly convergent.

Problem 20. *If $f_n \to f$ (weak) and $\|f_n\| \to \|f\|$, then $f_n \to f$ (strong).*

21. Semicontinuity of norm. The misbehavior of the example that shows the weak discontinuity of norm (Problem 20) is at the top, so to speak: norm fails to be upper semicontinuous. Definition: a real-valued function $\quad$ on a topological space is upper semicontinuous if

$$\limsup_n \varphi(x_n) \leqq \varphi(x)$$

whenever $x_n \to x$ (sequence or net); similarly φ is lower semicontinuous if

$$\varphi(x) \leqq \liminf_n \varphi(x_n)$$

13

whenever $x_n \to x$. (Here is how to remember which way the inequalities must point: always $\liminf_n \varphi(x_n) \leqq \limsup_n \varphi(x_n)$, so that if φ is both lower and upper semicontinuous, then liminf and limsup are forced to be equal, which is a characteristic property of continuity.) Misbehavior at the bottom cannot occur.

Problem 21. *Norm is weakly lower semicontinuous.*

Explicitly: if $f_n \to f$ (weak), then $\|f\| \leqq \liminf_n \|f_n\|$. Equivalently: for every $\varepsilon > 0$, there exists an n_0 such that $\|f\| \leqq \|f_n\| + \varepsilon$ whenever $n \geqq n_0$.

22

22. Weak separability. Since the strong closure of every set is included in its weak closure (see Solution 19), it follows that if a Hilbert space is separable (that is, strongly separable), then it is weakly separable. What about the converse?

Problem 22. *Is every weakly separable Hilbert space separable?*

23

23. Weak compactness of the unit ball.

Problem 23. *The closed unit ball in a Hilbert space is weakly compact.*

The result is sometimes known as the Tychonoff–Alaoglu theorem. It is as hard as it is important. It is very important.

24

24. Weak metrizability of the unit ball. Compactness is good, but even compact sets are better if they are metric. Once the unit ball is known to be weakly compact, it is natural to ask if it is weakly metrizable also.

Problem 24. *Is the weak topology of the unit ball in a separable Hilbert space metrizable?*

25

25. Weak closure of the unit sphere.

Problem 25. *What is the weak closure of the unit sphere (i.e., of the set of all unit vectors)? If a set is weakly dense in a Hilbert space, does it follow that its intersection with the unit ball is weakly dense in the unit ball?*

26

26. Weak metrizability and separability.

Problem 26. *If the weak topology of the unit ball in a Hilbert space H is metrizable, must H be separable?*

27

27. Uniform boundedness. The celebrated "principle of uniform boundedness" (true for all Banach spaces) is the assertion that a pointwise bounded collection of bounded linear functionals is bounded. The assumption and the conclusion can be expressed in the terminology appropriate to a Hilbert space **H** as follows. The assumption of pointwise boundedness for a subset **T** of **H** could also be called *weak boundedness*; it means that for each f in **H** there exists a positive constant $\alpha(f)$ such that $|(f, g)| \leq \alpha(f)$ for all g in **T**. The desired conclusion means that there exists a positive constant β such that $|(f, g)| \leq \beta \|f\|$ for all f in **H** and all g in **T**; this conclusion is equivalent to $\|g\| \leq \beta$ for all g in **T**. It is clear that every bounded subset of a Hilbert space is weakly bounded. The principle of uniform boundedness (for vectors in a Hilbert space) is the converse: every weakly bounded set is bounded. The usual proof of the general principle is a mildly involved category argument. A standard reference for a general treatment of the principle of uniform boundedness is [39, p. 49].

Problem 27. *Find an elementary proof of the principle of uniform boundedness for Hilbert space.*

(In this context a proof is "elementary" if it does not use the Baire category theorem.)

A frequently used corollary of the principle of uniform boundedness is the assertion that a weakly convergent sequence must be bounded. The proof is completely elementary: since convergent sequences of numbers are bounded, it follows that a weakly convergent sequence of vectors is weakly bounded. Nothing like this is true for nets, of course. One easy generalization of the sequence result that is available is that every weakly compact set is bounded. Reason: for each f, the map $g \mapsto (f, g)$ sends the g's in a weakly compact set onto a compact and therefore bounded set of numbers, so that a weakly compact set is weakly bounded.

28. Weak metrizability of Hilbert space. Some of the preceding results, notably the weak compactness of the unit ball and the principle of uniform boundedness, show that for bounded sets the weak topology is well behaved. For unbounded sets it is not.

28

Problem 28. *The weak topology of an infinite-dimensional Hilbert space is not metrizable.*

The shortest proof of this is tricky.

29. Linear functionals on l^2. If

29

$$\langle \alpha_1, \alpha_2, \alpha_3, \cdots \rangle \in l^2 \text{ and } \langle \beta_1, \beta_2, \beta_3, \cdots \rangle \in l^2,$$

then

$$\langle \alpha_1 \beta_1, \alpha_2 \beta_2, \alpha_3 \beta_3, \cdots \rangle \in l^1.$$

The following assertion is a kind of converse; it says that l^2 sequences are the only ones whose product with every l^2 sequence is in l^1.

Problem 29. *If* $\sum_n |\alpha_n \beta_n| < \infty$ *whenever* $\sum_n |\alpha_n|^2 < \infty$, *then*

$$\sum_n |\beta_n|^2 < \infty.$$

30. Weak completeness. A sequence $\{g_n\}$ of vectors in a Hilbert space is a *weak Cauchy sequence* if (surely this definition is guessable) the numerical sequence $\{(f, g_n)\}$ is a Cauchy sequence for each f in the space. *Weak Cauchy nets* are defined exactly the same way: just replace "sequence" by "net" throughout. To say of a Hilbert space, or a subset of one, that it is *weakly complete* means that every weak Cauchy net has a weak limit (in the set under consideration). If the conclusion is known to hold for sequences only, the space is called *sequentially weakly complete*.

Problem 30. (a) *No infinite-dimensional Hilbert space is weakly complete.*
(b) *Which Hilbert spaces are sequentially weakly complete?*

CHAPTER 4

Analytic Functions

31. Analytic Hilbert spaces. Analytic functions enter Hilbert space theory   **31**
in several ways; one of their roles is to provide illuminating examples. The
typical way to construct these examples is to consider a region D ("region"
means a non-empty open connected subset of the complex plane), let μ be
planar Lebesgue measure in D, and let $\mathbf{A}^2(D)$ be the set of all complex-valued
functions that are analytic throughout D and square-integrable with respect
to μ. The most important special case is the one in which D is the open unit
disc, $D = \{z : |z| < 1\}$; the corresponding function space will be denoted
simply by $\mathbf{A}^2$. No matter what D is, the set $\mathbf{A}^2(D)$ is a vector space with
respect to pointwise addition and scalar multiplication. It is also an inner-
product space with respect to the inner product defined by

$$(f, g) = \int_D f(z)g(z)^* \, d\mu(z).$$

Problem 31. *Is the space $\mathbf{A}^2(D)$ of square-integrable analytic functions on a region D a Hilbert space, or does it have to be completed
before it becomes one?*

32. Basis for $\mathbf{A}^2$.   **32**

Problem 32. *If $e_n(z) = \sqrt{(n + 1)/\pi} \cdot z^n$ for $|z| < 1$ and $n = 0, 1, 2, \cdots$,
then the e_n's form an orthonormal basis for $\mathbf{A}^2$. If $f \in \mathbf{A}^2$, with Taylor
series $\sum_{n=0}^{\infty} \alpha_n z^n$, then $\alpha_n = \sqrt{(n + 1)/\pi}(f, e_n)$ for $n = 0, 1, 2, \cdots$.*

33. Real functions in $\mathbf{H}^2$. Except for size (dimension) one Hilbert space is   **33**
very like another. To make a Hilbert space more interesting than its
neighbors, it is necessary to enrich it by the addition of some external
structure. Thus, for instance, the spaces $\mathbf{A}^2(D)$ are of interest because of

17

the analytic properties of their elements. Another important Hilbert space, known as $\mathbf{H}^2$ (H is for Hardy this time), endowed with some structure not usually found in a Hilbert space, is defined as follows.

Let C be the unit circle (that means circumference) in the complex plane, $C = \{z : |z| = 1\}$, and let μ be Lebesgue measure (the extension of arc length) on the Borel sets of C, normalized so that $\mu(C) = 1$ (instead of $\mu(C) = 2\pi$). If $e_n(z) = z^n$ for $|z| = 1$ $(n = 0, \pm 1, \pm 2, \cdots)$, then, by elementary calculus, the functions e_n form an orthonormal set in $\mathbf{L}^2(\mu)$; it is an easy consequence of standard approximation theorems (e.g., the Weierstrass theorem on approximation by polynomials) that the e_n's form an orthonormal basis for $\mathbf{L}^2$. (Finite linear combinations of the e_n's are called trigonometric polynomials.) The space $\mathbf{H}^2$ is, by definition, the subspace of $\mathbf{L}^2$ spanned by the e_n's with $n \geq 0$; equivalently $\mathbf{H}^2$ is the orthogonal complement in $\mathbf{L}^2$ of $\{e_{-1}, e_{-2}, e_{-3}, \cdots\}$. A related space, playing a role dual to that of $\mathbf{H}^2$, is the span of the e_n's with $n \leq 0$; it will be denoted by $\mathbf{H}^{2*}$.

Fourier expansions with respect to the orthonormal basis $\{e_n : n = 0, \pm 1, \pm 2, \cdots\}$ are formally similar to the Laurent expansions that occur in analytic function theory. The analogy motivates calling the functions in $\mathbf{H}^2$ the *analytic* elements of $\mathbf{L}^2$; the elements of $\mathbf{H}^{2*}$ are called *co-analytic*. A subset of $\mathbf{H}^2$ (a linear manifold but not a subspace) of considerable technical significance is the set $\mathbf{H}^\infty$ of bounded functions in $\mathbf{H}^2$; equivalently, $\mathbf{H}^\infty$ is the set of all those functions f in $\mathbf{L}^\infty$ for which $\int f e_n{}^* \, d\mu = 0$ $(n = -1, -2, -3, \cdots)$. Similarly $\mathbf{H}^1$ is the set of all those elements f of $\mathbf{L}^1$ for which these same equations hold. What gives $\mathbf{H}^1$, $\mathbf{H}^2$, and $\mathbf{H}^\infty$ their special flavor is the structure of the semigroup of non-negative integers within the additive group of all integers.

It is customary to speak of the elements of spaces such as $\mathbf{H}^1$, $\mathbf{H}^2$, and $\mathbf{H}^\infty$ as functions, and this custom was followed in the preceding paragraph. The custom is not likely to lead its user astray, as long as the qualification "almost everywhere" is kept in mind at all times. Thus "bounded" means "essentially bounded", and, similarly, all statements such as "$f = 0$" or "f is real" or "$|f| = 1$" are to be interpreted, when asserted, as holding almost everywhere.

Some authors define the Hardy spaces so as to make them honest function spaces (consisting of functions analytic on the unit disc). In that approach (see Problem 35) the almost everywhere difficulties are still present, but they are pushed elsewhere; they appear in questions (which must be asked and answered) about the limiting behavior of the functions on the boundary.

Independently of the approach used to study them, the functions in $\mathbf{H}^2$ are anxious to behave like analytic functions. The following statement is evidence in that direction.

Problem 33. *If f is a real function in $\mathbf{H}^2$, then f is a constant.*

18

34. Products in H^2. The deepest statements about the Hardy spaces have to   **34**
do with their multiplicative structure. The following one is an easily accessible
sample.

Problem 34. *The product of two functions in H^2 is in H^1.*

A kind of converse of this statement is true: it says that every function
in H^1 is the product of two functions in H^2. (See [75, p. 52].) The direct
statement is more useful in Hilbert space theory than the converse, and the
techniques used in the proof of the direct statement are nearer to the ones
appropriate to this book.

35. Analytic characterization of H^2. If $f \in H^2$, with Fourier expansion   **35**
$f = \sum_{n=0}^{\infty} \alpha_n e_n$, then $\sum_{n=0}^{\infty} |\alpha_n|^2 < \infty$, and therefore the radius of convergence of the power series $\sum_{n=0}^{\infty} \alpha_n{}^2 z^n$ is greater than or equal to 1. It follows
from the usual expression for the radius of convergence in terms of the coef-
ficients that the power series $\sum_{n=0}^{\infty} \alpha_n z^n$ defines an analytic function $\tilde{f}$ in
the open unit disc D. The mapping $f \mapsto \tilde{f}$ (obviously linear) establishes a
one-to-one correspondence between H^2 and the set $\tilde{H}^2$ of those functions
analytic in D whose series of Taylor coefficients is square-summable.

Problem 35. *If φ is an analytic function in the open unit disc, $\varphi(z) =$
$\sum_{n=0}^{\infty} \alpha_n z^n$, and if $\varphi_r(z) = \varphi(rz)$ for $0 < r < 1$ and $|z| = 1$, then $\varphi_r \in H^2$
for each r; the series $\sum_{n=0}^{\infty} |\alpha_n|^2$ converges if and only if the norms $\|\varphi_r\|$
are bounded.*

Many authors define H^2 to be $\tilde{H}^2$; for them, that is, H^2 consists of
analytic functions in the unit disc with square-summable Taylor series,
or, equivalently, with bounded concentric L^2 norms. If φ and ψ are two
such functions, with $\varphi(z) = \sum_{n=0}^{\infty} \alpha_n z^n$ and $\psi(z) = \sum_{n=0}^{\infty} \beta_n z^n$, then the
inner product (φ, ψ) is defined to be $\sum_{n=0}^{\infty} \alpha_n \beta_n{}^*$. In view of the one-to-
one correspondence $f \mapsto \tilde{f}$ between H^2 and $\tilde{H}^2$, it all comes to the same
thing. If $f \in H^2$, its image $\tilde{f}$ in $\tilde{H}^2$ may be spoken of as the *extension* of f
into the interior (cf. Solution 40). Since H^∞ is included in H^2, this concept
makes sense for elements of H^∞ also; the set of all their extensions will be
denoted by $\tilde{H}^\infty$.

36. Functional Hilbert spaces. Many of the popular examples of Hilbert   **36**
spaces are called function spaces, but they are not. If a measure space has
a non-empty set of measure zero (and this is usually the case), then the L^2
space over it consists not of functions, but of equivalence classes of
functions modulo sets of measure zero, and there is no natural way to
identify such equivalence classes with representative elements. There is,
however, a class of examples of Hilbert spaces whose elements are bona
fide functions; they will be called functional Hilbert spaces. A *functional*

19

Hilbert space is a Hilbert space **H** of complex-valued functions on a (non-empty) set X; the Hilbert space structure of **H** is related to X in two ways (the only two natural ways it could be). It is required that (1) if f and g are in **H** and if α and β are scalars, then $(\alpha f + \beta g)(x) = \alpha f(x) + \beta g(x)$ for each x in X, i.e., the evaluation functionals on **H** are linear, and (2) to each x in X there corresponds a positive constant γ_x, such that $|f(x)| \leqq \gamma_x \|f\|$ for all f in **H**, i.e., the evaluation functionals on **H** are bounded. The usual sequence spaces are trivial examples of functional Hilbert spaces (whether the length of the sequences is finite or infinite); the role of X is played by the index set. More typical examples of functional Hilbert spaces are the spaces $\mathbf{A}^2$ and $\tilde{\mathbf{H}}^2$ of analytic functions.

There is a trivial way of representing every Hilbert space as a functional one. Given **H**, write $X = \mathbf{H}$, and let $\tilde{\mathbf{H}}$ be the set of all those functions f on X ($= \mathbf{H}$) that are bounded conjugate-linear functionals. There is a natural correspondence $f \mapsto \tilde{f}$ from **H** to $\tilde{\mathbf{H}}$, defined by $\tilde{f}(g) = (f, g)$ for all g in X. By the Riesz representation theorem the correspondence is one-to-one; since (f, g) depends linearly on f, the correspondence is linear. Write, by definition, $(\tilde{f}, \tilde{g}) = (f, g)$ (whence, in particular, $\|\tilde{f}\| = \|f\|$); it follows that $\tilde{\mathbf{H}}$ is a Hilbert space. Since $|\tilde{f}(g)| = |(f, g)| \leqq \|f\| \cdot \|g\| = \|\tilde{f}\| \cdot \|g\|$, it follows that $\tilde{\mathbf{H}}$ is a functional Hilbert space. The correspondence $f \mapsto \tilde{f}$ between **H** and $\tilde{\mathbf{H}}$ is a Hilbert space isomorphism.

Problem 36. *Give an example of a Hilbert space of functions such that the vector operations are pointwise, but not all the evaluation functionals are bounded.*

An early and still useful reference for functional Hilbert spaces is [5].

37

37. Kernel functions. If **H** is a functional Hilbert space, over X say, then the linear functional $f \mapsto f(y)$ on **H** is bounded for each y in X, and, consequently, there exists, for each y in X, an element K_y of **H** such that $f(y) = (f, K_y)$ for all f. The function K on $X \times X$, defined by $K(x, y) = K_y(x)$, is called the *kernel function* or the *reproducing kernel* of **H**.

The most trivial examples of functional Hilbert spaces are obtained by modifying the standard inner product in $\mathbf{C}^n$ ($n = 1, 2, 3, \cdots$). In other words, start with $X = \{1, \cdots, n\}$, and define the "standard" inner product of two complex-valued functions f and g on X by $(f, g) = \sum_j f(j)g(j)^*$; to "modify" it, consider a linear transformation A on $\mathbf{C}^n$, and define $(f, g)_A$ to be (Af, g). This definition yields a bona fide inner product if and only if A is positive and invertible.

If $\mathbf{H}_A$ is the vector space $\mathbf{C}^n$ with inner product defined by the positive linear transformation A, then $\mathbf{H}_A$ is a functional Hilbert space; what is its kernel function? For a convenient notation to express the answer in, consider the standard orthonormal basis $\{e_1, \cdots, e_n\}$ in $\mathbf{C}^n$ (where $e_j(i) = \delta_{ij}$, the

Kronecker delta). If the kernel function of $\mathbf{H}_A$ is K, then

$$f(j) = (f, e_j) = (f, K_j)_A = (Af, K_j) = (f, AK_j)$$

whenever $f \in \mathbf{H}_A$ and $j = 1, \cdots, n$. (Since A is positive, it is Hermitian.) Consequence: $AK_j = e_j$, so that $K_j = A^{-1}e_j$, and it follows that

$$K(i, j) = K_j(i) = (A^{-1}e_j, e_i).$$

In other words, the function K is the matrix of A^{-1} with respect to the standard basis.

Note that the Hermitian character of the function K persists in the general case, in this sense:

$$K(x, y) = K_y(x) = (K_y, K_x) = (K_x, K_y)^* = (K_x(y))^* = (K(y, x))^*.$$

Problem 37. *If $\{e_j\}$ is an orthonormal basis for a functional Hilbert space* **H**, *then the kernel function K of* **H** *is given by*

$$K(x, y) = \sum_j e_j(x)e_j(y)^*.$$

What are the kernel functions of $\mathbf{A}^2$ and of $\tilde{\mathbf{H}}^2$?

The kernel functions of $\mathbf{A}^2$ and of $\tilde{\mathbf{H}}^2$ are known, respectively, as the *Bergman kernel* and the *Szegő kernel*.

38. Conjugation in functional Hilbert spaces. If f is an element of a functional Hilbert space **H**, the complex conjugate f^* may fail to belong to **H**; the spaces $\mathbf{H}^2$ and $\tilde{\mathbf{H}}^2$ yield examples. Call a functional Hilbert space *self-conjugate* if it is closed under the formation of complex conjugates; an example is the sequence space l^2. A more sophisticated example is the set of all square-integrable complex harmonic functions in, say, the unit disc. (The quickest way to describe complex harmonic functions is to say that they are the functions of the form $u + iv$, where each of u and v is the real part of some analytic function. Other classical definitions refer to the solutions of Laplace's equation, or, alternatively, to the mean value property.)

The definition of functional Hilbert spaces requires a strong connection between the unitary geometry of the space and the values of the functions the space consists of. Is the postulated connection strong enough to extend to complex conjugation? What does the question mean? Possible interpretation: is conjugation isometric?

Problem 38. *If f is an element of a self-conjugate functional Hilbert space, does it follow that $\|f^*\| = \|f\|$?*

Whenever the answer is yes for all f in the space, then a routine polarization argument shows that $(f^*, g^*) = (f, g)^*$ for all f and g.

39   **39. Continuity of extension.**

Problem 39. *The extension mapping* $f \mapsto \tilde{f}$ *(from* $\mathbf{H}^2$ *to* $\tilde{\mathbf{H}}^2$*) is continuous not only in the Hilbert space sense, but also in the sense appropriate to analytic functions. That is: if* $f_n \to f$ *in* $\mathbf{H}^2$*, then* $\tilde{f}_n(z) \to \tilde{f}(z)$ *for* $|z| < 1$*, and, in fact, the convergence is uniform on each disc* $\{z : |z| \leqq r\}$*,* $0 < r < 1$*.*

40   **40. Radial limits.**

Problem 40. *If an element* f *of* $\mathbf{H}^2$ *is such that the corresponding analytic function* $\tilde{f}$ *in* $\tilde{\mathbf{H}}^2$ *is bounded, then* f *is bounded, (i.e.,* $f \in \mathbf{H}^\infty$*).*

41   **41. Bounded approximation.**

Problem 41. *If* $f \in \mathbf{H}^\infty$*, does it follow that* $\tilde{f}$ *is bounded?*

42   **42. Multiplicativity of extension.**

Problem 42. *Is the mapping* $f \mapsto \tilde{f}$ *multiplicative?*

43   **43. Dirichlet problem.**

Problem 43. *To each real function* u *in* $\mathbf{L}^2$ *there corresponds a unique real function* v *in* $\mathbf{L}^2$ *such that* $(v, e_0) = 0$ *and such that* $u + iv \in \mathbf{H}^2$*. Equivalently, to each real* u *in* $\mathbf{L}^2$ *there corresponds a unique* f *in* $\mathbf{H}^2$ *such that* (f, e_0) *is real and such that* $\operatorname{Re} f = u$*.*

The relation between u and v is expressed by saying that they are *conjugate functions*; alternatively, v is the *Hilbert transform* of u.

CHAPTER 5

Infinite Matrices

44. Column-finite matrices. Many problems about operators on finite-dimensional spaces can be solved with the aid of matrices; matrices reduce qualitative geometric statements to explicit algebraic computations. Not much of matrix theory carries over to infinite-dimensional spaces, and what does is not so useful, but it sometimes helps.

Suppose that $\{e_j\}$ is an orthonormal basis for a Hilbert space **H**. If A is an operator on **H**, then each Ae_j has a Fourier expansion,

$$Ae_j = \sum_i \alpha_{ij} e_i;$$

the entries of the matrix that arises this way are given by

$$\alpha_{ij} = (Ae_j, e_i).$$

The index set is arbitrary here; it does not necessarily consist of positive integers. Familiar words (such as row, column, diagonal) can nevertheless be used in their familiar senses. Note that if, as usual, the first index indicates rows and the second one columns, then the matrix is formed by writing the coefficients in the expansion of Ae_j as the j column.

The correspondence from operators to matrices (induced by a fixed basis) has the usual algebraic properties. The zero matrix and the unit matrix are what they ought to be, the linear operations on matrices are the obvious ones, adjoint corresponds to conjugate transpose, and operator multiplication corresponds to the matrix product defined by the familiar formula

$$\gamma_{ij} = \sum_k \alpha_{ik} \beta_{kj}.$$

There are several ways of showing that these sums do not run into convergence trouble; here is one. Since $\alpha_{ik} = (e_k, A^*e_i)$, it follows that for each

23

fixed i the family $\{\alpha_{ik}\}$ is square-summable; since, similarly, $\beta_{kj} = (Be_j, e_k)$, it follows that for each fixed j the family $\{\beta_{kj}\}$ is square-summable. Conclusion (via the Schwarz inequality): for fixed i and j the family $\{\alpha_{ik}\beta_{kj}\}$ is (absolutely) summable.

It follows from the preceding paragraph that each row and each column of the matrix of each operator is square-summable. These are necessary conditions on a matrix in order that it arise from an operator; they are not sufficient. (Example: the diagonal matrix whose n-th diagonal term is n.) A sufficient condition of the same kind is that the family of all entries be square-summable; if, that is, $\sum_i \sum_j |\alpha_{ij}|^2 < \infty$, then there exists an operator A such that $\alpha_{ij} = (Ae_j, e_i)$. (Proof: since $|\sum_j \alpha_{ij}(f, e_j)|^2 \leq \sum_j |\alpha_{ij}|^2 \cdot \|f\|^2$ for each i and each f, it follows that $\|\sum_i (\sum_j \alpha_{ij}(f, e_j))e_i\|^2 \leq \sum_i \sum_j |\alpha_{ij}|^2 \cdot \|f\|^2$.) This condition is not necessary. (Example: the unit matrix.) There are no elegant and usable necessary and sufficient conditions. It is perfectly possible, of course, to write down in matricial terms the condition that a linear transformation is everywhere defined and bounded, but the result is neither elegant nor usable. This is the first significant way in which infinite matrix theory differs from the finite version: every operator corresponds to a matrix, but not every matrix corresponds to an operator, and it is hard to say which ones do.

As long as there is a fixed basis in the background, the correspondence from operators to matrices is one-to-one; as soon as the basis is allowed to vary, one operator may be assigned many matrices. An enticing game is to choose the basis so as to make the matrix as simple as possible. Here is a sample theorem, striking but less useful than it looks.

Problem 44. *Every operator has a column-finite matrix. More precisely, if A is an operator on a Hilbert space $\mathbf{H}$, then there exists an orthonormal basis $\{e_j\}$ for $\mathbf{H}$ such that, for each j, the matrix entry (Ae_j, e_i) vanishes for all but finitely many i's.*

Reference: [141].

45 **45. Schur test.** While the algebra of infinite matrices is more or less reasonable, the analysis is not. Questions about norms and spectra are likely to be recalcitrant. Each of the few answers that is known is considered a respectable mathematical accomplishment. The following result (due in substance to Schur [129]) is an example.

Problem 45. *If $\alpha_{ij} \geq 0$, if $p_i > 0$ and $q_j > 0$ $(i, j = 0, 1, 2, \cdots)$, and if β and γ are positive numbers such that*

$$\sum_i \alpha_{ij}p_i \leq \beta q_j \qquad (j = 0, 1, 2, \cdots),$$

$$\sum_j \alpha_{ij}q_j \leq \gamma p_i \qquad (i = 0, 1, 2, \cdots),$$

then there exists an operator A (on a separable infinite-dimensional Hilbert space, of course) with $\|A\|^2 \leqq \beta\gamma$ and with matrix $\langle\alpha_{ij}\rangle$ (with respect to a suitable orthonormal basis).

For a related result, and a pertinent reference, see Problem 173.

46. Hilbert matrix.

46

Problem 46. *There exists an operator A (on a separable infinite-dimensional Hilbert space) with $\|A\| \leqq \pi$ and with matrix $\langle 1/(i+j+1)\rangle$ $(i,j = 0, 1, 2, \cdots)$.*

The matrix is named after Hilbert; the norm of the matrix is in fact equal to π ([67, p. 226]).

47. Exponential Hilbert matrix.

47

A matrix whose $\langle i, j\rangle$ entry is a function of $i + j$ only is called a *Hankel matrix*. Thus, for instance, the Hilbert matrix (see Problem 46) is the Hankel matrix that corresponds to the function φ defined by $\varphi(x) = 1/(x + 1)$ (i.e., $\alpha_{ij} = \varphi(i + j)$, $i, j = 0, 1, 2, \cdots$; the main assertion of Problem 46 is that the matrix is "bounded" (meaning that it is the matrix of some operator). The same question, and other sharper ones, can be asked for other functions φ. A pleasant function is given by $\varphi(x) = 2^{-(x+1)}$. In that case all questions have a simple answer.

Problem 47. *The matrix $\langle 2^{-(i+j+1)}\rangle$ is bounded. What is its norm?*

48. Positivity of the Hilbert matrix.

48

The exponential Hilbert matrix (Problem 47) is Hermitian and its spectrum is positive (Solution 47); consequence: the corresponding operator is positive. The classical Hilbert matrix (Problem 46) is also Hermitian; its spectrum, however, is not quite so easily visible.

Problem 48. *Is the Hilbert matrix positive?*

49. Series of vectors.

49

If $\{\alpha_n\}$ is a sequence of complex numbers and $\{f_n\}$ is a sequence of vectors in a Hilbert space **H**, then the series $\sum_n \alpha_n f_n$ sometimes converges in **H** and sometimes does not. If, for instance, $\{f_n\}$ is an orthonormal sequence, then a necessary and sufficient condition for the convergence of $\sum_n \alpha_n f_n$ is that the sequence α be in l^2. If, for another example, **H** is the 1-dimensional vector space of complex numbers, then a necessary and sufficient condition that $\sum_n |\alpha_n f_n| < \infty$ for *every* sequence $\{f_n\}$ in l^2 is, again, that α be in l^2 (Problem 29).

For an interesting concrete question not covered by either of these examples consider this one: if the functions f_n in $\mathbf{L}^2(0, 1)$ are defined by $f_n(x) = x^n$, $n = 1, 2, \cdots$, and if $\sum_n |\alpha_n|^2 < \infty$, does it follow that the series $\sum_n \alpha_n f_n$ converges in L^2? It turns out that a general question along these lines has an elegant and usable answer.

Problem 49. *Under what conditions on a sequence $\{f_n\}$ of vectors in a Hilbert space* **H** *does the series* $\sum_n \alpha_n f_n$ *converge in* **H** *for every sequence* α *in* l^2?

CHAPTER 6

Boundedness and Invertibility

50. Boundedness on bases. Boundedness is a useful and natural condition, but it is a very strong condition on a linear transformation. The condition has a profound effect throughout operator theory, from its mildest algebraic aspects to its most complicated topological ones. To avoid certain obvious mistakes, it is important to know that boundedness is more than just the conjunction of an infinite number of conditions, one for each element of a basis. If A is an operator on a Hilbert space $\mathbf{H}$ with an orthonormal basis $\{e_1, e_2, e_3, \cdots\}$, then the numbers $\|Ae_n\|$ are bounded; if, for instance, $\|A\| \leq 1$, then $\|Ae_n\| \leq 1$ for all n; and, of course, if $A = 0$, then $Ae_n = 0$ for all n. The obvious mistakes just mentioned are based on the assumption that the converses of these assertions are true.

> **Problem 50.** (a) *Give an example of an unbounded linear transformation that is bounded on a basis.* (b) *Is there such an example that annihilates a basis?* (c) *Is there such an example that is bounded on each basis?* (d) *Give examples of operators of arbitrarily large norms that are bounded by 1 on a basis.* (e) *Could all the operators in such an example be normal?*

51. Uniform boundedness of linear transformations. Sometimes linear transformations between two Hilbert spaces play a role even when the center of the stage is occupied by operators on one Hilbert space. Much of the two-space theory is an easy adaptation of the one-space theory.

If $\mathbf{H}$ and $\mathbf{K}$ are Hilbert spaces, a linear transformation A from $\mathbf{H}$ into $\mathbf{K}$ is *bounded* if there exists a positive number α such that $\|Af\| \leq \alpha\|f\|$ for all f in $\mathbf{H}$; the *norm* of A, in symbols $\|A\|$, is the infimum of all such values of α. Given a bounded linear transformation A, the inner product (Af, g) makes sense whenever f is in $\mathbf{H}$ and g is in $\mathbf{K}$; the inner product is formed in $\mathbf{K}$.

For fixed g the inner product defines a bounded linear functional of f, and, consequently, it is identically equal to $(f, \tilde{g})$ for some $\tilde{g}$ in $\mathbf{H}$. The mapping from g to $\tilde{g}$ is the *adjoint* of A; it is a bounded linear transformation A^* from $\mathbf{K}$ into $\mathbf{H}$. By definition

$$(Af, g) = (f, A^*g)$$

whenever $f \in \mathbf{H}$ and $g \in \mathbf{K}$; here the left inner product is formed in $\mathbf{K}$ and the right one in $\mathbf{H}$. The algebraic properties of this kind of adjoint can be stated and proved the same way as for the classical kind. An especially important (but no less easily proved) connection between A and A^* is that the orthogonal complement of the range of A is equal to the kernel of A^*; since $A^{**} = A$, this assertion remains true with A and A^* interchanged.

All these algebraic statements are trivialities; the generalization of the principle of uniform boundedness from linear functionals to linear transformations is somewhat subtler. The generalization can be formulated almost exactly the same way as the special case: a pointwise bounded collection of bounded linear transformations is uniformly bounded. The assumption of pointwise boundedness can be formulated in a "weak" manner and a "strong" one. A set $\mathbf{Q}$ of linear transformations (from $\mathbf{H}$ into $\mathbf{K}$) is *weakly bounded* if for each f in $\mathbf{H}$ and each g in $\mathbf{K}$ there exists a positive constant $\alpha(f, g)$ such that $|(Af, g)| \leq \alpha(f, g)$ for all A in $\mathbf{Q}$. The set $\mathbf{Q}$ is *strongly bounded* if for each f in $\mathbf{H}$ there exists a positive constant $\beta(f)$ such that $\|Af\| \leq \beta(f)$ for all A in $\mathbf{Q}$. It is clear that every bounded set is strongly bounded and every strongly bounded set is weakly bounded. The principle of uniform boundedness for linear transformations is the best possible converse.

Problem 51. *Every weakly bounded set of bounded linear transformations is bounded.*

52. Invertible transformations. A bounded linear transformation A from a Hilbert space $\mathbf{H}$ to a Hilbert space $\mathbf{K}$ is *invertible* if there exists a bounded linear transformation B (from $\mathbf{K}$ into $\mathbf{H}$) such that $AB = 1$ ($=$the identity operator on $\mathbf{K}$) and $BA = 1$ ($=$the identity operator on $\mathbf{H}$). If A is invertible, then A is a one-to-one mapping of $\mathbf{H}$ onto $\mathbf{K}$. In the sense of pure set theory the converse is true: if A maps $\mathbf{H}$ one-to-one onto $\mathbf{K}$, then there exists a unique mapping A^{-1} from $\mathbf{K}$ to $\mathbf{H}$ such that $AA^{-1} = 1$ and $A^{-1}A = 1$; the mapping A^{-1} is linear. It is not obvious, however, that the linear transformation A^{-1} must be bounded; it is conceivable that A could be invertible as a set-theoretic mapping but not invertible as an operator. To guarantee that A^{-1} is bounded it is customary to strengthen the condition that A be one-to-one. The proper strengthening is to require that A be bounded from below, i.e., that there exist a positive number δ such that $\|Af\| \geq \delta\|f\|$ for every f in $\mathbf{H}$. (It is trivial to verify that if A is bounded from below, then A is indeed one-to-one.) If that strengthened condition is satisfied, then the other usual condition (onto) can be weakened: the requirement that the range of A be equal to $\mathbf{K}$ can be replaced by the requirement that the range of A be dense in $\mathbf{K}$. In sum: A is invertible if and only if it is bounded from below and has a dense range (see [50, p. 38]). Observe that the linear transformations A and A^* are invertible together; if they are invertible, then each of A^{-1} and A^{*-1} is the adjoint of the other.

It is perhaps worth a short digression to discuss the possibility of the range of an operator not being closed, and its consequences. If, for instance, A is defined on l^2 by $A \langle \xi_1, \xi_2, \xi_3, \cdots \rangle = \langle \xi_1, \frac{1}{2}\xi_2, \frac{1}{3}\xi_3, \cdots \rangle$, then the range of A consists of all vectors

$$\langle \eta_1, \eta_2, \eta_3, \cdots \rangle \quad \text{with} \quad \sum_n n^2 |\eta_n|^2 < \infty.$$

Since this range contains all finitely non-zero sequences, it is dense in l^2; since, however, it does not contain the sequence $\langle 1, \frac{1}{2}, \frac{1}{3}, \cdots \rangle$, it is not closed. Another example: for f in $\mathbf{L}^2(0, 1)$, define $(Af)(x) = xf(x)$. These operators are, of course, not bounded from below; if they were, their ranges would be closed.

Operators with non-closed ranges can be used to give a very simple example of two subspaces whose vector sum is not closed; cf. [50, p. 110]. Let A be an operator on a Hilbert space $\mathbf{H}$; the construction itself takes place in the direct sum $\mathbf{H} \oplus \mathbf{H}$. Let $\mathbf{M}$ be the "x-axis", i.e., the set of all vectors (in $\mathbf{H} \oplus \mathbf{H}$) of the form $\langle f, 0 \rangle$, and let $\mathbf{N}$ be the "graph" of A, i.e., the set of all vectors of the form $\langle f, Af \rangle$. It is trivial to verify that both $\mathbf{M}$ and $\mathbf{N}$ are subspaces of $\mathbf{H} \oplus \mathbf{H}$. When does $\langle f, g \rangle$ belong to $\mathbf{M} + \mathbf{N}$? The answer is if and only if it has the form $\langle u, 0 \rangle + \langle v, Av \rangle = \langle u + v, Av \rangle$; since u and v are arbitrary, a vector in $\mathbf{H} \oplus \mathbf{H}$ has that form if and only if its second coordinate belongs to the range $\mathbf{R}$ of the operator A. (In other words, $\mathbf{M} + \mathbf{N} = \mathbf{H} \oplus \mathbf{R}$.) Is $\mathbf{M} + \mathbf{N}$ closed? This means: if $\langle f_n, g_n \rangle \to \langle f, g \rangle$, where $f_n \in \mathbf{H}$ and $g_n \in \mathbf{R}$, does it follow that $f \in \mathbf{H}$? (trivially yes), and does it follow that $g \in \mathbf{R}$? (possibly no). Conclusion: $\mathbf{M} + \mathbf{N}$ is closed in $\mathbf{H} \oplus \mathbf{H}$ if and only if $\mathbf{R}$ is closed in $\mathbf{H}$. Since A can be chosen so that $\mathbf{R}$ is not closed, the vector sum of two subspaces need not be closed either.

The theorems and the examples seem to indicate that set-theoretic invertibility and operatorial invertibility are indeed distinct; it is one of the pleasantest and most useful facts about operator theory that they are the same after all.

Problem 52. *If* $\mathbf{H}$ *and* $\mathbf{K}$ *are Hilbert spaces, and if* A *is a bounded linear transformation that maps* $\mathbf{H}$ *one-to-one onto* $\mathbf{K}$, *then* A *is invertible.*

The corresponding statement about Banach spaces is usually proved by means of the Baire category theorem. The result is a special case of the so-called closed graph theorem; see Problem 58.

53. Diminishable complements. A *complement* of a subspace $\mathbf{M}$ in a   **53**
Hilbert space $\mathbf{H}$ is a subspace $\mathbf{N}$ such that $\mathbf{M} \cap \mathbf{N} = 0$ and $\mathbf{M} \vee \mathbf{N} = \mathbf{H}$.

Problem 53. (a) *It is possible for a subspace* **M** *to have a diminishable complement* **N**, *in the sense that there exists another complement* N_0, *with* $N_0 \subset N, N_0 \neq N$. (b) *Can a complement* **N** *be infinitely diminishable, in the sense that there exists a complement* N_0 *with* $N_0 \subset N$ *and* $\dim(N \cap N_0^\perp)$ *infinite?*

The statement (a) is a generalization of the existence of non-closed vector sums, i.e., the existence of subspaces **M** and **N** such that $M \vee N \neq M + N$. Indeed, if **N** is a diminishable complement of **M**, then $M \vee N$ cannot possibly be equal to $M + N$. The reason is that the condition $M \cap N = 0$ implies the uniqueness of the representation of a vector in the form $f + g$ ($f \in M, g \in N$); if any part of **N** is discarded, the rest cannot form a conspiracy with **M** to recapture it.

54. Dimension in inner-product spaces. The statement that any two maximal orthonormal sets have the same cardinal number is true for every inner product space; the standard proof makes no use of completeness. Orthogonal dimension can therefore be defined for every inner product space, word for word as for Hilbert spaces (the common cardinal number of all maximal orthonormal sets), but although it makes sense for "bad" inner-product spaces, its properties are likely to be bad too.

Problem 54. *There exists a linear manifold* **G** *dense in a Hilbert space* **H** *such that* $\dim G \neq \dim H$.

55. Total orthonormal sets. A total orthonormal set in an inner-product space (not necessarily a Hilbert space) is maximal. A possible proof of the converse goes as follows. Assume that $\{e_j\}$ is a maximal orthonormal set; given an arbitrary vector f, form the Fourier expansion $\sum_j (f, e_j)e_j$; since the difference $f - \sum_j (f, e_j)e_j$ is orthogonal to each e_j, use maximality to infer the desired conclusion. The crucial point in this proof is the formation of $\sum_j (f, e_j)e_j$. Bessel's inequality guarantees that the terms are not too large to be added, but completeness is needed to guarantee that the sum exists in the space. (Cf. Solution 32.) Is this use of completeness an awkwardness in the proof, or is it in the nature of things?

Problem 55. *Is every maximal orthonormal set in an inner-product space total?*

56. Preservation of dimension. An important question about operators is what do they do to the geometry of the underlying space. It is familiar from the study of finite-dimensional vector spaces that a linear transformation can lower dimension: the transformation 0, for an extreme example, collapses every space to a 0-dimensional one. If, however, a linear transformation on a finite-dimensional vector space is one-to-one (i.e., its kernel is 0), then it cannot lower dimension; since the same can be said about the inverse transformation (from the range back to the domain), it follows that dimension is preserved. The following assertion is, in a sense, the generalization of this finite-dimensional result to arbitrary Hilbert spaces.

Problem 56. *If there exists a one-to-one bounded linear transformation from a Hilbert space* **H** *into a Hilbert space* **K**, *then* dim **H** $\leq$ dim **K**. *If the image of* **H** *is dense in* **K**, *then equality holds.*

57. Projections of equal rank.   57

Problem 57. *If* P *and* Q *are projections such that* $\|P - Q\| < 1$, *then* P *and* Q *have the same rank.*

This is a special case of Problem 130.

58. Closed graph theorem. The *graph* of a linear transformation A (not   **58**
necessarily bounded) between inner product spaces **H** and **K** (not necessarily complete) is the set of all those ordered pairs $\langle f, g \rangle$ (elements of **H** $\oplus$ **K**) for which $Af = g$. (The terminology is standard. It is curious that it should be so, but it is. According to a widely adopted approach to the foundations of mathematics, a function, by definition, is a set of ordered pairs satisfying a certain univalence condition. According to that approach, the graph of A is A, and it is hard to see what is accomplished by giving it another name. Nevertheless most mathematicians cheerfully accept the unnecessary word; at the very least it serves as a warning that the same object is about to be viewed from a different angle.) A linear transformation is called *closed* if its graph is a closed set.

Problem 58. *A linear transformation from a Hilbert space into a Hilbert space is closed if and only if it is bounded.*

The assertion is known as the *closed graph theorem* for Hilbert spaces; its proof for Banach spaces is usually based on a category argument ([39, p. 57]). The theorem does not make the subject of closed but unbounded linear transformations trivial. Such transformations occur frequently in the applications of functional analysis; what the closed graph theorem says is that they can occur on incomplete inner-product spaces only (or non-closed linear manifolds in Hilbert spaces).

59. Range inclusion and factorization. If an operator A is "left divisible" by   **59**
an operator B, that is, if there exists an operator X such that $A = BX$, then ran $A \subset$ ran B. Is the converse true?

Problem 59. *If* ran $A \subset$ ran B, *does it follow that A is left divisible by B?*

The corresponding question about right divisibility is different, and, as it happens, easier to answer. If A is right divisible by B, then A is "majorized" by B; that is, there exists a constant α such that $\|Af\| \leqq \alpha\|Bf\|$ for all f. (If $A = XB$, then $\|X\|$ will do for α.) The converse is true; the proof is a straightforward geometric construction. Reference: [38].

60 **60. Unbounded symmetric transformations.** A linear transformation A (not necessarily bounded) on an inner-product space **H** (not necessarily complete) is called *symmetric* if $(Af, g) = (f, Ag)$ for all f and g in **H**. It is advisable to use this neutral term (rather than "Hermitian" or "self-adjoint"), because in the customary approach to Hermitian operators $(A = A^*)$ boundedness is an assumption necessary for the very formulation of the definition. Is there really a distinction here?

Problem 60. (a) *Is a symmetric linear transformation on an inner-product space* **H** *necessarily bounded?* (b) *What if* **H** *is a Hilbert space?*

CHAPTER 7

Multiplication Operators

61. Diagonal operators. Operator theory, like every other part of mathe- **61**
matics, cannot be properly studied without a large stock of concrete examples.
The purpose of several of the problems that follow is to build a stock of concrete operators, which can then be examined for the behavior of their
norms, inverses, and spectra.

Suppose, for a modest first step, that **H** is a Hilbert space and that $\{e_j\}$ is a
family of vectors that constitute an orthonormal basis for **H**. An operator A
is called a *diagonal operator* if Ae_j is a scalar multiple of e_j, say $Ae_j = \alpha_j e_j$,
for each j; the family $\{\alpha_j\}$ may properly be called the *diagonal* of A.

It is sometimes convenient to use the symbol

$$\operatorname{diag}\langle \alpha_0, \alpha_1, \alpha_2, \cdots \rangle$$

to denote the diagonal operator with the indicated diagonal terms. The
definition of a diagonal operator depends, of course, on the basis $\{e_j\}$, but in
most discussions of diagonal operators a basis is (perhaps tacitly) fixed in
advance, and then never mentioned again. Alternatively, diagonal operators
can be characterized in invariant terms as normal operators whose eigenvectors span the space. (The proof of the characterization is an easy exercise.)
Usually diagonal operators are associated with an orthonormal *sequence*;
the emphasis is on both the cardinal number ($\aleph_0$) and the order (ω) of the
underlying index set. That special case makes possible the use of some convenient language (e.g., "the first element of the diagonal") and the use of some
convenient techniques (e.g., constructions by induction).

Problem 61. *A necessary and sufficient condition that a family $\{\alpha_j\}$ be the
diagonal of a diagonal operator is that it be bounded; if it is bounded, then
the equations $Ae_j = \alpha_j e_j$ uniquely determine an operator A, and $\|A\| =
\sup_j |\alpha_j|$.*

33

62   **62. Multiplications on l^2.** Each sequence $\{\alpha_n\}$ of complex scalars induces a linear transformation A that maps l^2 into the vector space of all (not necessarily square-summable) sequences; by definition

$$A\langle \xi_1, \xi_2, \xi_3, \cdots \rangle = \langle \alpha_1\xi_1, \alpha_2\xi_2, \alpha_3\xi_3, \cdots \rangle.$$

Half of Problem 61 implies that if A is an operator (i.e., a bounded linear transformation of l^2 into itself), then the sequence $\{\alpha_n\}$ is bounded. What happens if the boundedness assumption on A is dropped?

> **Problem 62.** *Can an unbounded sequence of scalars induce a (possibly unbounded) transformation of l^2 into itself?*

The emphasis is that all l^2 is in the domain of the transformation, i.e., that if $\langle \xi_1, \xi_2, \xi_3, \cdots \rangle \in l^2$, then $\langle \alpha_1\xi_1, \alpha_2\xi_2, \alpha_3\xi_3, \cdots \rangle \in l^2$. The question should be compared with Problem 29. That problem considered sequences that multiply l^2 into l^1 (and concluded that they must belong to l^2); this one considers sequences that multiply l^2 into l^2 (and asks whether they must belong to l^∞). See Problem 66 for the generalization to $\mathbf{L}^2$.

63   **63. Spectrum of a diagonal operator.** The set of all bounded sequences $\{\alpha_n\}$ of complex numbers is an algebra (pointwise operations), with unit ($\alpha_n = 1$ for all n), with a conjugation ($\{\alpha_n\} \to \{\alpha_n{}^*\}$), and with a norm ($\|\{\alpha_n\}\| = \sup_n |\alpha_n|$). A bounded sequence $\{\alpha_n\}$ will be called *invertible* if it has an inverse in this algebra, i.e., if there exists a bounded sequence $\{\beta_n\}$ such that $\alpha_n\beta_n = 1$ for all n. A necessary and sufficient condition for this to happen is that $\{\alpha_n\}$ be bounded away from 0, i.e., that there exist a positive number δ such that $|\alpha_n| \geq \delta$ for all n.

If $\mathbf{H}$ is a Hilbert space with an orthonormal basis $\{e_n\}$, then it is easy to verify that the correspondence $\{\alpha_n\} \mapsto A$, where A is the operator on $\mathbf{H}$ such that $Ae_n = \alpha_n e_n$ for all n, is an isomorphism (an embedding) of the sequence algebra into the algebra of operators on $\mathbf{H}$. The correspondence preserves not only the familiar algebraic operations, but also conjugation; that is, if $\{\alpha_n\} \to A$, then $\{\alpha_n{}^*\} \to A^*$. The correspondence preserves the norm also (see Problem 61).

> **Problem 63.** *A diagonal operator with diagonal $\{\alpha_n\}$ is an invertible operator if and only if the sequence $\{\alpha_n\}$ is an invertible sequence. Consequence: the spectrum of a diagonal operator is the closure of the set of its diagonal terms.*

The result has the following useful corollary: every non-empty compact subset of the complex plane is the spectrum of some operator (and, in fact, of some diagonal operator). Proof: find a sequence of complex numbers dense in the prescribed compact set, and form a diagonal operator with that sequence as its diagonal.

64. Norm of a multiplication. Diagonal operators are special cases of a **64** general measure-theoretic construction. Suppose that X is a measure space with measure μ. If φ is a complex-valued bounded (i.e., essentially bounded) measurable function on X, then the *multiplication operator* (or just *multiplication*, for short) induced by φ is the operator A on $\mathbf{L}^2(\mu)$ defined by

$$(Af)(x) = \varphi(x)f(x)$$

for all x in X. (Here, as elsewhere in measure theory, two functions are identified if they differ on a set of measure zero only. This applies to the bounded φ's as well as to the square-integrable f's.) If X is the set of all positive integers and μ is the counting measure (the measure of every set is the number of elements in it), then multiplication operators reduce to diagonal operators.

Problem 64. *What, in terms of the multiplier φ, is the norm of the multiplication induced by φ?*

65. Boundedness of multipliers. Much of the theory of diagonal operators **65** extends to multiplication operators on measure spaces, but the details become a little fussy at times. A sample is the generalization of the assertion that if a sequence is the diagonal of a diagonal operator, then it is bounded.

Problem 65. *If an operator A on $\mathbf{L}^2$ (for a σ-finite measure) is such that $Af = \varphi \cdot f$ for all f in $\mathbf{L}^2$ (for some function φ), then φ is measurable and bounded.*

66. Boundedness of multiplications. Each complex-valued measurable **66** function φ induces a linear transformation A that maps $\mathbf{L}^2$ into the vector space of all (not necessarily square-integrable) measurable functions; by definition $(Af)(x) = \varphi(x)f(x)$. Half of Problem 65 implies that if A is an operator (i.e., a bounded linear transformation of $\mathbf{L}^2$ into itself), then the function φ is bounded. What happens if the boundedness assumption on A is dropped?

Problem 66. *Can an unbounded function induce a (possibly unbounded) transformation of $\mathbf{L}^2$ (for a σ-finite measure) into itself?*

This is the generalization to measures of Problem 62.

67. Spectrum of a multiplication. Some parts of the theory of diagonal **67** operators extend to multiplication operators almost verbatim, as follows. The set of all bounded measurable functions (identified modulo sets of measure zero) is an algebra (pointwise operations), with unit ($\varphi(x) = 1$ for all x), with a conjugation ($\varphi \mapsto \varphi^*$), and with a norm ($\|\varphi\|_\infty$). A bounded measurable function is *invertible* if it has an inverse in this algebra, i.e., if there exists a bounded measurable function ψ such that $\varphi(x)\psi(x) = 1$ for almost every x. A necessary and sufficient condition for this to happen is that φ be bounded away from 0 almost everywhere, i.e., that there exist a positive number δ such that $|\varphi(x)| \geqq \delta$ for almost every x.

The correspondence $\varphi \mapsto A$, where A is the multiplication operator defined by $(Af)(x) = \varphi(x)f(x)$, is an isomorphism (an embedding) of the function algebra into the algebra of operators on L^2. The correspondence preserves not only the familiar algebraic operations, but also the conjugation; that is, if $\varphi \to A$, then $\varphi^* \to A^*$. If the measure is σ-finite, the correspondence preserves the norm also (see Solution 64).

The role played by the range of a sequence is played, in the general case, by the *essential range* of a function φ; by definition, that is the set of all complex numbers λ such that for each neighborhood N of λ the set $\varphi^{-1}(N)$ has positive measure.

Problem 67. *The multiplication operator on* L^2 *(for a σ-finite measure) induced by φ is an invertible operator if and only if φ is an invertible function. Consequence: the spectrum of a multiplication is the essential range of the multiplier.*

68 **68. Multiplications on functional Hilbert spaces.** If a function φ multiplies L^2 into itself, then φ is necessarily bounded (Solution 66), and therefore multiplication by φ is necessarily an operator on L^2. Are the analogues of these assertions true for functional Hilbert spaces?

Problem 68. *Suppose that H is a functional Hilbert space, over a set X say, and suppose that φ is a complex-valued function on X such that $\varphi \cdot f \in H$ whenever $f \in H$. (a) If $Af = \varphi \cdot f$, is the linear transformation A bounded? (b) If $Af = \varphi \cdot f$ and if A is bounded, is the function φ bounded?*

69 **69. Multipliers of functional Hilbert spaces.** Suppose that H is a functional Hilbert space over a set X. A function φ on X is a *multiplier* of H if $\varphi \cdot f \in H$ for every f in H. Solution 68 says that every multiplier is bounded. It is frequently interesting and important to determine all multipliers of a functional Hilbert space.

For l^2, the easiest infinite-dimensional space, it is easy to prove that a necessary and sufficient condition that a function (i.e., a sequence) be a multiplier is that it be bounded. In a certain sense the space l^2 has too many multipliers: most of them do not belong to the space.

The space A^2 behaves differently: for it a necessary and sufficient condition that a function be a multiplier is that it be bounded and belong to the space. In a certain sense the space has too few multipliers: most of the functions in the space are not among them.

If X is finite and if H consists of all functions on X, then the set of multipliers of H is neither too large nor too small: it consists exactly of the elements of H. Can this happen for infinite-dimensional spaces?

Problem 69. *Construct an infinite-dimensional functional Hilbert space* **H** *such that the multipliers of* **H** *are exactly the elements of* **H**.

To say that every element of **H** is a multiplier is the same as to say that **H** is closed under multiplication, i.e., that **H** is an algebra. The constant function 1 is a multiplier of every **H**; hence, to say that every multiplier of **H** belongs to **H** is the same as to say that $1 \in$ **H**. If $1 \in$ **H**, then, of course, the algebra **H** has a unit, but trivial examples show that the converse is not true. Thus, the construction of an infinite-dimensional functional Hilbert space that is an algebra with unit (under pointwise functional multiplication) is not quite, but almost, what the problem asks for.

CHAPTER 8

Operator Matrices

70. Commutative operator determinants. An orthonormal basis serves to express a Hilbert space as the direct sum of one-dimensional subspaces. Some of the matrix theory associated with orthonormal bases deserves to be extended to more general direct sums. Suppose, to be specific, that $\mathbf{H} = \mathbf{H}_1 \oplus \mathbf{H}_2 \oplus \mathbf{H}_3 \oplus \cdots$. (Uncountable direct sums work just as well, and finite ones even better.) If the direct sum is viewed as an "internal" one, so that the $\mathbf{H}_i$'s are subspaces of $\mathbf{H}$, then the elements f of $\mathbf{H}$ are sums

$$f = f_1 + f_2 + f_3 + \cdots,$$

with f_i in $\mathbf{H}_i$. If A is an operator on $\mathbf{H}$, then

$$Af = Af_1 + Af_2 + Af_3 + \cdots.$$

Each Af_j, being an element of $\mathbf{H}$, has a decomposition:

$$Af_j = g_{1j} + g_{2j} + g_{3j} + \cdots,$$

with g_{ij} in $\mathbf{H}_i$. The g_{ij}'s depend, of course, on f_j, and the dependence is linear and continuous. It follows that

$$g_{ij} = A_{ij} f_j,$$

where A_{ij} is a bounded linear transformation from $\mathbf{H}_j$ to $\mathbf{H}_i$. The construction is finished: corresponding to each A on $\mathbf{H}$ there is a matrix $\langle A_{ij} \rangle$, whose entry in row i and column j is the projection onto the i component of the restriction of A to $\mathbf{H}_j$.

The correspondence from operators to matrices (induced by a fixed direct decomposition) has all the right algebraic properties. If $A = 0$, then $A_{ij} = 0$ for all i and j; if $A = 1$ (on $\mathbf{H}$), then $A_{ij} = 0$ when $i \neq j$ and $A_{ii} = 1$ (on $\mathbf{H}_i$). The linear operations on operator matrices are the obvious ones. The matrix

38

of A^* is the adjoint transpose of the matrix of A; that is, the matrix of A^* has the entry A_{ji}^* in row i and column j. The multiplication of operators corresponds to the matrix product defined by $\sum_k A_{ik} B_{kj}$. There is no convergence trouble here, but there may be commutativity trouble; the order of the factors must be watched with care.

The theory of operator matrices does not become trivial even if the number of direct summands is small (say two) and even if all the direct summands are identical. The following situation is the one that occurs most frequently: a Hilbert space **H** is given, the role of what was **H** in the preceding paragraph is played now by the direct sum **H** $\oplus$ **H**, and operators on that direct sum are expressed as two-by-two matrices whose entries are operators on **H**.

Problem 70. *If A, B, C, and D are pairwise commutative operators on a Hilbert space, then a necessary and sufficient condition that the operator matrix*

$$\begin{pmatrix} A & B \\ C & D \end{pmatrix}$$

be invertible is that the formal determinant $AD - BC$ be invertible.

71. Operator determinants. There are many situations in which the invertibility of an operator matrix

$$\begin{pmatrix} A & B \\ C & D \end{pmatrix}$$

plays a central role but in which the entries are not commutative; any special case is worth knowing.

Problem 71. *If C and D commute, and if D is invertible, then a necessary and sufficient condition that*

$$\begin{pmatrix} A & B \\ C & D \end{pmatrix}$$

be invertible is that $AD - BC$ be invertible. Construct examples to show that if the assumption that D is invertible is dropped, then the condition becomes unnecessary and insufficient.

For finite matrices more is known (cf. [130]): if C and D commute, then

$$\begin{pmatrix} A & B \\ C & D \end{pmatrix}$$

and $AD - BC$ have the same determinant. The proof for the general case can be arranged so as to yield this strengthened version for the finite-dimensional case.

72   **72. Operator determinants with a finite entry.** If A, B, and D are operators on a Hilbert space **H**, then the operator matrix

$$M = \begin{pmatrix} A & B \\ 0 & D \end{pmatrix}$$

induces (is) an operator on $\mathbf{H} \oplus \mathbf{H}$, and (cf. Problem 71) if both A and D are invertible, then M is invertible. The converse (if M is invertible, then A and D are) is not true (see Problem 71 again).

Operator matrices define operators on direct sums of Hilbert spaces whether the direct summands are identical or not. In at least one special case of interest the converse that was false in the preceding paragraph becomes true.

Problem 72. *If* **H** *and* **K** *are Hilbert spaces, with* dim **H** $< \infty$, *and if*

$$M = \begin{pmatrix} A & B \\ 0 & D \end{pmatrix}$$

is an invertible operator on $\mathbf{H} \oplus \mathbf{K}$, *then both* A *and* D *are invertible. Consequence: the spectrum of* M *is the union of the spectra of* A *and* D.

Note that A operates on **H**, D operates on **K**, and B maps **K** into **H**.

CHAPTER 9

Properties of Spectra

73. Spectra and conjugation. It is often useful to ask of a point in the spectrum of an operator how it got there. To say that λ is in the spectrum of A means that $A - \lambda$ is not invertible. The question reduces therefore to this: why is a non-invertible operator not invertible? There are several possible ways of answering the question; they have led to several (confusingly overlapping) classifications of spectra.

Perhaps the simplest approach to the subject is to recall that if an operator is bounded from below and has a dense range, then it is invertible. Consequence: if spec A is the spectrum of A, if $\Pi(A)$ is the set of complex numbers λ such that $A - \lambda$ is not bounded from below, and if $\Gamma(A)$ is the set of complex numbers λ such that the closure of the range of $A - \lambda$ is a proper subspace of $\mathbf{H}$ (i.e., distinct from $\mathbf{H}$), then

$$\text{spec } A = \Pi(A) \cup \Gamma(A).$$

The set $\Pi(A)$ is called the *approximate point spectrum* of A; a number λ belongs to $\Pi(A)$ if and only if there exists a sequence $\{f_n\}$ of unit vectors such that $\|(A - \lambda)f_n\| \to 0$. An important subset of the approximate point spectrum is the *point spectrum* $\Pi_0(A)$; a number λ belongs to it if and only if there exists a unit vector f such that $Af = \lambda f$ (i.e., $\Pi_0(A)$ is the set of all eigenvalues of A). The set $\Gamma(A)$ is called the *compression spectrum* of A. Schematically: think of the spectrum as the union of two overlapping discs (Π and Γ), one of which (Π) is divided into two parts (Π_0 and $\Pi - \Pi_0$) by a diameter perpendicular to the overlap. The result is a partition of the spectrum into five parts, each one of which may be sometimes present and sometimes absent. The born taxonomist may amuse himself by trying to see which one of the 2^5 a priori possibilities is realizable, but he would be well advised to postpone the attempt until he has seen several more examples of operators than have appeared in this book so far.

This is a good opportunity to comment on a sometimes confusing aspect of the nomenclature of operator theory. There is something called the spectral theorem for normal operators (see Problem 123), and there are things called spectra for all operators. The study of the latter might be called spectral theory, and sometimes it is. In the normal case the spectral theorem gives information about spectral theory, but, usually, that information can be bought cheaper elsewhere. Spectral theory in the present sense of the phrase is one of the easiest aspects of operator theory.

There is no consensus on which concepts and symbols are most convenient in this part of operator theory. Apparently every book introduces its own terminology, and the present one is no exception. A once popular approach was to divide the spectrum into three disjoint sets, namely, the point spectrum Π_0, the *residual spectrum* $\Gamma - \Pi_0$, and the *continuous spectrum* $\Pi - (\Gamma \cup \Pi_0)$. (The sets Π and Γ may overlap; examples will be easy to construct a little later.) As for symbols: the spectrum is often σ (or Σ) instead of spec.

The best way to master these concepts is, of course, through illuminating special examples, but a few general facts should come first; they help in the study of the examples. The most useful things to know are the relations of spectra to the algebra and topology of the complex plane. Perhaps the easiest algebraic questions concern conjugation.

> **Problem 73.** *What happens to the point spectrum, the compression spectrum, and the approximate point spectrum when an operator is replaced by its adjoint?*

74

74. Spectral mapping theorem. An assertion such as that if A is an operator and p is a polynomial, then spec $p(A) = p(\text{spec } A)$ (see [50, p. 53]) is called a *spectral mapping theorem*; other instances of it have to do with functions other than polynomials, such as inversion, conjugation, and wide classes of analytic functions ([39, p. 569]).

> **Problem 74.** *Is the spectral mapping theorem for polynomials true with* Π_0, *or* Π, *or* Γ *in place of* spec? *What about the spectral mapping theorem for inversion* $(p(z) = 1/z$ *when* $z \neq 0)$, *applied to invertible operators, with* Π_0, *or* Π, *or* Γ?

75

75. Similarity and spectrum. Two operators A and B are *similar* if there exists an invertible operator P such that $P^{-1}AP = B$.

> **Problem 75.** *Similar operators have the same spectrum, the same point spectrum, the same approximate point spectrum, and the same compression spectrum.*

76. Spectrum of a product. If A and B are operators, and if at least one of   **76**
them is invertible, then AB and BA are similar. (For the proof, apply $BA = A^{-1}(AB)A$ in case A is invertible or $AB = B^{-1}(BA)B$ in case B is.) This
implies (Problem 75) that if at least one of A and B is invertible, then AB
and BA have the same spectrum. In the finite-dimensional case more is
known: with no invertibility assumptions, AB and BA always have the same
characteristic polynomial. If neither A nor B is invertible, then, in the
infinite-dimensional case, the two products need not have the same
spectrum (many examples occur below), but their spectra cannot differ
by much. Here is the precise assertion.

Problem 76. *The non-zero elements of* spec AB *and* spec BA *are the*
same.

77. Closure of approximate point spectrum.   **77**

Problem 77. *Is the approximate point spectrum always closed?*

78. Boundary of spectrum.

Problem 78. *The boundary of the spectrum of an operator is included in*   **78**
the approximate point spectrum.

CHAPTER 10

Examples of Spectra

79

79. Residual spectrum of a normal operator. The time has come to consider special cases. The first result is that for normal operators, the most amenable large class known, the worst spectral pathology cannot occur.

Problem 79. *If A is normal, then $\Gamma(A) = \Pi_0(A)$ (and therefore* spec $A = \Pi(A)$). *Alternative formulation: the residual spectrum of a normal operator is always empty.*

Recall that the residual spectrum of A is $\Gamma(A) - \Pi_0(A)$.

80

80. Spectral parts of a diagonal operator. The spectrum of a diagonal operator was determined (Problem 63) as the closure of its diagonal; the determination of the fine structure of the spectrum requires another look.

Problem 80. *For each diagonal operator, find its point spectrum, compression spectrum, and approximate point spectrum.*

81

81. Spectral parts of a multiplication.

Problem 81. *For each multiplication, find its point spectrum, compression spectrum, and approximate point spectrum.*

82

82. Unilateral shift. The most important single operator, which plays a vital role in all parts of Hilbert space theory, is called the *unilateral shift*. Perhaps the simplest way to define it is to consider the Hilbert space l^2 of square-summable sequences; the unilateral shift is the operator U on l^2 defined by

$$U\langle \xi_0, \xi_1, \xi_2, \cdots \rangle = \langle 0, \xi_0, \xi_1, \xi_2, \cdots \rangle.$$

(The unilateral shift has already occurred in this book, although it was not named until now; see Solution 71.) Linearity is obvious. As for boundedness, it is true with room to spare. Norms are not only kept within reasonable bounds, but they are preserved exactly; the unilateral shift is an isometry. The range of U is not l^2 but a proper subspace of l^2, the subspace of vectors with vanishing first coordinate. The existence of an isometry whose range is not the whole space is characteristic of infinite-dimensional spaces.

If e_n is the vector $\langle \xi_0, \xi_1, \xi_2, \cdots \rangle$ for which $\xi_n = 1$ and $\xi_i = 0$ whenever $i \neq n$ ($n = 0, 1, 2, \cdots$), then the e_n's form an orthonormal basis for l^2. The effect of U on this basis is described by

$$Ue_n = e_{n+1} \qquad (n = 0, 1, 2, \cdots).$$

These equations uniquely determine U, and in most of the study of U they may be taken as its definition.

A familiar space that comes equipped with an orthonormal basis indexed by non-negative integers is $\mathbf{H}^2$ (see Problem 33). Since, in that space, $e_n(z) = z^n$, the effect of shifting forward by one index is the same as the effect of multiplication by e_1. In other words, the unilateral shift is the same as the multiplication operator on $\mathbf{H}^2$ defined by

$$(Uf)(z) = zf(z).$$

To say that it is the "same", and, in fact, to speak of "the" unilateral shift is a slight abuse of language, a convenient one that will be maintained throughout the sequel. Properly speaking the unilateral shift is a unitary equivalence class of operators, but no confusion will result from regarding it as one operator with many different manifestations.

Problem 82. *What is the spectrum of the unilateral shift, and what are its parts (point spectrum, compression spectrum, and approximate point spectrum)? What are the answers to the same questions for the adjoint of the unilateral shift?*

83. Structure of the set of eigenvectors. The unilateral shift shows that the set of eigenvalues of an operator can be much richer than finite-dimensional experience suggests. What about the set of eigenvectors? A dull answer to the question is given by scalars; every vector is an eigenvector, or every non-zero vector is one, depending on how the definition of eigenvector was formulated. Is there an interesting answer? **83**

Problem 83. *Is there a non-scalar operator whose set of eigenvectors has a non-empty interior?*

84. Bilateral shift. A close relative of the unilateral shift is the *bilateral shift*. **84** To define it, let $\mathbf{H}$ be the Hilbert space of all two-way (bilateral) square-summable sequences. The elements of **H** are most conveniently written in the form

$$\langle \cdots, \xi_{-2}, \xi_{-1}, (\xi_0), \xi_1, \xi_2, \cdots \rangle;$$

the term in parentheses indicates the one corresponding to the index 0. The bilateral shift is the operator W on **H** defined by

$$W\langle \cdots, \xi_{-2}, \xi_{-1}, (\xi_0), \xi_1, \xi_2, \cdots \rangle = \langle \cdots, \xi_{-3}, \xi_{-2}, (\xi_{-1}), \xi_0, \xi_1, \cdots \rangle.$$

Linearity is obvious, and boundedness is true with room to spare; the bilateral shift, like the unilateral one, is an isometry. Since the range of the bilateral shift is the entire space **H**, it is even unitary.

If e_n is the vector $\langle \cdots. \xi_{-1}, (\xi_0), \xi_1, \cdots \rangle$ for which $\xi_n = 1$ and $\xi_i = 0$ whenever $i \neq n$ ($n = 0, \pm 1, \pm 2, \cdots$), then the e_n's form an orthonormal basis for **H**. The effect of W on this basis is described by

$$W e_n = e_{n+1} \qquad (n = 0, \pm 1, \pm 2, \cdots).$$

Problem 84. *What is the spectrum of the bilateral shift, and what are its parts (point spectrum, compression spectrum, and approximate point spectrum)? What are the answers to the same questions for the adjoint of the bilateral shift?*

85

85. Spectrum of a functional multiplication. Every operator studied so far has been a multiplication, either in the legitimate sense (on an L^2) or in the extended sense (on a functional Hilbert space). The latter kind is usually harder to study; it does, however, have the advantage of having a satisfactory characterization in terms of its spectrum.

Problem 85. *A necessary and sufficient condition that an operator A on a Hilbert space **H** be representable as a multiplication on a functional Hilbert space is that the eigenvectors of A* span **H**.*

Caution: as the facts for multiplications on L^2 spaces show (cf. Solution 81) this characterization is applicable to functional Hilbert spaces only. The result seems to be due to P. R. Halmos and A. L. Shields.

CHAPTER 11

Spectral Radius

86. Analyticity of resolvents. Suppose that A is an operator on a Hilbert space **H**. If λ does not belong to the spectrum of A, then the operator $A - \lambda$ is invertible; write $\rho(\lambda) = (A - \lambda)^{-1}$. (When it is necessary to indicate the dependence of the function ρ on the operator A, write $\rho = \rho_A$.) The function ρ is called the *resolvent* of A. The domain of ρ is the complement of the spectrum of A; its values are operators on **H**.

The definition of the resolvent is very explicit; this makes it seem plausible that the resolvent is a well-behaved function. To formulate this precisely, consider, quite generally, functions φ whose domains are open sets in the complex plane and whose values are operators on **H**. Such a function φ will be called *analytic* if, for each f and g in **H**, the numerical function $\lambda \mapsto (\varphi(\lambda)f, g)$ (with the same domain as φ) is analytic in the usual sense. (To distinguish this concept from other closely related ones, it is sometimes called *weak* analyticity.) In case the function ψ defined by $\psi(\lambda) = \varphi(1/\lambda)$ can be assigned a value at the origin so that it becomes analytic there, then (just as for numerical functions) φ will be called analytic at ∞, and φ is assigned at ∞ the value of ψ at 0.

Problem 86. *The resolvent of every operator is analytic at each point of its domain, and at ∞; its value at ∞ is (the operator) 0.*

For a detailed study of resolvents, see [39, VII, 3].

87. Non-emptiness of spectra. Does every operator have a non-empty spectrum? The question was bound to arise sooner or later. Even the finite-dimensional case shows that the question is non-trivial. To say that every finite matrix has an eigenvalue is the same as to say that the characteristic polynomial of every finite matrix has at least one zero, and that is no more and no less general than to say that every polynomial equation (with complex coefficients) has at least one (complex) zero. In other words, the finite-dimensional case of the general question about spectra is as deep as the fundamental theorem of algebra, whose proof is usually based on the theory of complex analytic functions. It should not be too surprising now that the theory of such functions enters the study of operators in every case (whether the dimension is finite or infinite).

Problem 87. *Every operator has a non-empty spectrum.*

88   **88. Spectral radius.** The *spectral radius* of an operator A, in symbols $r(A)$, is defined by

$$r(A) = \sup\{|\lambda| : \lambda \in \operatorname{spec} A\}.$$

Clearly $0 \leq r(A) \leq \|A\|$; the spectral mapping theorem implies also that $r(A^n) = (r(A))^n$ for every positive integer n. It frequently turns out that the spectral radius of an operator is easy to compute even when it is hard to find the spectrum; the tool that makes it easy is the following assertion.

Problem 88. *For each operator A,*

$$r(A) = \lim_n \|A^n\|^{1/n},$$

in the sense that the indicated limit always exists and has the indicated value.

It is an easy consequence of this result that if A and B are commutative operators, then

$$r(AB) \leq r(A)r(B).$$

It is a somewhat less easy consequence, but still a matter of no more than a little fussy analysis with inequalities, that if A and B commute, then

$$r(A + B) \leq r(A) + r(B).$$

If no commutativity assumptions are made, then two-dimensional examples, such as

$$A = \begin{pmatrix} 0 & 0 \\ 1 & 0 \end{pmatrix}, \qquad B = \begin{pmatrix} 0 & 1 \\ 0 & 0 \end{pmatrix},$$

show that neither the submultiplicative nor the subadditive property persists.

89   **89. Weighted shifts.** A *weighted shift* is the product of a shift (one-sided or two) and a compatible diagonal operator. More explicitly, suppose that $\{e_n\}$ is an orthonormal basis ($n = 0, 1, 2, \cdots$, or else $n = 0, \pm 1, \pm 2, \cdots$), and suppose that $\{\alpha_n\}$ is a bounded sequence of complex numbers (the set of n's being the same as before). A weighted shift is an operator of the form SP, where S is a shift $(Se_n = e_{n+1})$ and P is a diagonal operator with diagonal $\{\alpha_n\}$ $(Pe_n = \alpha_n e_n)$. Not everything about weighted shifts is known, but even the little that is makes them almost indispensable in the construction of examples and counterexamples. It is sometimes convenient to use the symbol

$$\text{shift}\langle \alpha_0, \alpha_1, \alpha_2, \cdots \rangle$$

to denote the weighted shift with the indicated weights.

Problem 89. *If P and Q are diagonal operators, with diagonals $\{\alpha_n\}$ and $\{\beta_n\}$, and if $|\alpha_n| = |\beta_n|$ for all n, then the weighted shifts $A = SP$ and $B = SQ$ are unitarily equivalent.*

A discussion of two weighted shifts should, by rights, refer to two orthonormal bases, but the generality gained that way is shallow. If $\{e_n\}$ and $\{f_n\}$ are orthonormal bases, then there exists a unitary operator U such that $Ue_n = f_n$ for all n, and U can be carried along gratis with any unitary equivalence proof.

The result about the unitary equivalence of weighted shifts has two useful consequences. First, the weighted shift with weights α_n is unitarily equivalent to the weighted shift with weights $|\alpha_n|$. Since unitarily equivalent operators are "abstractly identical", there is never any loss of generality in restricting attention to weighted shifts whose weights are non-negative; this is what really justifies the use of the word "weight". Second, if A is a weighted shift and if α is a complex number of modulus 1, then, since αA is a weighted shift, whose weights have the same moduli as the corresponding weights of A, it follows that A and αA are unitarily equivalent. In other words, to within unitary equivalence, a weighted shift is not altered by multiplication by a number of modulus 1. This implies, for instance, that the spectrum of a weighted shift has circular symmetry: if λ is in the spectrum and if $|\alpha| = 1$, then $\alpha\lambda$ is in the spectrum.

90. Similarity of weighted shifts. Is the converse of Problem 89 true? **90** Suppose, in other words, that A and B are weighted shifts, with weights $\{\alpha_n\}$ and $\{\beta_n\}$; if A and B are unitarily equivalent, does it follow that $|\alpha_n| = |\beta_n|$ for all n? The answer can be quite elusive, but with the right approach it is easy. The answer is no; the reason is that, for bilateral shifts, a translation of the weights produces a unitarily equivalent shift. That is: if $Ae_n = \alpha_n e_{n+1}$ and $Be_n = \alpha_{n+1}e_{n+1}$ $(n = 0, \pm 1, \pm 2, \cdots)$, then A and B are unitarily equivalent. If, in fact, W is the bilateral shift $(We_n = e_{n+1}, n = 0, \pm 1, \pm 2, \cdots)$, then $W^*AW = B$; if, however, the sequence $\{|\alpha_n|\}$ is not constant, then there is at least one n such that $|\alpha_n| \neq |\alpha_{n+1}|$.

Unilateral shifts behave differently. If some of the weights are allowed to be zero, the situation is in part annoying and in part trivial. In the good case (no zero weights), the kernel of the adjoint A^* of a unilateral weighted shift is

49

spanned by e_0, the kernel of A^{*2} is spanned by e_0 and e_1, and, in general, the kernel of A^{*n} is spanned by $e_0, \cdots, e_{n-1}$ ($n = 1, 2, 3, \cdots$). If A and B are unitarily equivalent weighted shifts, then A^{*n} and B^{*n} are unitarily equivalent; if, say, $A = U^*BU$, then U must send ker A^{*n} onto ker B^{*n}. This implies that the span of $\{e_0, \cdots, e_{n-1}\}$ is invariant under U, and from this, in turn, it follows that U is a diagonal operator. Since the diagonal entries of a unitary diagonal matrix have modulus 1, it follows that, for each n, the effect of A on e_n can differ from that of B by a factor of modulus 1 only.

This settles the unitary equivalence theory for weighted shifts with non-zero weights; what about similarity?

Problem 90. *If A and B are unilateral weighted shifts, with non-zero weights $\{\alpha_n\}$ and $\{\beta_n\}$, then a necessary and sufficient condition that A and B be similar is that the sequence of quotients*

$$\left| \frac{\alpha_0 \cdots \alpha_n}{\beta_0 \cdots \beta_n} \right|$$

be bounded away from 0 and from ∞.

Similarity is a less severe restriction than unitary equivalence; questions about similarity are usually easier to answer. By a modification of the argument for one-sided shifts, a modification whose difficulties are more notational than conceptual, it is possible to get a satisfactory condition, like that in Problem 90, for the similarity of two-sided shifts; this was done by R. L. Kelley.

91   91. Norm and spectral radius of a weighted shift.

Problem 91. *Express the norm and the spectral radius of a weighted shift in terms of its weights.*

92   92. Power norms. The power norms of an operator A, that is the numbers $\|A^k\|$, constitute an interesting sequence.

Problem 92. *For which sequences $\{p_k\}$ of positive numbers does there exist an operator A such that $p_k = \|A^k\|$, $k = 0, 1, 2, \cdots$?*

93   93. Eigenvalues of weighted shifts. The exact determination of the spectrum and its parts for arbitrary weighted shifts is a non-trivial problem. Here is a useful fragment.

Problem 93. *Find all the eigenvalues of all unilateral weighted shifts (with non-zero weights) and of their adjoints.*

The possible presence of 0 among the weights is not a genuine difficulty but a nuisance. A unilateral weighted shift, one of whose weights vanishes, becomes thereby the direct sum of a finite-dimensional operator and another weighted shift. The presence of an infinite number of zero weights can cause some interesting trouble (cf. Problem 98), but the good problems about shifts have to do with non-zero weights.

94. Approximate point spectrum of a weighted shift. For each operator A there exists a complex number of modulus $r(A)$ in the approximate point spectrum of A (Problem 78). If A is a weighted shift, then the circular symmetry of each part of the spectrum (Problem 89) implies that every complex number of modulus $r(A)$ belongs to the approximate point spectrum of A. Consequence: the approximate point spectrum of a weighted shift always includes a circle (possibly degenerate). Question: does it ever include more?

94

> **Problem 94.** *Is there a weighted shift whose approximate point spectrum has interior points?*

95. Weighted sequence spaces. The expression "weighted shift" means one thing, but it could just as well have meant something else. What it does mean is to modify the ordinary shift on the ordinary sequence space l^2 by attaching weights to the transformation; what it could have meant is to modify by attaching weights to the space.

95

To get an explicit description of the alternative, let $p = \{p_0, p_1, p_2, \cdots\}$ be a sequence of strictly positive numbers, and let $l^2(p)$ be the set of complex sequences $\langle \xi_0, \xi_1, \xi_2, \cdots \rangle$ with $\sum_{n=0}^{\infty} p_n |\xi_n|^2 < \infty$. With respect to the coordinatewise linear operations and the inner product defined by

$$(\langle \xi_0, \xi_1, \xi_2, \cdots \rangle, \langle \eta_0, \eta_1, \eta_2, \cdots \rangle) = \sum_{n=0}^{\infty} p_n \xi_n \eta_n^*,$$

the set $l^2(p)$ is a Hilbert space; it may be called a weighted sequence space. (All this is unilateral; the bilateral case can be treated similarly.) When is the shift an operator on this space? When, in other words, is it true that if $f = \langle \xi_0, \xi_1, \xi_2, \cdots \rangle \in l^2(p)$, then $Sf = \langle 0, \xi_0, \xi_1, \xi_2, \cdots \rangle \in l^2(p)$, and, as f varies over $l^2(p)$, $\|Sf\|$ is bounded by a constant multiple of $\|f\|$? The answer is easy. An obviously necessary condition is that there exist a positive constant α such that $\|e_{n+1}\| \leq \alpha \|e_n\|$, where e_n, of course, is the vector whose coordinate with index n is 1 and all other coordinates are 0. Since $\|e_n\|^2 = p_n$, this condition says that the sequence $\{p_{n+1}/p_n\}$ is bounded. It is almost obvious that this necessary condition is also sufficient. If $p_{n+1}/p_n \leq \alpha^2$ for all n, then

$$\|Sf\|^2 = \sum_{n=1}^{\infty} p_n |\xi_{n-1}|^2 = \sum_{n=1}^{\infty} \frac{p_n}{p_{n-1}} p_{n-1} |\xi_{n-1}|^2$$

$$\leq \alpha^2 \sum_{n=0}^{\infty} p_n |\xi_n|^2 = \alpha^2 \|f\|^2.$$

Every question about weighted shifts on the ordinary sequence space can be re-asked about the ordinary shift on weighted sequence spaces; here is a sample.

Problem 95. *If $p = \{p_n\}$ is a sequence of positive numbers such that $\{p_{n+1}/p_n\}$ is bounded, what, in terms of $\{p_n\}$, is the spectral radius of the shift on $l^2(p)$?*

96

96. One-point spectrum. The proof in Problem 63 (every non-empty compact subset of the plane is the spectrum of some operator) is not sufficiently elastic to yield examples of all the different ways spectral parts can behave. That proof used diagonal operators, which always have eigenvalues; from that proof alone it is not possible to infer the existence of operators whose point spectrum is empty. Multiplication operators come to the rescue. If D is a bounded region, if $\varphi(z) = z$ for z in D, and if A is the multiplication operator induced by φ on the $\mathbf{L}^2$ space of planar Lebesgue measure in D, then the spectrum of A is the closure $\bar{D}$, but the point spectrum of A is empty. Similar techniques show the existence of operators A with $\Pi_0(A) = \varnothing$ and spec $A = [0, 1]$, say; just use linear Lebesgue measure in $[0, 1]$. Whenever a compact set M in the plane is the support of a measure (on the Borel sets) that gives zero weight to each single point, then M is the spectrum of an operator with no eigenvalues. (To say that M is the support of μ means that if N is an open set with $\mu(M \cap N) = 0$, then $M \cap N = \varnothing$.) It is a routine exercise in topological measure theory to prove that every non-empty, compact, perfect set (no isolated points) in the plane is the support of a measure (on the Borel sets) that gives zero weight to each single point. (The proof is of no relevance to Hilbert space theory.) It follows that every such set is the spectrum of an operator with no eigenvalues. What about sets that are not perfect?

A very satisfactory answer can be given in terms of the appropriate analytic generalization of the algebraic concept of nilpotence. An operator is nilpotent if some positive integral power of it is zero (and the least such power is the index of nilpotence); an operator A is *quasinilpotent* if $\lim_n \|A^n\|^{1/n} = 0$. It is obvious that nilpotence implies quasinilpotence. The spectral mapping theorem implies that if A is nilpotent, then spec $A = \{0\}$. The expression for the spectral radius in terms of norms implies that if A is quasinilpotent, then spec $A = \{0\}$, and that, moreover, the converse is true. A nilpotent operator always has a non-trivial kernel, and hence a non-empty point spectrum; for quasinilpotent operators that is not so.

Problem 96. *Construct a quasinilpotent operator whose point spectrum is empty.*

Observe that on finite-dimensional spaces such a construction is clearly impossible.

97. Analytic quasinilpotents. Polynomials in an operator make sense; what **97** about the obvious extension to "infinite polynomials", or, more precisely, power series? There are several possible approaches to defining expressions such as $f(A)$, where f is an analytic function and A is an operator. The simplest of them has to do with operators that have the smallest possible spectrum, that is, quasinilpotent operators. If A is such an operator, and if f is a function that is analytic in a neighborhood of 0, then f has a power series expansion, $f(z) = \sum_{n=0}^{\infty} \alpha_n z^n$, convergent when, say, $|z| < \varepsilon$; in that case write $f(A) = \sum_{n=0}^{\infty} \alpha_n A^n$. This operator series converges (in the norm). Reason: since $\|A^n\|^{1/n} < \varepsilon$ for n sufficiently large, it follows that $\|A^n\| < \varepsilon^n$ for n sufficiently large.

The correspondence $f \mapsto f(A)$ between functions and operators respects the algebraic structures involved; sums and products of functions correspond to sums and products of operators. This phenomenon, or, more precisely, the homomorphism from functions f analytic at 0 to operators $f(A)$ (where A is a fixed quasinilpotent operator) is called a functional calculus; it is an extension of the trivial functional calculus for polynomials.

An operator A is called *algebraic* if there exists a non-zero polynomial p such that $p(A) = 0$; by a natural extension of language it seems proper to say that A is *analytic* if there exists a non-zero function f analytic at 0 such that $f(A) = 0$. Every nilpotent operator is algebraic; is the generalization to "infinite polynomials" true?

Problem 97. *Is every quasinilpotent operator analytic?*

98. Spectrum of a direct sum. The spectrum of the direct sum of two operators is the union of their spectra, and the same is true of the point spectrum, the **98** approximate point spectrum, and the compression spectrum. The extension of this result from two direct summands to any finite number is a trivial induction. What happens if the number of summands is infinite? A possible clue to the answer is the behavior of diagonal operators on infinite-dimensional spaces. Such an operator is an infinite direct sum, each summand of which is an operator on a one-dimensional space, and its spectrum is the closure of the union of their spectra (Problem 63).

Problem 98. *Is the spectrum of a direct sum of operators always the closure of the union of their spectra?*

CHAPTER 12

Norm Topology

99. Metric space of operators. If the distance between two operators A and B is defined to be $\|A - B\|$, the set $\mathbf{B(H)}$ of all operators on a Hilbert space $\mathbf{H}$ becomes a metric space. Some of the standard metric and topological questions about that space have more interesting answers than others. Thus, for instance, it is not more than minimum courtesy to ask whether or not the space is complete. The answer is yes. The proof is the kind of routine analysis every mathematician has to work through at least once in his life; it offers no surprises. The result, incidentally, has been tacitly used already. In Solution 86, the convergence of the series $\sum_{n=0}^{\infty} A^n$ was inferred from the assumption $\|A\| < 1$. The alert reader should have noted that the justification of this inference is in the completeness result just mentioned. (It takes less alertness to notice that the very concept of convergence refers to some topology.)

So much for completeness; what about separability? If the underlying Hilbert space is not separable, it is not to be expected that the operator space is, and, indeed, it is easy to prove that it is not. That leaves one more natural question along these lines.

Problem 99. *If a Hilbert space $\mathbf{H}$ is separable, does it follow that the metric space $\mathbf{B(H)}$ of operators on it is separable?*

100. Continuity of inversion. Soon after the introduction of a topology on an algebraic structure, such as the space of operators on a Hilbert space, it is customary and necessary to ask about the continuity of the pertinent algebraic operations. In the present case it turns out that all the elementary algebraic operations (linear combination, conjugation, multiplication) are continuous in all their variables simultaneously, and the norm of an operator is also a continuous function of its argument. The proofs are boring.

The main algebraic operation not mentioned above is inversion. Since not every operator is invertible, the question of the continuity of inversion makes sense on only a subset of the space of operators.

Problem 100. *The set of invertible operators is open. Is the mapping $A \mapsto A^{-1}$ of that set onto itself continuous?*

The statement that the set of invertible operators is open does not answer all questions about the geometry of that set. It does not say, for instance, whether or not invertible operators can completely surround a singular ($=$non-invertible) one. In more technical language: are there any isolated singular operators? The answer is no; the set of singular operators is (arcwise) connected. Reason: if A is singular, so is tA for all scalars t; the mapping $t \mapsto tA$ is a continuous curve that joins the operator 0 to the operator A. Is the open set of invertible operators connected also? That question is much harder; see Problem 141.

Here is an amusing puzzle about exponentiation and invertibility with a perhaps unexpected solution: determine all operators A and B such that B is invertible and $A^n \to B$ as $n \to \infty$. (A trivial example is given by $A = B = 1$.)

101. Interior of conjugate class. What are the topological properties of similarity? The question includes several precise subquestions. One of them is about conjugate classes. (They are the equivalence classes of the relation of similarity. In more detail: the conjugate class of an operator A is the set of all operators similar to A.)

Conjugate classes do not have to be closed sets. (Example: the standard finite-dimensional similarity theory implies that all the matrices $\begin{pmatrix} 0 & \alpha \\ 0 & 0 \end{pmatrix}$, with $\alpha \neq 0$, are similar; their closure contains $\begin{pmatrix} 0 & 0 \\ 0 & 0 \end{pmatrix}$.) It is, however, quite possible for a conjugate class to be closed; for an example, consider $A = 0$. Conjugate classes do not have to be open (consider $A = 0$ again); can they be?

Problem 101. *Is there an open conjugate class?*

102. Continuity of spectrum. The spectrum (restricted for a moment to operators on just one fixed Hilbert space) is a function whose domain consists of operators and whose range consists of compact sets of complex numbers. It would be quite reasonable to try to define what it means for a function of this kind to be continuous. Is the spectrum continuous? The following example is designed to prove that however the question is interpreted, the answer is always no.

Problem 102. *If $k = 1, 2, 3, \cdots$ and if $k = \infty$, let A_k be the two-sided weighted shift such that $A_k e_n$ is e_{n+1} or $(1/k)e_{n+1}$ according as $n \neq 0$ or $n = 0$. (Put $1/\infty = 0$.) What are the spectra of the operators A_k ($k = 1, 2, 3, \cdots, \infty$)?*

103

103. Semicontinuity of spectrum. The example of Problem 102 shows that there exists an operator with a large spectrum in every neighborhood of which there are operators with relatively small spectra. Could it happen the other way? Is there a small spectrum with arbitrarily near large spectra? The answer turns out to be no; the relevant concept is that of an upper semicontinuous function. (Cf. Problem 21.)

Upper semicontinuity for a set-valued function such as spec has at least two definitions, a metric one and a sequential one. Metric definition: to each open set Λ_0 that includes spec A there corresponds a positive number ε such that if $\|A - B\| < \varepsilon$, then spec $B \subset \Lambda_0$. The sequential definition has to be preceded by an auxiliary concept: if $\{\Lambda_n\}$ is a sequence of sets of complex numbers, then $\limsup_n \Lambda_n$ is the set of all limit points of those sequences $\{\lambda_n\}$ for which $\lambda_n \in \Lambda_n$ for each n. Sequential definition of upper semicontinuity: whenever $A_n \to A$, it follows that

$$\limsup_n \text{spec } A_n \subset \text{spec } A.$$

The two definitions are equivalent. The proofs needed for this assertion are straightforward; in outline form they go like this. If spec is known to be metrically upper semicontinuous, and if $A_n \to A$ and $\lambda \notin$ spec A, then separate λ and spec A by disjoint open sets, and infer that λ cannot belong to $\limsup_n$ spec A_n. If, conversely, spec is known to be sequentially upper semicontinuous and Λ_0 is an open set that includes spec A, then the negation of metric upper semicontinuity leads to a sequence $\{A_n\}$ with $A_n \to A$ that contradicts the assumption.

Problem 103. *Spectrum is upper semicontinuous.*

This is a standard result. One standard reference is [74, p. 167]; another is [112, p. 35]. The semicontinuity of related functions is discussed in [63].

104

104. Continuity of spectral radius. Since the spectrum is upper semicontinuous (Problem 103), so is the spectral radius. That is: to each operator A and to each positive number δ there corresponds a positive number ε such that if $\|A - B\| < \varepsilon$, then $r(B) < r(A) + \delta$. (The proof is immediate from

Problem 103.) The spectrum is not continuous (Problem 102); what about the spectral radius?

Problem 104. *Is it true that to each operator A and to each positive number δ there corresponds a positive number ε such that if $\|A - B\| < \varepsilon$, then $|r(A) - r(B)| < \delta$? Equivalently: if $A_n \to A$, does it follow that $r(A_n) \to r(A)$?*

This is hard. Note that the example in Problem 102 gives no information; in that case the spectral radius is equal to 1 for each term of the sequence and also for the limit.

105. Normal continuity of spectrum. Does it make sense to speak of the continuity (not just semicontinuity) of a set-valued function such as spec? The most convenient answer is in sequential terms. If $\{\Lambda_n\}$ is a sequence of sets of complex numbers, define $\liminf_n \Lambda_n$ to be the set of all limits of those *convergent* sequences $\{\lambda_n\}$ for which $\lambda_n \in \Lambda_n$ for each n. (Recall that $\limsup_n \Lambda_n$ was defined as the set of all limit points of not necessarily convergent sequences of this kind; see Problem 103.) The sequence $\{\Lambda_n\}$ is called convergent ($\Lambda_n \to \Lambda$) if and only if

$$\liminf_n \Lambda_n = \limsup_n \Lambda_n;$$

in that case the common value deserves to be called $\lim_n \Lambda_n$. To say that spec is continuous at a particular operator A means that $A_n \to A$ implies

$$\text{spec } A_n \to \text{spec } A.$$

Since upper semicontinuity is true for all A, a necessary and sufficient condition for continuity at A is that

$$\text{spec } A \subset \liminf_n \text{spec } A_n.$$

The determination of the set of all points of continuity of spec is a nontrivial problem. Some interesting subsets of it were studied by Newburgh in his seminal paper [102]. (Example: if spec A is totally disconnected, then A is a point of continuity of spec. Special case: if A is of finite rank, or, more generally, if A is compact, then A is a point of continuity of spec. Cf. Solution 106.) A general characterization is given in [27].

Another useful question asks whether the restriction of spec to various sets of operators is continuous. (Sample result: the restriction of spec to every commutative set is continuous.) Normal operators frequently behave not only algebraically but also topologically better than others; what happens if spec is restricted to them?

Problem 105. *Is the restriction of spec to the set of normal operators continuous?*

106

106. Quasinilpotent perturbations of spectra. The discontinuity of the spectrum shows that if an operator is perturbed by a "small" summand, the spectrum can undergo a large change. Does the converse behave any better? That is: does a perturbation that produces only a small change in the spectrum have to be small? If small is interpreted in the sense of norm, the answer is easily seen to be no. If, for instance, $A = \begin{pmatrix} \alpha & 0 \\ 0 & \beta \end{pmatrix}$ and $B = \begin{pmatrix} 0 & 1 \\ 0 & 0 \end{pmatrix}$, then spec$(A + B) =$ spec $A = \{\alpha, \beta\}$, but B is not small. The operator A can in fact be perturbed by arbitrarily large summands without changing its spectrum at all: spec$(A + nB) =$ spec A for all n.

There is a sense in which the operator B in the preceding example is small: its spectrum is small. Is that the clue?

Problem 106. *If A and B are operators such that* spec$(A + nB) =$ spec A *for $n = 0, 1, 2, \cdots$, does it follow that B is quasinilpotent?*

The answer is clearly yes in some degenerate cases. If, for instance, A is a scalar, and if spec$(A + nB) =$ spec A for just one non-zero value of n, then spec $B = 0$.

CHAPTER 13

Operator Topologies

107. Topologies for operators. A Hilbert space has two useful topologies (weak and strong); the space of operators on a Hilbert space has several. The metric topology induced by the norm is one of them; to distinguish it from the others, it is usually called the *norm topology* or the *uniform topology*. The next two are natural outgrowths for operators of the strong and weak topologies for vectors. A subbase for the *strong* operator topology is the collection of all sets of the form

$$\{A: \|(A - A_0)f\| < \varepsilon\};$$

correspondingly a base is the collection of all sets of the form

$$\{A: \|(A - A_0)f_i\| < \varepsilon, i = 1, \cdots, k\}.$$

Here k is a positive integer, $f_1, \cdots, f_k$ are vectors, and ε is a positive number. A subbase for the *weak* operator topology is the collection of all sets of the form

$$\{A: |((A - A_0)f, g)| < \varepsilon\},$$

where f and g are vectors and $\varepsilon > 0$; as above (as always) a base is the collection of all finite intersections of such sets. The corresponding concepts of convergence (for sequences and nets) are easy to describe: $A_n \to A$ strongly if and only if $A_n f \to Af$ strongly for each f (i.e., $\|(A_n - A)f\| \to 0$ for each f), and $A_n \to A$ weakly if and only if $A_n f \to Af$ weakly for each f (i.e., $(A_n f, g) \to (Af, g)$ for each f and g). For a slightly different and often very efficient definition of the strong and weak operator topologies see Problems 224 and 225.

The easiest questions to settle are the ones about comparison. The weak topology is smaller (weaker) than the strong topology, and the strong topology is smaller than the norm topology. In other words, every weak open set is a strong open set, and every strong open set is norm open. In still other words: every weak neighborhood of each operator includes a strong neighborhood of that operator, and every strong neighborhood includes a metric neighborhood. Again: norm convergence implies strong convergence, and strong convergence implies weak convergence. These facts are immediate from the definitions. In the presence of uniformity on the unit sphere, the implications are reversible.

Problem 107. *If* $(A_n f, g) \to (Af, g)$ *uniformly for* $\|g\| = 1$, *then* $\|A_n f - Af\| \to 0$, *and if* $\|A_n f - Af\| \to 0$ *uniformly for* $\|f\| = 1$, *then* $\|A_n - A\| \to 0$.

The plethora of topologies yields many questions. A few of those questions are interesting and have useful answers, but even those few are more like routine operator extensions of the corresponding vector questions and answers (in Chapters 2 and 3) than inspiring novelties. In most cases the operator proofs are not ingenious corollaries of the vector facts but repetitions of them, with some extra notational complications. Here are a few sample questions, with curt answers and no proofs.

(1) Is $\mathbf{B}_1(\mathbf{H})$ (the closed unit ball in $\mathbf{B}(\mathbf{H})$) compact? (A topological term such as "compact", with no indication of topology, always refers to the norm.) Answer: yes if and only if the underlying Hilbert space $\mathbf{H}$ is finite-dimensional; cf. Problem 16. (The answer is a special case of the corresponding theorem about arbitrary Banach spaces.)

(2) If $\mathbf{H}$ is separable, is $\mathbf{B}_1(\mathbf{H})$ separable? (No; see Solution 99.) Is it weakly separable?, weakly metrizable?, strongly separable?, strongly metrizable? (Yes, to all parts; cf. Problems 17 and 24.)

(3) Is $\mathbf{B}_1(\mathbf{H})$ weakly compact? Yes; imitate Solution 23. Is $\mathbf{B}_1(\mathbf{H})$ strongly compact? No; see, for instance, Solution 115. Associated with these questions there is a small verbal misfortune. The space $\mathbf{B}(\mathbf{H})$ is a Banach space, and, as such, it has a conjugate space, which induces a topology in $\mathbf{B}(\mathbf{H})$ called, regrettably, the weak topology. This weak topology is defined in terms of the set of *all* bounded linear functionals on $\mathbf{B}(\mathbf{H})$. The definition of the weak topology for operators above uses only some linear functionals on $\mathbf{B}(\mathbf{H})$, namely the ones given by inner products $(A \mapsto (Af, g))$. There is a big difference between the two.

The Banach space weak topology is rarely used in the operator context; when it is, a small additional effort must be made to avoid confusion. In an attempt to do so, one school of thought abandons the grammatically appropriate form of "weak" in favor of a parenthetical WOT, standing for weak operator topology. People who do that use SOT also (strong), but they stop short of NOT (norm). None of these acronyms will ever be seen below; "weak" for operators will always refer to the topology defined by inner products.

60

108. Continuity of norm. In the study of topological algebraic structures **108** (such as the algebra of operators on a Hilbert space, endowed with one of the appropriate operator topologies) the proof that something is continuous is usually dull; the interesting problems arise in proving that something is not continuous. Thus, for instance, it is true that the linear operations on operators ($\alpha A + \beta B$) are continuous in all variables simultaneously, and the proof is a matter of routine. (Readers who have never been through this routine are urged to check it before proceeding.) Here is a related question that is easy but not quite so mechanical.

Problem 108. *Which of the three topologies (uniform, strong, weak) makes the norm (i.e., the function $A \mapsto \|A\|$) continuous?*

109. Semicontinuity of operator norm. The norm is not weakly continuous; **109** for vectors this was discussed in Problem 20, and for operators in Solution 108 (where it was shown that the norm is not even strongly sequentially continuous). Half of continuity is, however, available.

Problem 109. *The norm (of an operator) is weakly lower semicontinuous (and a fortiori strongly lower semicontinuous).*

The assertion means that if $\{A_n\}$ is a net converging weakly to A, then $\|A\| \leq \liminf_n \|A_n\|$. Equivalently: for every $\varepsilon > 0$ there exists an n_0 such that $\|A\| - \|A_n\| \leq \varepsilon$ whenever $n \geq n_0$.

110. Continuity of adjoint. **110**

Problem 110. *Which of the three topologies (uniform, strong, weak) makes the adjoint (i.e., the mapping $A \mapsto A^*$) continuous?*

111. Continuity of multiplication. The most useful, and most recalcitrant, **111** questions concern products. Since a product (unlike the norm and the adjoint) is a function of two variables, a continuity statement about products has a "joint" and a "separate" interpretation. It is usual, when nothing is said to the contrary, to interpret such statements in the "joint" sense, i.e., to interpret them as referring to the mapping that sends an ordered pair $\langle A, B \rangle$ onto the product AB.

Problem 111. *Multiplication is continuous with respect to the uniform topology and discontinuous with respect to the strong and weak topologies.*

The proof is easy, but the counterexamples are hard; the quickest ones depend on unfair trickery.

112

112. Separate continuity of multiplication. Although multiplication is not jointly continuous with respect to either the strong topology or the weak, it is separately continuous in each of its arguments with respect to both topologies. A slightly more precise formulation runs as follows.

Problem 112. *Each of the mappings* $A \mapsto AB$ *(for fixed B) and* $B \mapsto AB$ *(for fixed A) is both strongly and weakly continuous.*

113

113. Sequential continuity of multiplication. Separate continuity (strong and weak) of multiplication is a feeble substitute for joint continuity; another feeble (but sometimes usable) substitute is joint continuity in the sequential sense.

Problem 113. (a) *If* $\{A_n\}$ *and* $\{B_n\}$ *are sequences of operators that strongly converge to A and B, respectively, then* $A_n B_n \to AB$ *strongly.* (b) *Does the assertion remain true if "strongly" is replaced by "weakly" in both hypothesis and conclusion?*

114

114. Weak sequential continuity of squaring. There is a familiar argument that can be used frequently to show that if the linear operations and squaring are continuous, then multiplication is continuous also. Since the argument depends on the identity $ab = \frac{1}{4}((a + b)^2 - (a - b)^2)$, it can be used only when multiplication is commutative. Operator multiplication is not commutative; does that mean that there is a hope for the weak sequential continuity of squaring, at least for some operators?

Problem 114. *For which operators A (on an infinite-dimensional Hilbert space) is squaring weakly sequentially continuous? In other words, for which A is it true that if* $A_n \to A$ *weakly, then* $A_n^2 \to A^2$ *weakly (n = 1, 2, 3, ···)?*

115

115. Weak convergence of projections. Are the weak and the strong operator topologies the same? The answer is no, and proofs of that answer can be deduced from much of what precedes; note, for instance, the different ways that sequential continuity of multiplication behaves. In one respect, however, weak and strong are alike: if a net $\{P_n\}$ of projections converges weakly to a projection P, then it converges strongly to P. Proof: for each f,

$$\|P_n f\|^2 = (P_n f, f) \to (Pf, f) = \|Pf\|^2,$$

and therefore Problem 20 is applicable. How much does the restriction to projections simplify matters?

Problem 115. *Is every weakly convergent sequence of projections strongly convergent?*

CHAPTER 14

Strong Operator Topology

116. Strong normal continuity of adjoint. Even though the adjoint is not strongly continuous, it has an important continuous part.

    Problem 116. *The restriction of the adjoint to the set of normal operators is strongly continuous.*

117. Strong bounded continuity of multiplication. The crux of the proof that multiplication is strongly sequentially continuous (Solution 113) is boundedness. That is: if $\{A_n\}$ and $\{B_n\}$ are nets that converge strongly to A and B, respectively, and if $\{\|A_n\|\}$ is bounded, then $\{A_n B_n\}$ converges strongly to AB. Is this result symmetric with respect to the interchange of right and left?

    Problem 117. *If $\{A_n\}$ and $\{B_n\}$ are nets that converge strongly to 0, and if $\{\|B_n\|\}$ is bounded, does it follow that $\{A_n B_n\}$ converges strongly to 0?*

118. Strong operator versus weak vector convergence.

    Problem 118. *If $\{f_n\}$ is a sequence of vectors and $\{A_n\}$ is a sequence of operators such that $f_n \to f$ weakly and $A_n \to A$ strongly, does it follow that $A_n f_n \to Af$ weakly?*

119. Strong semicontinuity of spectrum. The spectrum of an operator varies upper semicontinuously (Problem 103). If, that is, an operator is replaced by one that is near it in the norm topology, then the spectrum can increase only a little. What if the strong topology is used in place of the norm topology?

Problem 119. *Is the spectrum strongly upper semicontinuous? What can be said about the spectral radius?*

120 **120. Increasing sequences of Hermitian operators.** A bounded increasing sequence of Hermitian operators is weakly convergent (to a necessarily Hermitian operator). To see this, suppose that $\{A_n\}$ is an increasing sequence of Hermitian operators (i.e., $(A_n f, f) \leq (A_{n+1} f, f)$ for all n and all f), bounded by α (i.e., $(A_n f, f) \leq \alpha \|f\|^2$ for all n and all f). If $\psi_n(f) = (A_n f, f)$, then each ψ_n is a quadratic form. The assumptions imply that the sequence $\{\psi_n\}$ is convergent and hence (Solution 1) that the limit ψ is a quadratic form. It follows that $\psi(f) = (Af, f)$ for some (necessarily Hermitian) operator A; polarization justifies the conclusion that $A_n \to A$ (weakly).

Does the same conclusion follow with respect to the strong and the uniform topologies?

Problem 120. *Is a bounded increasing sequence of Hermitian operators necessarily strongly convergent? uniformly convergent?*

121 **121. Square roots.** The assertion that a positive operator has a unique positive square root is an easy consequence of the spectral theorem. In some approaches to spectral theory, however, the existence of square roots is proved first, and the spectral theorem is based on that result. The following assertion shows how to get square roots without the spectral theorem.

Problem 121. *If A is an operator such that $0 \leq A \leq 1$, and if a sequence $\{B_n\}$ is defined recursively by the equations*

$$B_0 = 0 \quad and \quad B_{n+1} = \tfrac{1}{2}((1 - A) + B_n^2), \qquad n = 0, 1, 2, \cdots,$$

then the sequence $\{B_n\}$ is strongly convergent. If $\lim_n B_n = B$, then $(1 - B)^2 = A$.

122 **122. Infimum of two projections.** If E and F are projections with ranges **M** and **N**, then it is sometimes easy and sometimes hard to find, in terms of E and F, the projections onto various geometric constructs formed with **M** and **N**. Things are likely to be easy if E and F commute. Thus, for instance, if $\mathbf{M} \subset \mathbf{N}$, then it is easy to find the projection with range $\mathbf{N} \cap \mathbf{M}^\perp$, and if $\mathbf{M} \perp \mathbf{N}$, then it is easy to find the projection with range $\mathbf{M} \vee \mathbf{N}$. In the absence of such special assumptions, the problems become more interesting.

Problem 122. *If E and F are projections with ranges **M** and **N**, find the projection $E \wedge F$ with range $\mathbf{M} \cap \mathbf{N}$.*

The problem is to find an "expression" for the projection described. Although most mathematicians would read the statement of such a problem with sympathetic understanding, it must be admitted that rigorously speaking it does not really mean anything. The most obvious way to make it precise is to describe certain classes of operators by the requirement that they be closed under some familiar algebraic and topological operations, and then try to prove that whenever E and F belong to such a class, then so does $E \wedge F$. The most famous and useful classes pertinent here are the von Neumann algebras (called "rings of operators" by von Neumann). A *von Neumann algebra* is an algebra of operators (i.e., a collection closed under addition and multiplication, and closed under multiplication by arbitrary scalars), self-adjoint (i.e., closed under adjunction), containing 1, and strongly closed (i.e., closed with respect to the strong operator topology). For von Neumann algebras, then, the problem is this: prove that if a von Neumann algebra contains two projections E and F, then it contains $E \wedge F$.

Reference: [150, vol. 2, p. 55].

CHAPTER 15

Partial Isometries

123. Spectral mapping theorem for normal operators. Normal operators constitute the most important tractable class of operators known; the most important statement about them is the spectral theorem. Students of operator theory generally agree that the finite-dimensional version of the spectral theorem has to do with diagonal forms. (Every finite normal matrix is unitarily equivalent to a diagonal one.) The general version, applicable to infinite-dimensional spaces, does not have a universally accepted formulation. Sometimes bounded operator representations of function algebras play the central role, and sometimes Stieltjes integrals with unorthodox multiplicative properties. There is a short, simple, and powerful statement that does not attain maximal generality (it applies to only one operator at a time, not to algebras of operators), but that does have all classical formulations of the spectral theorem as easy corollaries, and that has the advantage of being a straightforward generalization of the familiar statement about diagonal forms. That statement will be called the spectral theorem in what follows; it says that *every normal operator is unitarily equivalent to a multiplication*. The statement can be proved by exactly the same techniques as are usually needed for the spectral theorem; see [56], [40, pp. 911–912].

The multiplication version of the spectral theorem has a technical drawback: the measures that it uses may fail to be σ-finite. This is not a tragedy, for two reasons. In the first place, the assumption of σ-finiteness in the treatment of multiplications is a matter of convenience, not of necessity (see [131]). In the second place, non-σ-finite measures need to be considered only when the underlying Hilbert space is not separable; the pathology of measures accompanies the pathology of operators. In the sequel when reference is made to the spectral theorem, the reader may choose one of two courses: treat the general case and proceed with the caution it requires, or restrict attention to the separable case and proceed with the ease that the loss of generality permits.

In some contexts some authors choose to avoid a proof that uses the spectral theorem even if the alternative is longer and more involved. This sort of ritual circumlocution is common to many parts of mathematics; it is the fate of many big theorems to be more honored in evasion than in use. The reason is not just mathematical mischievousness. Often a long but "elementary" proof gives more insight, and leads to more fruitful generalizations, than a short proof whose brevity is made possible by a powerful but overly specialized tool.

This is not to say that use of the spectral theorem is to be avoided at all costs. Powerful general theorems exist to be used, and their willful avoidance can lose insight at least as often as gain it. Thus, for example, the spectral theorem yields an immediate and perspicuous proof that every positive operator has a positive square root (because every positive measurable function has one); the approximation trickery of Problem 121 is fun, and has its uses, but it is not nearly so transparent. A non-spectral treatment of a related property of square roots is in Problem 124. For another example, consider the assertion that a Hermitian operator whose spectrum consists of the two numbers 0 and 1 is a projection. To prove it, let A be the operator, and write $B = A - A^2$. Clearly B is Hermitian, and, by the spectral mapping theorem, spec $B = \{0\}$. This implies that $\|B\| = r(B) = 0$ and hence that $B = 0$. (It is true for all normal operators that the norm is equal to the spectral radius, but for Hermitian operators it is completely elementary; see [50, p. 55].) Compare this with the proof via the spectral theorem: if φ is a function whose range consists of the two numbers 0 and 1, then $\varphi^2 = \varphi$. For a final example, try to prove, without using the spectral theorem, that every normal operator with a real spectrum (i.e., with spectrum included in the real line) is Hermitian.

The spectral theorem makes possible a clear and efficient description of the so-called *functional calculus*. If A is a normal operator and if F is a bounded Borel measurable function on spec A, then the functional calculus yields an operator $F(A)$. To define $F(A)$ represent A as a multiplication, with multiplier φ, say, on a measure space X; the operator $F(A)$ is then the multiplication induced by the composite function $F \circ \varphi$. In order to be sure that this makes sense, it is necessary to know that φ maps almost every point of X into spec A, i.e., that if the domain of φ is altered by, at worst, a set of measure zero, then the range of φ comes to be included in its essential range. The proof goes as follows. By definition, every point in the complement of spec A has a neighborhood whose inverse image under φ has measure zero. Since the plane is a Lindelöf space, it follows that the complement of spec A is covered by a countable collection of neighborhoods with that property, and hence that the inverse image of the entire complement of spec A has measure zero.

67

The mapping $F \mapsto F(A)$ has many pleasant properties. Its principal property is that it is an algebraic homomorphism that preserves conjugation also (i.e., $F^*(A) = (F(A))^*$); it follows, for instance, that if $F(\lambda) = |\lambda|^2$, then $F(A) = A^*A$. The functions F that occur in the applications of the functional calculus are not always continuous (e.g., characteristic functions of Borel sets are of importance), but continuous functions are sometimes easier to handle. The problem that follows is a spectral mapping theorem; it is very special in that it refers to normal operators only, but it is very general in that it allows all continuous functions.

Problem 123. *If A is a normal operator and if F is a continuous function on spec A, then spec $F(A) = F(\text{spec } A)$.*

Something like $F(A)$ might seem to make sense sometimes even for nonnormal A's, but the result is not likely to remain true. Suppose, for instance, that $F(\lambda) = \lambda^*\lambda (= |\lambda|^2)$, and define $F(A)$, for every operator A, as A^*A. There is no hope for the statement spec $F(A) = F(\text{spec } A)$; for a counterexample, contemplate the unilateral shift.

124 **124. Decreasing squares.** If A is a positive contraction, i.e., $0 \leq A \leq 1$, then $A^2 \leq A$; if, conversely, A is a Hermitian operator such that $A^2 \leq A$, then $0 \leq A \leq 1$. Proof: if φ is a measurable function and $0 \leq \varphi \leq 1$, then $\varphi^2 \leq \varphi$; if, conversely, $\varphi > 1$ on a set of positive measure, then it is false that $\varphi^2 \leq \varphi$. This is a typical use of the spectral theorem to prove an algebraic fact about Hermitian operators; some people have found the possibility of getting along without the spectral theorem a little more than commonly elusive in this case.

Problem 124. *Prove without using the spectral theorem that, for Hermitian operators, $0 \leq A \leq 1$ if and only if $A^2 \leq A$.*

125 **125. Polynomially diagonal operators.** If an operator A is diagonal (see Problem 61), then all the powers of A are diagonal, and so is any sum of such powers. Is the converse true? (Note that if A is diagonal, then it is normal, and, consequently, every function of it is normal.)

Problem 125. *If A is a normal operator on a separable Hilbert space, and if $1 + A + \cdots + A^n$ is diagonal for some positive integer n, does it follow that A is diagonal?*

126 **126. Continuity of the functional calculus.** A functional calculus is a mapping from functions and operators to operators. For a fixed value of the second argument (the operator), it has good properties as a function of the first argument (the function); see, for instance, Problem 123. What properties can it have as a function of the second argument?

For a special example, fix a polynomial p and consider the mapping $X \mapsto p(X)$, defined for all operators X. Assertion: the mapping is norm continuous. Proof: obvious, since addition and multiplication are norm continuous. Does the functional calculus for normal operators remain continuous in this sense if polynomials are replaced by more general functions? The answer is sometimes no. If, for instance, F is the characteristic function of $(0, \infty)$ (considered as a subset of the complex plane), and if A_n is the scalar $1/n$, then $A_n \to 0$, but $F(A_n)$ $(=1)$ does not converge to $F(0)$ $(=0)$. What is the source of the trouble? Is it just that the function F in this example is not continuous?

Problem 126. *If F is a continuous function on the complex plane, is the mapping $A \mapsto F(A)$, defined for all normal operators, continuous?*

127. Partial isometries. An *isometry* is a linear transformation U (from a **127** Hilbert space into itself, or from one Hilbert space into another) such that $\|Uf\| = \|f\|$ for all f. An isometry is a distance-preserving transformation: $\|Uf - Ug\| = \|f - g\|$ for all f and g. A necessary and sufficient condition that a linear transformation U be an isometry is that $U^*U = 1$. Indeed: the conditions (1) $\|Uf\|^2 = \|f\|^2$, (2) $(U^*Uf, f) = (f, f)$, (3) $(U^*Uf, g) = (f, g)$, and (4) $U^*U = 1$ are mutually equivalent. (To pass from (2) to (3), polarize.) Caution: the conditions $U^*U = 1$ and $UU^* = 1$ are not equivalent. The latter condition is satisfied in case U^* is an isometry; in that case U is called a *co-isometry*.

It is sometimes convenient to consider linear transformations U that act isometrically on a subset (usually a linear manifold, but not necessarily a subspace) of a Hilbert space; this just means that $\|Uf\| = \|f\|$ for all f in that subset. A *partial isometry* is a linear transformation that is isometric on the orthogonal complement of its kernel. There are two large classes of examples of partial isometries that are in a sense opposite extreme cases; they are the isometries (and, in particular, the unitary operators), and the projections. The definition of partial isometries is deceptively simple, and these examples continue the deception; the structure of partial isometries can be quite complicated. In any case, however, it is easy to verify that a partial isometry U is bounded, in fact if U is not 0, then $\|U\| = 1$.

The orthogonal complement of the kernel of a partial isometry is frequently called its *initial space*. The initial space of a partial isometry U turns out to be equal to the set of all those vectors f for which $\|Uf\| = \|f\|$. (What needs proof is that if $\|Uf\| = \|f\|$, then $f \perp \ker U$. Write $f = g + h$, with $g \in \ker U$ and $h \perp \ker U$; then $\|f\| = \|Uf\| = \|Ug + Uh\| = \|Uh\| = \|h\|$; since $\|f\|^2 = \|g\|^2 + \|h\|^2$, it follows that $g = 0$.) The range of a partial isometry is equal to the image of the initial space and is necessarily closed. (Since U is isometric on the initial space, the image is a complete metric space.) For partial isometries, the range is sometimes called the *final space*.

Problem 127. *A bounded linear transformation U is a partial isometry if and only if U^*U is a projection.*

Corollary 1. *If U is a partial isometry, then the initial space of U is the range of U^*U.*

Corollary 2. *The adjoint of a partial isometry is a partial isometry, with initial space and final space interchanged.*

Corollary 3. *A bounded linear transformation U is a partial isometry if and only if $U = UU^*U$.*

128

128. Maximal partial isometries. It is natural to define a (partial) order for partial isometries as follows: if U and V are partial isometries, write $U \leq V$ in case V agrees with U on the initial space of U. This implies that the initial space of U is included in the initial space of V. (Cf. the characterization of initial spaces given in Problem 127.) It follows that if $U \leq V$ with respect to the present order, then $U^*U \leq V^*V$ with respect to the usual order for operators. (The "usual" order, usually considered for Hermitian operators only, is the one according to which $A \leq B$ if and only if $(Af, f) \leq (Bf, f)$ for all f. Note that $U^*U \leq V^*V$ in this sense is equivalent to $\|Uf\| \leq \|Vf\|$ for all f.) The converse is not true; if all that is known about the partial isometries U and V is that $U^*U \leq V^*V$, then, to be sure, the initial space of U is included in the initial space of V, but it cannot be concluded that U and V necessarily agree on the smaller initial space.

If $U^*U = 1$, i.e., if U is an isometry, then the only partial isometry that can dominate U is U itself: an isometry is a maximal partial isometry. Are there any other maximal partial isometries? One way to get the answer is to observe that if $U \leq V$, then the final space of U (i.e., the initial space of U^*) is included in the final space of V (the initial space of V^*), and, moreover, V^* agrees with U^* on the initial space of U^*. In other words, if $U \leq V$, then $U^* \leq V^*$, and hence, in particular, $UU^* \leq VV^*$. This implies that if $UU^* = 1$, i.e., if U is a co-isometry, then, again, U is maximal. If a partial isometry U is neither an isometry nor a co-isometry, then both U and U^* have non-zero kernels. In that case it is easy to enlarge U to a partial isometry that maps a prescribed unit vector in ker U onto a prescribed unit vector in ker U^* (and, of course, agrees with U on ker$^\perp U$). Conclusion: a partial isometry is maximal if and only if either it or its adjoint is an isometry.

The easy way to be a maximal partial isometry is to be unitary. If U is unitary on **H** and if **M** is a subspace of **H**, then a necessary and sufficient condition that **M** reduce U is that $U\mathbf{M} = \mathbf{M}$. If U is merely a partial isometry, then it can happen that $U\mathbf{M} = \mathbf{M}$ but **M** does not reduce U, and it can happen that **M** reduces U but $U\mathbf{M} \neq \mathbf{M}$. What if U is a maximal partial isometry?

Problem 128. *Discover the implication relations between the statements* "$UM = M$" *and* "M *reduces* U" *when* U *is a maximal partial isometry.*

129. Closure and connectedness of partial isometries. Some statements about **129** partial isometries are slightly awkward just because 0 must be counted as one of them. The operator 0 is an isolated point of the set of partial isometries; it is the only partial isometry in the interior of the unit ball. For this reason, for instance, the set of all partial isometries is obviously not connected. What about the partial isometries on the boundary of the unit ball?

Problem 129. *The set of all non-zero partial isometries is closed but not connected (with respect to the norm topology of operators).*

130. Rank, co-rank, and nullity. If U is a partial isometry, write $\rho(U) =$ **130** dim ran U, $\rho'(U) = $ dim ran$^{\perp} U$, and $v(U) = $ dim ker U. (That U is a partial isometry is not really important in these definitions; similar definitions can be made for arbitrary operators.) These three cardinal numbers, called the *rank*, the *co-rank*, and the *nullity* of U, respectively, are not completely independent of one another; they are such that both $\rho + \rho'$ and $\rho + v$ are equal to the dimension of the underlying Hilbert space. (Caution: subtraction of infinite cardinal numbers is slippery; it does not follow that $\rho' = v$.) It is easy to see that if ρ, ρ', and v are any three cardinal numbers such that $\rho + \rho' = \rho + v$, then there exist partial isometries with rank ρ, co-rank ρ', and nullity v. (Symmetry demands the consideration of $v'(U) = $ dim ker$^{\perp} U$, the *co-nullity* of U, but there is no point in it; since U is isometric on ker$^{\perp} U$ it follows that $v' = \rho$.)

Recall that if U is a partial isometry, then so is U^*; the initial space of U^* is the final space of U, and vice versa. It follows that $v(U^*) = \rho'(U)$ and $\rho'(U^*) = v(U)$.

One reason that the functions ρ, ρ', and v are useful is that they are continuous. To interpret this statement, use the norm topology for the space **P** of partial isometries (on a fixed Hilbert space), and use the discrete topology for cardinal numbers. With this explanation the meaning of the continuity assertion becomes unambiguous: if U is sufficiently near to V, then U and V have the same rank, the same co-rank, and the same nullity. The following assertion is a precise quantitative formulation of the result.

Problem 130. *If* U *and* V *are partial isometries such that* $\|U - V\| < 1$, *then* $\rho(U) = \rho(V)$, $\rho'(U) = \rho'(V)$, *and* $v(U) = v(V)$.

For each fixed ρ, ρ', and v let $\mathbf{P}(\rho, \rho', v)$ be the set of partial isometries (on a fixed Hilbert space) with rank ρ, co-rank ρ', and nullity v. Clearly the sets of the form $\mathbf{P}(\rho, \rho', v)$ constitute a partition of the space **P** of all partial isometries; it is a consequence of the statement of Problem 130 that each set

$P(\rho, \rho', v)$ is both open and closed. It follows that the set of all isometries ($v = 0$) is both open and closed, and so is the set of all unitary operators ($\rho' = v = 0$).

131

131. Components of the space of partial isometries. If φ is a measurable function on a measure space, such that $|\varphi| = 1$ almost everywhere, then there exists a measurable real-valued function θ on that space such that $\varphi = e^{i\theta}$ almost everywhere. This is easy to prove. What it essentially says is that a measurable function always has a measurable logarithm. The reason is that the exponential function has a Borel measurable inverse (in fact many of them) on the complement of the origin in the complex plane. (Choose a continuous logarithm on the complement of the negative real axis, and extend it by requiring one-sided continuity on, say, the upper half plane.)

In the language of the functional calculus, the result of the preceding paragraph can be expressed as follows: if U is a unitary operator, then there exists a Hermitian operator A such that $U = e^{iA}$. If $U_t = e^{itA}$, $0 \leq t \leq 1$, then $t \mapsto U_t$ is a continuous curve of unitary operators joining 1 ($= U_0$) to U ($= U_1$). Conclusion: the set of all unitary operators is arcwise connected. In the notation of Problem 130, the open-closed set $P(\rho, 0, 0)$ (on a Hilbert space of dimension ρ) is connected; it is a component of the set P of all partial isometries. Question: what are the other components? Answer: the sets of the form $P(\rho, \rho', v)$.

Problem 131 *Each pair of partial isometries (on the same Hilbert space) with the same rank, co-rank, and nullity, can be joined by a continuous curve of partial isometries with the same rank, co-rank, and nullity.*

132

132. Unitary equivalence for partial isometries. If A is a *contraction* (that means $\|A\| \leq 1$), then $1 - AA^*$ is positive. It follows that there exists a unique positive operator whose square is $1 - AA^*$; call it A'. Assertion: the operator matrix

$$M(A) = \begin{pmatrix} A & A' \\ 0 & 0 \end{pmatrix}$$

is a partial isometry. Proof (via Problem 127): check that $MM^*M = M$. Consequence: every contraction can be extended to a partial isometry.

Problem 132. *If A and B are unitarily equivalent contractions, then $M(A)$ and $M(B)$ are unitarily equivalent, and conversely.*

There are many ways that a possibly "bad" operator A can be used to manufacture a "good" one. Samples: $A + A^*$ and

$$\begin{pmatrix} 0 & A \\ A^* & 0 \end{pmatrix}.$$

72

None of these ways yields sufficiently many usable unitary invariants for A. It is usually easy to prove that if A and B are unitarily equivalent, then so are the various constructs in which they appear. It is, however, usually false that if the constructs are unitarily equivalent, then the original operators themselves are. The chief interest of the assertion of Problem 132 is that, for the special partial isometry construct it deals with, the converse happens to be true.

The result is that the unitary equivalence problem for an apparently very small class of operators (partial isometries) is equivalent to the problem for the much larger class of invertible contractions. The unitary equivalence problem for invertible contractions is, in turn, trivially equivalent to the unitary equivalence problem for arbitrary operators. The reason is that by a translation $(A \mapsto A + \alpha)$ and a change of scale $(A \mapsto \beta A)$ every operator becomes an invertible contraction, and translations and changes of scale do not affect unitary equivalence. The end product of all this is a reduction of the general unitary equivalence problem to the special case of partial isometries.

133. Spectrum of a partial isometry. What conditions must a set of complex   **133**
numbers satisfy in order that it be the spectrum of some partial isometry? Since a partial isometry is a contraction, its spectrum is necessarily a subset of the closed unit disc. If the spectrum of a partial isometry does not contain the origin, i.e., if a partial isometry is invertible, then it is unitary, and, therefore, its spectrum is a subset of the unit circle (perimeter). Since every non-empty compact subset of the unit circle is the spectrum of some unitary operator (cf. Problem 63), the problem of characterizing the spectra of invertible partial isometries is solved. What about the non-invertible ones?

Problem 133. *What conditions must a set of complex numbers satisfy in order that it be the spectrum of some non-unitary partial isometry?*

CHAPTER 16

Polar Decomposition

134   **134. Polar decomposition.** Every complex number is the product of a non-negative number and a number of modulus 1; except for the number 0, this polar decomposition is unique. The generalization to finite matrices says that every complex matrix is the product of a positive matrix and a unitary one. If the given matrix is invertible, and if the order of the factors is specified (UP or PU), then, once again, this polar decomposition is unique. It is possible to get a satisfactory uniqueness theorem for every matrix, but only at the expense of changing the kind of factors admitted; this is a point at which partial isometries can profitably enter the study of finite-dimensional vector spaces. In the infinite-dimensional case, partial isometries are unavoidable. It is not true that every operator on a Hilbert space is equal to a product UP, with U unitary and P positive, and it does not become true even if U is required to be merely isometric. (The construction of concrete counterexamples may not be obvious now, but it will soon be an easy by-product of the general theory.) The correct statements are just as easy for transformations between different spaces as for operators on one space.

> **Problem 134.** *If A is a bounded linear transformation from a Hilbert space $\mathbf{H}$ to a Hilbert space $\mathbf{K}$, then there exists a partial isometry U (from $\mathbf{H}$ to $\mathbf{K}$) and there exists a positive operator P (on $\mathbf{H}$) such that $A = UP$. The transformations U and P can be found so that $\ker U = \ker P$, and this additional condition uniquely determines them.*

The representation of A as the product of the unique U and P satisfying the stated conditions is called the *polar decomposition* of A, or, more accurately, the right-handed polar decomposition of A. The corresponding left-handed theory ($A = PU$) follows by a systematic exploitation of adjoints.

74

Corollary 1. *If $A = UP$ is the polar decomposition of A, then $U^*A = P$.*

Corollary 2. *If $A = UP$ is the polar decomposition of A, then a necessary and sufficient condition that U be an isometry is that A be one-to-one, and a necessary and sufficient condition that U be a co-isometry is that the range of A be dense.*

135. Maximal polar representation. **135**

Problem 135. *Every bounded linear transformation is the product of a maximal partial isometry and a positive operator.*

136. Extreme points. **136**
The closed unit ball in the space of operators is convex. For every interesting convex set, it is of interest to determine the extreme points.

Problem 136. *What are the extreme points of the closed unit ball in the space of operators on a Hilbert space?*

137. Quasinormal operators. **137**
The condition of normality can be weakened in various ways; the most elementary of these leads to the concept of quasinormality. An operator A is called *quasinormal* if A commutes with A^*A. It is clear that every normal operator is quasinormal. The converse is obviously false. If, for instance, A is an isometry, then $A^*A = 1$ and therefore A commutes with A^*A, but if A is not unitary, then A is not normal. (For a concrete example consider the unilateral shift.)

Problem 137. *An operator with polar decomposition UP is quasinormal if and only if $UP = PU$.*

Quasinormal operators (under another name) were first introduced and studied in [18].

138. Mixed Schwarz inequality. **138**
If A is a positive operator, then (Af, g) defines an inner product (not necessarily strictly positive); it follows that A satisfies the Schwarz-like inequality

$$|(Af, g)|^2 \leqq (Af, f) \cdot (Ag, g).$$

(To feel comfortable about this relation it helps to recall that the positiveness of A implies $(\sqrt{A})^2 = A$.) If A is not positive, then the inner product expression (Af, g) still defines something (sesquilinear, yes, positive and symmetric, probably no), but since the equation $(\sqrt{A})^2 = A$ is not available, it is not clear what, if any, Schwarz-like inequality prevails. A possible guess is to replace A on the right (majorant) side by $\sqrt{A^*A}$.

(Reason: if z is a complex number, then $\sqrt{z^*z} = |z|$.) Objection: why $\sqrt{A^*A}$? Why not $\sqrt{AA^*}$?

Problem 138. *Is it true for every operator A that $|(Af, g)|^2 \leqq (\sqrt{A^*Af}, f) \cdot (\sqrt{A^*Ag}, g)$? What if $\sqrt{AA^*}$ is used in place of $\sqrt{A^*A}$? What if they are both used, one in each factor?*

139

139. Quasinormal weighted shifts. The unilateral shift is an example of a quasinormal operator that is not normal. There is a tempting generalization nearby that is frequently a rich source of illuminating examples.

Problem 139. *Which weighted shifts are quasinormal?*

140

140. Density of invertible operators. It sometimes happens that a theorem is easy to prove for invertible operators but elusive in the general case. This makes it useful to know that every finite (square) matrix is the limit of invertible matrices. In the infinite-dimensional case the approximation technique works, with no difficulty, for normal operators. (Invoke the spectral theorem to represent the given operator as a multiplication, and, by changing the small values of the multiplier, approximate it by operators that are bounded from below.) If, however, the space is infinite-dimensional and the operator is not normal, then there is trouble.

Problem 140. *The set of all operators that have either a left or a right inverse is dense, but the set of all operators that have both a left and a right inverse (i.e., the set of all invertible operators) is not. What about the set of left-invertible operators?*

141

141. Connectedness of invertible operators.

Problem 141. *The set of all invertible operators is connected.*

CHAPTER 17

Unilateral Shift

142. Reducing subspaces of normal operators. One of the principal achievements of the spectral theorem is to reduce the study of a normal operator to subspaces with various desirable properties. The following assertion is one way to say that the spectral theorem provides many reducing subspaces.

> **Problem 142.** *If A is a normal operator on an infinite-dimensional Hilbert space* **H**, *then* **H** *is the direct sum of a countably infinite collection of subspaces that reduce A, all with the same infinite dimension.*

143. Products of symmetries. A *symmetry* is a unitary involution, i.e., an operator Q such that $Q^*Q = QQ^* = Q^2 = 1$. It may be pertinent to recall that if an operator possesses any two of the properties "unitary", "involutory", and "Hermitian", then it possesses the third; the proof is completely elementary algebraic manipulation.

> **Problem 143.** *Discuss the assertion: every unitary operator is the product of a finite number of symmetries.*

144. Unilateral shift versus normal operators. The main point of Problem 142 is to help solve Problem 143 (and, incidentally, to provide a non-trivial application of the spectral theorem). The main point of Problem 143 is to emphasize the role of certain shift operators. Shifts (including the simple unilateral and bilateral ones introduced before) are a basic tool in operator theory. The unilateral shift, in particular, has many curious properties, both algebraic and analytic. The techniques for discovering and proving these properties are frequently valuable even when the properties themselves have no visible immediate application. Here are three sample questions.

Problem 144. (a) *Is the unilateral shift the product of a finite number of normal operators?* (b) *What is the norm of the real part of the unilateral shift?* (c) *How far is the unilateral shift from the set of normal operators?*

The last question takes seriously the informal question: "How far does the unilateral shift miss being normal?" The question can be asked for every operator and the answer is a unitary invariant that may occasionally be useful.

145. Square root of shift.

Problem 145. *Does the unilateral shift have a square root? In other words, if U is the unilateral shift, does there exist an operator V such that $V^2 = U$?*

146. Commutant of the bilateral shift. The *commutant* of an operator (or of a set of operators) is the set of all operators that commute with it (or with each operator in the set). The commutant is one of the most useful things to know about an operator. One of the most important purposes of the so-called multiplicity theory is to discuss the commutants of normal operators. In some special cases the determination of the commutant is accessible by relatively elementary methods; a case in point is the bilateral shift.

The bilateral shift W can be viewed as multiplication by e_1 on $\mathbf{L}^2$ of the unit circle (cf. Problem 84). Here $e_n(z) = z^n$ $(n = 0, \pm 1, \pm 2, \cdots)$ whenever $|z| = 1$, and $\mathbf{L}^2$ is formed with normalized Lebesgue measure.

Problem 146. *The commutant of the bilateral shift is the set of all multiplications.*

Corollary. *Each reducing subspace of the bilateral shift is determined by a Borel subset M of the circle as the set of all functions (in $\mathbf{L}^2$) that vanish outside M.*

Both the main statement and the corollary have natural generalizations that can be bought at the same price. The generalizations are obtained by replacing the unit circle by an arbitrary bounded Borel set X in the complex plane and replacing Lebesgue measure by an arbitrary finite Borel measure in X. The generalization of the bilateral shift is the multiplication induced by e_1 (where $e_1(z) = z$ for all z in X).

147. Commutant of the unilateral shift. The unilateral shift is the restriction of the bilateral shift to $\mathbf{H}^2$. If the bilateral shift is regarded as a multiplication, then its commutant can be described as the set of all multiplications on the same $\mathbf{L}^2$ (Problem 146). The wording suggests a superficially plausible conjecture: perhaps the commutant of the unilateral shift consists of the restrictions to $\mathbf{H}^2$ of all multiplications. On second thought this is absurd: $\mathbf{H}^2$ need not be invariant under a multiplication, and, consequently, the restriction of a multiplication to $\mathbf{H}^2$ is not necessarily an operator on $\mathbf{H}^2$. If, however, the multiplier itself is in $\mathbf{H}^2$ (and hence in $\mathbf{H}^\infty$), then $\mathbf{H}^2$ is invariant under the induced multiplication (cf. Problem 34), and the conjecture makes sense.

Problem 147. *The commutant of the unilateral shift is the set of all restrictions to $\mathbf{H}^2$ of multiplications by multipliers in $\mathbf{H}^\infty$.*

Corollary. *The unilateral shift is irreducible, in the sense that its only reducing subspaces are 0 and $\mathbf{H}^2$.*

Just as for the bilateral shift, the main statement has a natural generalization. Replace the unit circle by an arbitrary bounded Borel subset X of the complex plane, and replace Lebesgue measure by an arbitrary finite Borel measure μ in X. The generalization of $\mathbf{H}^2$, sometimes denoted by $\mathbf{H}^2(\mu)$, is the span in $\mathbf{L}^2(\mu)$ of the functions e_n, $n = 0, 1, 2, \cdots$, where $e_n(z) = z^n$ for all z in X. The generalization of the unilateral shift is the restriction to $\mathbf{H}^2(\mu)$ of the multiplication induced by e_1.

The corollary does not generalize so smoothly as the main statement. The trouble is that the structure of $\mathbf{H}^2(\mu)$ within $\mathbf{L}^2(\mu)$ depends strongly on X and μ; it can, for instance, happen that $\mathbf{H}^2(\mu) = \mathbf{L}^2(\mu)$.

The characterization of the commutant of the unilateral shift yields a curious alternative proof of, and corresponding insight into, the assertion that U has no square root (Solution 145). Indeed, if $V^2 = U$, then V commutes with U, and therefore V is the restriction to $\mathbf{H}^2$ of the multiplication induced by a function φ in $\mathbf{H}^\infty$. Apply V^2 to e_0, apply U to e_0, and infer that $(\varphi(z))^2 = z$ almost everywhere. This implies that $(\tilde\varphi(z))^2 = z$ in the unit disc (see Solution 42), i.e., that the function $\tilde e_1$ has an analytic square root; the contradiction has arrived.

148. Commutant of the unilateral shift as limit.   148

Problem 148. *Every operator that commutes with the unilateral shift is the limit (strong operator topology) of a sequence of polynomials in the unilateral shift.*

149. Characterization of isometries. What can an isometry look like? Some   149
isometries are unitary, and some are not; an example of the latter kind is the unilateral shift. Since a direct sum (finite or infinite) of isometries is an isometry, a mixture of the two kinds is possible. More precisely, the direct sum of a unitary operator and a number of copies (finite or infinite) of the unilateral shift is an isometry. (There is no point in forming direct sums of unitary operators—they are no more unitary than the summands.) The useful theorem along these lines is that that is the only way to get isometries. It follows

that the unilateral shift is more than just an example of an isometry, with interesting and peculiar properties; it is in fact one of the fundamental building blocks out of which all isometries are constructed.

Problem 149. *Every isometry is either unitary, or a direct sum of one or more copies of the unilateral shift, or a direct sum of a unitary operator and some copies of the unilateral shift.*

An isometry for which the unitary direct summand is absent is called *pure*.

150

150. Distance from shift to unitary operators.

Problem 150. *How far is the unilateral shift from the set of unitary operators?*

151

151. Square roots of shifts.

If U is an isometry on a Hilbert space $\mathbf{H}$, and if there exists a unit vector e_0 in $\mathbf{H}$ such that the vectors $e_0, Ue_0, U^2e_0, \cdots$ form an orthonormal basis for $\mathbf{H}$, then (obviously) U is unitarily equivalent to the unilateral shift, or, by a slight abuse of language, U is the unilateral shift. This characterization of the unilateral shift can be reformulated as follows: U is an isometry on a Hilbert space $\mathbf{H}$ for which there exists a one-dimensional subspace $\mathbf{N}$ such that the subspaces $\mathbf{N}, U\mathbf{N}, U^2\mathbf{N}, \cdots$ are pairwise orthogonal and span $\mathbf{H}$. If there is such a subspace $\mathbf{N}$, then it must be equal to the *co-range* $(U\mathbf{H})^\perp$. In view of this comment another slight reformulation is possible: the unilateral shift is an isometry U of co-rank 1 on a Hilbert space $\mathbf{H}$ such that the subspaces $(U\mathbf{H})^\perp, U(U\mathbf{H})^\perp, U^2(U\mathbf{H})^\perp, \cdots$ span $\mathbf{H}$. (Since U is an isometry, it follows that they must be pairwise orthogonal.) Most of these remarks are implicit in Solution 149.

A generalization lies near at hand. Consider an isometry U on a Hilbert space $\mathbf{H}$ such that the subspaces $(U\mathbf{H})^\perp, U(U\mathbf{H})^\perp, U^2(U\mathbf{H})^\perp, \cdots$ are pairwise orthogonal and span $\mathbf{H}$, but make no demands on the value of the co-rank. Every such isometry may be called a shift (a unilateral shift). The co-rank of a shift (also called its *multiplicity*) constitutes a complete set of unitary invariants for it; the original unilateral shift is determined (to within unitary equivalence) as the shift of multiplicity 1 (the *simple* unilateral shift).

Unilateral shifts of higher multiplicities are just as important as the simple one. Problem 149 shows that they are exactly the pure isometries. They play a vital role in the study of all operators, not only isometries. Here, to begin their study, is a puzzle that makes contact with Problem 145 and with a curious infinite-dimensional manifestation of Sylvester's law of nullity.

Problem 151. *Which unilateral shifts have square roots?*

If U is the simple unilateral shift, then it is quite easy to see that U^2 is a shift of multiplicity 2. (More generally, shifts of multiplicity n, where n is a positive integer, can be obtained by forming the n-th power of U. The study of shifts of infinite multiplicity is much harder and much more important than that of the finite ones; they cannot be obtained from U in such a simple manner.) This remark shows (see Problem 145) that the answer to Problem 151 is neither "all" nor "none".

152. Shifts as universal operators. The most important aspect of shifts is that they, or rather their adjoints, turn out to be universal operators.

152

A *part* of an operator is a restriction of it to an invariant subspace. Each part of an isometry is an isometry; the study of the parts of unilateral shifts does not promise anything new. What about parts of the adjoints of unilateral shifts? If U is a unilateral shift, then $\|U\| = \|U^*\| = 1$, and it follows that if A is a part of U^*, then $\|A\| \leq 1$. Since, moreover, $U^{*n} \to 0$ in the strong topology (cf. Solution 110), it follows that $A^n \to 0$ (strong). The miraculous and useful fact is that these two obviously necessary conditions are also sufficient; cf. [43] and [32, 33].

Problem 152. *Every contraction whose powers tend strongly to 0 is unitarily equivalent to a part of the adjoint of a unilateral shift.*

153. Similarity to parts of shifts. For many purposes similarity is just as good as unitary equivalence. When is an operator A similar to a part of the adjoint of a shift U? Since similarity need not preserve norm, there is no obvious condition that $\|A\|$ must satisfy. There is, however, a measure of size that similarity does preserve, namely the spectral radius; since $r(U^*) = 1$, it follows that $r(A) \leq 1$. It is easy to see that this necessary condition is not sufficient. The reason is that one of the necessary conditions for unitary equivalence ($A^n \to 0$ strongly, cf. Problem 152) is necessary for similarity also. (That is: if $A^n \to 0$ strongly, and if $B = S^{-1}AS$, then $B^n \to 0$ strongly.) Since there are many operators A such that $r(A) \leq 1$ but A^n does not tend to 0 in any sense (example: 1), the condition on the spectral radius is obviously not sufficient. There is a condition on the spectral radius alone that is sufficient for similarity to a part of the adjoint of a shift, but it is quite a bit stronger than $r(A) \leq 1$; it is, in fact, $r(A) < 1$.

153

Problem 153. *Every operator whose spectrum is included in the interior of the unit disc is similar to a contraction whose powers tend strongly to 0.*

Corollary 1. *Every operator whose spectrum is included in the interior of the unit disc is similar to a part of the adjoint of a unilateral shift.*

Corollary 2. *Every operator whose spectrum is included in the interior of the unit disc is similar to a strict contraction.*

(A *strict* contraction is an operator A with $\|A\| < 1$.)

Corollary 3. *Every quasinilpotent operator is similar to operators with arbitrarily small norms.*

These simple but beautiful and general results are in [119].

Corollary 4. *The spectral radius of every operator A is the infimum of the numbers $\|S^{-1}AS\|$ for all invertible operators S.*

154 **154. Similarity to contractions.** One of the most difficult open problems of operator theory is to determine exactly which operators are similar to contractions [59]. Corollary 2 of Problem 153 gives a sufficient condition for similarity to a strict contraction. What happens to that result when the strict inequalities in it, in both hypothesis and conclusion, are replaced by the corresponding weak inequalities?

Problem 154. *If $r(A) \leq 1$, is A similar to a contraction?*

155 **155. Wandering subspaces.** If A is an operator on a Hilbert space $\mathbf{H}$, a subspace $\mathbf{N}$ of $\mathbf{H}$ is called *wandering* for A if it is orthogonal to all its images under the positive powers of A. This concept is especially useful in the study of isometries. If U is an isometry and $\mathbf{N}$ is a wandering subspace for U, then $U^m\mathbf{N} \perp U^n\mathbf{N}$ whenever m and n are distinct positive integers. In other words, if f and g are in $\mathbf{N}$, then $U^mf \perp U^ng$. (Proof: reduce to the case $m > n$, and note that $(U^mf, U^ng) = (U^{*n}U^mf, g) = (U^{m-n}f, g)$.) If U is unitary, even more is true: in that case $U^m\mathbf{N} \perp U^n\mathbf{N}$ whenever m and n are any two distinct integers, positive, negative, or zero. (Proof: find k so that $m + k$ and $n + k$ are positive and note that $(U^mf, U^ng) = (U^{m+k}f, U^{n+k}g)$.)

Wandering subspaces are important because they are connected with invariant subspaces, in this sense: if U is an isometry, then there is a natural one-to-one correspondence between all wandering subspaces $\mathbf{N}$ and some invariant subspaces $\mathbf{M}$. The correspondence is given by setting $\mathbf{M} = \bigvee_{n=0}^{\infty} U^n\mathbf{N}$. (To prove that this correspondence is one-to-one, observe that $U\mathbf{M} = \bigvee_{n=1}^{\infty} U^n\mathbf{N}$, so that $\mathbf{N} = \mathbf{M} \cap (U\mathbf{M})^{\perp}$.) For at least one operator, namely the unilateral shift, the correspondence is invertible.

Problem 155. *If U is the (simple) unilateral shift and if $\mathbf{M}$ is a non-zero subspace invariant under U, then there exists a (necessarily unique) one-dimensional wandering subspace $\mathbf{N}$ such that $\mathbf{M} = \bigvee_{n=0}^{\infty} U^n\mathbf{N}$.*

The equation connecting $\mathbf{M}$ and $\mathbf{N}$ can be expressed by saying that every non-zero part of the simple unilateral shift is a shift. To add that dim $\mathbf{N} = 1$ is perhaps an unsurprising sharpening, but a useful and non-trivial one. In view of these comments, the following concise statement is just a reformulation of the problem: every non-zero part of the simple unilateral shift is (unitarily equivalent to) the simple unilateral shift. With almost no additional effort, and only the obviously appropriate changes in the statement, all these considerations extend to shifts of higher multiplicities.

156. Special invariant subspaces of the shift. One of the most recalcitrant **156** unsolved problems of Hilbert space theory is whether or not every operator has a non-trivial invariant subspace. A promising, interesting, and profitable thing to do is to accumulate experimental evidence by examining concrete special cases and seeing what their invariant subspaces look like. A good concrete special case to look at is the unilateral shift.

There are two kinds of invariant subspaces: the kind whose orthogonal complement is also invariant (the reducing subspaces), and the other kind. The unilateral shift has no reducing subspaces (Problem 147); the question remains as to how many of the other kind it has and what they look like.

The easiest way to obtain an invariant subspace of the unilateral shift U is to fix a positive integer k, and consider the span $\mathbf{M}_k$ of the e_n's with $n \geqq k$. After this elementary observation most students of the subject must stop and think; it is not at all obvious that any other invariant subspaces exist. A recollection of the spectral behavior of U is helpful here. Indeed, since each complex number λ of modulus less than 1 is a simple eigenvalue of U^* (Solution 82), with corresponding eigenvector $f_\lambda = \sum_{n=0}^\infty \lambda^n e_n$, it follows that the orthogonal complement of the singleton $\{f_\lambda\}$ is a non-trivial subspace invariant under U.

> **Problem 156.** If $\mathbf{M}_k(\lambda)$ is the orthogonal complement of $\{f_\lambda, \cdots, U^{k-1}f_\lambda\}$ $(k = 1, 2, 3, \cdots)$, then $\mathbf{M}_k(\lambda)$ is invariant under U, $\dim \mathbf{M}_k^\perp(\lambda) = k$, and $\bigvee_{k=1}^\infty \mathbf{M}_k^\perp(\lambda) = \mathbf{H}^2$.

Note that the spaces $\mathbf{M}_k$ considered above are the same as the spaces $\mathbf{M}_k(0)$.

157. Invariant subspaces of the shift. What are the invariant subspaces of the **157** unilateral shift? The spaces $\mathbf{M}_k$ and their generalizations $\mathbf{M}_k(\lambda)$ (see Problem 156) are examples. The lattice operations (intersection and span) applied to them yield some not particularly startling new examples, and then the well seems to run dry. New inspiration can be obtained by abandoning the sequential point of view and embracing the functional one; regard U as the restriction to $\mathbf{H}^2$ of the multiplication induced by e_1.

> **Problem 157.** A non-zero subspace $\mathbf{M}$ of $\mathbf{H}^2$ is invariant under U if and only if there exists a function φ in $\mathbf{H}^\infty$, of constant modulus 1 almost everywhere, such that $\mathbf{M}$ is the range of the restriction to $\mathbf{H}^2$ of the multiplication induced by φ.

This basic result is due to Beurling [13]. It has received considerable attention since then; cf. [89, 54, 71].

In more informal language, **M** can be described as the set of all *multiples* of φ (multiples by functions in $\mathbf{H}^2$, that is). Correspondingly it is suggestive to write $\mathbf{M} = \varphi \cdot \mathbf{H}^2$. For no very compelling reason, the functions such as φ (functions in $\mathbf{H}^\infty$, of constant modulus 1) are called *inner* functions.

Corollary 1. *If φ and ψ are inner functions such that $\varphi \cdot \mathbf{H}^2 \subset \psi \cdot \mathbf{H}^2$, then φ is divisible by ψ, in the sense that there exists an inner function θ such that $\varphi = \psi \cdot \theta$. If $\varphi \cdot \mathbf{H}^2 = \psi \cdot \mathbf{H}^2$, then φ and ψ are constant multiples of one another, by constants of modulus 1.*

The characterization in terms of inner functions does not solve all problems about invariant subspaces of the shift, but it does solve some. Here is a sample.

Corollary 2. *If* **M** *and* **N** *are non-zero subspaces invariant under the unilateral shift, then* $\mathbf{M} \cap \mathbf{N} \neq 0$.

Corollary 2 says that the lattice of invariant subspaces of the unilateral shift is about as far as can be from being complemented.

158 **158. F. and M. Riesz theorem.** It is always a pleasure to see a piece of current (soft) mathematics reach into the past to illuminate and simplify some of the work of the founding fathers on (hard) analysis; the characterization of the invariant subspaces of the unilateral shift does that. The elements of $\mathbf{H}^2$ are related to certain analytic functions on the unit disc (Problem 35), and, although they themselves are defined on the unit circle only, and only almost everywhere at that, they tend to imitate the behavior of analytic functions. A crucial property of an analytic function is that it cannot vanish very often without vanishing everywhere. An important theorem of F. and M. Riesz asserts that the elements of $\mathbf{H}^2$ exhibit the same kind of behavior; here is one possible formulation.

Problem 158. *A function in* $\mathbf{H}^2$ *vanishes either almost everywhere or almost nowhere.*

Corollary. *If f and g are in* $\mathbf{H}^2$ *and if $fg = 0$ almost everywhere, then $f = 0$ almost everywhere or $g = 0$ almost everywhere.*

Concisely: there are no zero-divisors in $\mathbf{H}^2$.

For a more general discussion of the F. and M. Riesz theorem, see [75, p. 47].

159. Reducible weighted shifts. Very little of the theory of reducing and **159** invariant subspaces of the bilateral and the unilateral shift is known for weighted shifts. There is, however, one striking fact that deserves mention; it has to do with the reducibility of two-sided weighted shifts. It is due to R. L. Kelley.

Problem 159. *If A is a bilateral weighted shift with strictly positive weights α_n, $n = 0, \pm 1, \pm 2, \cdots$, then a necessary and sufficient condition that A be reducible is that the sequence $\{\alpha_n\}$ be periodic.*

CHAPTER 18

Cyclic Vectors

160. Cyclic vectors. An operator A on a Hilbert space $\mathbf{H}$ has a *cyclic vector* f if the vectors $f, Af, A^2 f, \cdots$ span $\mathbf{H}$. Equivalently, f is a cyclic vector for A in case the set of all vectors of the form $p(A)f$, where p varies over all polynomials, is dense in $\mathbf{H}$. The simple unilateral shift has many cyclic vectors; a trivial example is e_0.

On finite-dimensional spaces the existence of a cyclic vector indicates something like multiplicity 1. If, to be precise, A is a finite diagonal matrix, then a necessary and sufficient condition that A have a cyclic vector is that the diagonal entries be distinct (i.e., that the eigenvalues be simple). Indeed, if the diagonal entries are $\lambda_1, \cdots, \lambda_n$, then

$$p(A)\langle \xi_1, \cdots, \xi_n \rangle = \langle p(\lambda_1)\xi_1, \cdots, p(\lambda_n)\xi_n \rangle$$

for every polynomial p. For $f = \langle \xi_1, \cdots, \xi_n \rangle$ to be cyclic, it is clearly necessary that $\xi_i \neq 0$ for each i; otherwise the i coordinate of $p(A)f$ is 0 for all p. If the λ's are not distinct, nothing is sufficient to make f cyclic. If, for instance, $\lambda_1 = \lambda_2$, then $\langle \xi_2^*, -\xi_1^*, 0, \cdots, 0 \rangle$ is orthogonal to $p(A)f$ for all p. If, on the other hand, the λ's are distinct, then $p(\lambda_1), \cdots, p(\lambda_n)$ can be prescribed arbitrarily, so that if none of the ξ's vanishes, then the $p(A)f$'s exhaust the whole space.

Some trace of the relation between the existence of cyclic vectors and multiplicity 1 is visible even for non-diagonal matrices. Thus, for instance, if A is a finite matrix, then the direct sum $A \oplus A$ cannot have a cyclic vector. Reason: by virtue of the Hamilton-Cayley equation, at most n of the matrices $1, A, A^2, \cdots$ are linearly independent (where n is the size of A), and consequently, no matter what f and g are, at most n of the vectors $A^j f \oplus A^j g$ are linearly independent; it follows that their span can never be $2n$-dimensional.

If A has multiplicity 1 in any sense, it is reasonable to expect that A^* also has; this motivates the conjecture that if A has a cyclic vector, then so does A^*. For finite matrices this is true. For a proof, note that, surely, if a matrix has a cyclic vector, then so does its complex conjugate, and recall that every matrix is similar to its transpose.

The methods of the preceding paragraphs are very parochially finite-dimensional; that indicates that the theory of cyclic vectors in infinite-dimensional spaces is likely to be refractory, and it is. There is, to begin with, a trivial difficulty with cardinal numbers. If there is a cyclic vector, then a countable set spans the space, and therefore the space is separable; in other words, in non-separable spaces there are no cyclic vectors. This difficulty can be got around; that is one of the achievements of the multiplicity theory of normal operators ([50, III]). For normal operators, the close connection between multiplicity 1 and the existence of cyclic vectors persists in infinite-dimensional spaces, and, suitably reinterpreted, even in spaces of uncountable dimension.

For non-normal operators, things are peculiar. It is possible for a direct sum $A \oplus A$ to have a cyclic vector, and it is possible for A to have a cyclic vector when A^* does not. These facts were first noticed by D. E. Sarason.

Problem 160. *If U is a unilateral shift of multiplicity not greater than $\aleph_0$, then U^* has a cyclic vector.*

It is obvious that the simple unilateral shift has a cyclic vector, but it is not at all obvious that its adjoint has one. It does, but that by itself does not imply anything shocking. The first strange consequence of the present assertion is that if U is the simple unilateral shift, then $U^* \oplus U^*$ (which is the adjoint of a unilateral shift of multiplicity 2) has a cyclic vector. The promised strange behavior becomes completely exposed with the remark that $U \oplus U$ cannot have a cyclic vector (and, all the more, the same is true for direct sums with more direct summands). To prove the negative assertion, consider a candidate

$$\langle\langle \xi_0, \xi_1, \xi_2, \cdots \rangle, \langle \eta_0, \eta_1, \eta_2, \cdots \rangle\rangle$$

for a cyclic vector of $U \oplus U$. If $\langle \alpha, \beta \rangle$ is an arbitrary non-zero vector orthogonal to $\langle \xi_0, \eta_0 \rangle$ in the usual two-dimensional complex inner product space, then the vector

$$\langle\langle \alpha, 0, 0, \cdots \rangle, \langle \beta, 0, 0, \cdots \rangle\rangle$$

is orthogonal to

$$(U \oplus U)^n \langle\langle \xi_0, \xi_1, \xi_2, \cdots \rangle, \langle \eta_0, \eta_1, \eta_2, \cdots \rangle\rangle$$

for all n ($=0, 1, 2, \cdots$), and that proves that the cyclic candidate fails. (Here is a slightly more sophisticated way of expressing the same proof: if an operator has a cyclic vector, then its co-rank is at most 1; the co-rank of $U \oplus U$ is 2.)

161   **161. Density of cyclic operators.** How large is the set of cyclic operators? (It is convenient to say that if an operator has a cyclic vector, then it is a *cyclic operator.*)

If the underlying space is finite-dimensional, a necessary and sufficient condition for an operator to be cyclic is that every eigenvalue be of multiplicity 1. (The relevant multiplicity is the geometric one, the dimension of the corresponding eigenspace.) It follows that in the finite-dimensional case the , set of cyclic operators is open. If the dimension is n, then the operators with n distinct eigenvalues constitute a dense set; it follows that the set of cyclic operators is dense, and hence is a "large" set in the sense of Baire category. (Since not every operator is cyclic, it follows also that the set of cyclic operators is not closed. Concrete example: $\begin{pmatrix} 1/n & 0 \\ 0 & -1/n \end{pmatrix} \rightarrow \begin{pmatrix} 0 & 0 \\ 0 & 0 \end{pmatrix}$ as $n \rightarrow \infty$.)

In the separable infinite-dimensional case the set of cyclic operators is not open. Concrete example: the cyclic operator $\text{diag}\langle 1, \frac{1}{2}, \frac{1}{3}, \cdots \rangle$ is the limit of the non-cyclic operators $\text{diag}\langle 1, \frac{1}{2}, \cdots, 1/n, 0, 0, 0, \cdots \rangle$.

Problem 161. *If* $\dim \mathbf{H} = \aleph_0$, *is the set of cyclic operators on* $\mathbf{H}$ *dense?*

162   **162. Density of non-cyclic operators.** If $\dim \mathbf{H} = \aleph_0$, the set of cyclic operators is not dense; what about its complement? Even special cases of the question are not completely trivial.

Problem 162. *Is the unilateral shift a limit of non-cyclic operators?*

163   **163. Cyclicity of a direct sum.** Operators constructed out of the unilateral shift exhibit various cyclicity properties: U, U^*, and $U^* \oplus U^*$ are cyclic, but $U \oplus U$ is not. At least one question remains.

Problem 163. *Is* $U \oplus U^*$ *cyclic?*

164   **164. Cyclic vectors of adjoints.** Even if an operator is cyclic, its adjoint does not have to be (remember $U \oplus U$), and even if both an operator and its adjoint are cyclic, they don't necessarily have the same cyclic vectors. (Example: $\begin{pmatrix} 0 & 0 \\ 1 & 0 \end{pmatrix}$ and $\langle 1, 0 \rangle$.) Does normality improve matters?

Problem 164. *If* f *is a cyclic vector for a normal operator* A, *does it follow that* f *is cyclic for* A^*?

165   **165. Cyclic vectors of a position operator.** It is always good to know *all* the cyclic vectors of an operator, but to find them all is rarely easy. Even for one of the most natural operators on an infinite-dimensional space the problem leads to some interesting analysis.

Problem 165. *What are all the cyclic vectors of the position operator on* $\mathbf{L}^2(0, 1)$?

Recall that if μ is a Borel measure with compact support in the complex plane, then the position operator A on $\mathbf{L}^2(\mu)$ is defined by $(Af)(z) = zf(z)$.

166. Totality of cyclic vectors. How many cyclic vectors does an operator have? One possible answer is none at all. Another conceivable answer is that, for some operators, every non-zero vector is cyclic; it is not known whether that is actually possible. Solution 165 describes examples for which the set of cyclic vectors is, nevertheless, quite large. Can it be medium-sized?

166

Problem 166. *Is there an operator for which the span of the set of cyclic vectors is a non-trivial subspace* (*that is, different from both* 0 *and the whole space*)?

167. Cyclic operators and matrices. What's special about a cyclic matrix (meaning the matrix, with respect to some orthonormal basis, of a cyclic operator)? Nothing much: the only matrices of size 2, for instance, that are *not* cyclic are the scalars. Despite this bad news, matrices can be helpful in the study of cyclic operators.

167

The pertinent concept is a slight generalization of triangularity. A matrix $\langle \alpha_{ij} \rangle$ is *triangular* (specifically, upper triangular) if every entry below the main diagonal is zero (that is, $\alpha_{ij} = 0$ whenever $i > j$). A matrix is *triangular* $+ 1$ if every entry more than one step below the main diagonal is zero (that is, $\alpha_{ij} = 0$ whenever $i > j + 1$). Extensions of this language (triangular $+ k$) are obviously possible, but are not needed now.

Problem 167. (a) *If an operator has a matrix that is triangular* $+1$ *and is such that none of the entries on the diagonal just below the main one is zero* ($\alpha_{ij} \neq 0$ *when* $i = j + 1$), *then it is cyclic.* (b) *Is the converse true?*

168. Dense orbits. To say that an operator A has a cyclic vector f is to say that finite linear combinations of the vectors $A^n f$ are dense. A much stronger property is conceivable; is it possible?

168

Problem 168. *Is there an operator A on a Hilbert space $\mathbf{H}$ and a vector f in $\mathbf{H}$ such that the orbit of f under A is dense in $\mathbf{H}$?*

The *orbit* of f is the set of all vectors of the form $A^n f$, $n = 0, 1, 2, \cdots$, with no scalar multiples and no sums allowed.

CHAPTER 19

Properties of Compactness

169 **169. Mixed continuity.** Corresponding to the strong (s) and weak (w) topologies for a Hilbert space **H**, there are four possible interpretations of continuity for a transformation from **H** into **H**: they are the ones suggested by the symbols (s → s), (w → w), (s → w), and (w → s). Thus, to say that A is continuous (s → w) means that the inverse image under A of each w-open set is s-open; equivalently it means that the direct image under A of a net s-convergent to f is a net w-convergent to Af. Four different kinds of continuity would be too much of a good thing; it is fortunate that three of them collapse into one.

> **Problem 169.** *For a linear transformation A the three kinds of continuity* (s → s), (w → w), *and* (s → w) *are equivalent (and hence each is equivalent to boundedness), and continuity* (w → s) *implies that A has finite rank.*

> **Corollary.** *The image of the closed unit ball under an operator on a Hilbert space is always strongly closed.*

It is perhaps worth observing that for linear transformations of finite rank all four kinds of continuity are equivalent; this is a trivial finite-dimensional assertion.

170 **170. Compact operators.** A linear transformation on a Hilbert space is called *compact* (also *completely continuous*) if its restriction to the unit ball is (w → s) continuous (see Problem 169). Equivalently, a linear transformation is compact if it maps each bounded weakly convergent net onto a strongly convergent net. Since weakly convergent sequences are bounded, it follows that a compact linear transformation maps every weakly convergent sequence onto a strongly convergent one.

The image of the closed unit ball under a compact linear transformation is strongly compact. (Proof: the closed unit ball is weakly compact.) This implies that the image of each bounded set is precompact (i.e., has a strongly compact closure). (Proof: a bounded set is included in some closed ball.) The converse implication is also true: if a linear transformation maps bounded sets onto precompact sets, then it maps the closed unit ball onto a compact set. To prove this, observe first that compact (and precompact) sets are bounded, and that therefore a linear transformation that maps bounded sets onto precompact sets is necessarily bounded itself. (This implies, incidentally, that every compact linear transformation is bounded.) It follows from the corollary to Problem 169 that the image of the closed unit ball is strongly closed; this, together with the assumption that that image is precompact, implies that that image is actually compact. (The converse just proved is not universally true for Banach spaces.) The compactness conditions, here treated as consequences of the continuity conditions used above to define compact linear transformations, can in fact be shown to be equivalent to those continuity conditions and are frequently used to define compact linear transformations. (See [39, p. 484].)

An occasionally useful property of compact operators is that they "attain their norm". Precisely said: if A is compact, then there exists a unit vector f such that $\|Af\| = \|A\|$. The reason is that the mapping $f \mapsto Af$ is (w $\to$ s) continuous on the unit ball, and the mapping $g \mapsto \|g\|$ is strongly continuous; it follows that $f \mapsto \|Af\|$ is weakly continuous on the unit ball. Since the unit ball is weakly compact, this function attains its maximum, so that $\|Af\| = \|A\|$ for some f with $\|f\| \leq 1$. If $A = 0$, then f can be chosen to have norm 1; if $A \neq 0$, then f necessarily has norm 1. Reason: since $f \neq 0$ and $1/\|f\| \geq 1$, it follows that

$$\|A\| \leq \frac{\|A\|}{\|f\|} = \frac{\|Af\|}{\|f\|} \leq \|A\|.$$

Problem 170. *The set* **K** *of all compact operators on a Hilbert space is a closed self-adjoint (two-sided) ideal.*

Here "closed" refers to the norm topology, "self-adjoint" means that if $A \in \mathbf{K}$, then $A^* \in \mathbf{K}$, and "ideal" means that linear combinations of operators in **K** are in **K** and that products with at least one factor in **K** are in **K**.

171. Diagonal compact operators. Is the identity operator compact? Since in finite-dimensional spaces the strong and the weak topologies coincide, the answer is yes for them. For infinite-dimensional spaces, the answer is no; the reason is that the image of the unit ball is the unit ball, and in an infinite- **171**

dimensional space the unit ball cannot be strongly compact (Problem 16).

The indistinguishability of the strong and the weak topologies in finite-dimensional spaces yields a large class of examples of compact operators, namely all operators of finite rank. Examples of a slightly more complicated structure can be obtained by exploiting the fact that the set of compact operator is closed.

Problem 171. *A diagonal operator with diagonal $\{\alpha_n\}$ is compact if and only if $\alpha_n \to 0$ as $n \to \infty$.*

Corollary. *A weighted shift with weights $\{\alpha_n : n = 0, 1, 2, \cdots\}$ is compact if and only if $\alpha_n \to 0$ as $n \to \infty$.*

172

172. Normal compact operators. It is easy to see that if a normal operator has the property that every non-zero element in its spectrum is isolated (i.e., is not a cluster point of the spectrum), then it is a diagonal operator. (For each non-zero eigenvalue λ of A, choose an orthonormal basis for the subspace $\{f : Af = \lambda f\}$; the union of all these little bases, together with a basis for the kernel of A, is a basis for the whole space.) If, moreover, each non-zero eigenvalue has finite multiplicity, then the operator is compact. (Compare Problem 171; note that under the assumed conditions the set of eigenvalues is necessarily countable.) The remarkable and useful fact along these lines goes in the converse direction.

Problem 172. *The spectrum of a compact normal operator is countable; all its non-zero elements are eigenvalues of finite multiplicity.*

Corollary. *Every compact normal operator is the direct sum of the operator 0 (on a space that can be anything from absent to non-separable) and a diagonal operator (on a separable space).*

A less sharp but shorter formulation of the corollary is this: every compact normal operator is diagonal.

173

173. Hilbert–Schmidt operators. Matrices have valuable "continuous" generalizations. The idea is to replace sums by integrals, and it works—up to a point. To see where it goes wrong, consider a measure space X with measure μ (σ-finite as usual), and consider a measurable function K on the product space $X \times X$. A function of two variables, such as K, is what a generalized matrix can be expected to be. Suppose that A is an operator on $\mathbf{L}^2(\mu)$ whose relation to K is similar to the usual relation of an operator to its matrix. In precise terms this means that if $f \in \mathbf{L}^2(\mu)$, then

$$(Af)(x) = \int K(x, y)f(y)d\mu(y)$$

for almost every x. Under these conditions A is called an *integral operator* and K is called its *kernel*.

92

In the study of a Hilbert space $\mathbf{H}$, to say "select an orthonormal basis" is a special case of saying "select a particular way of representing $\mathbf{H}$ as $\mathbf{L}^2$". Many phenomena in $\mathbf{L}^2$ spaces are the natural "continuous" generalizations of more familiar phenomena in sequence spaces. One simple fact about sequence spaces is that every operator on them has a matrix, and this is true whether the sequences (families) that enter are finite or infinite. (It is the reverse procedure that goes wrong in the infinite case. From operators to matrices all is well; it is from matrices to operators that there is trouble.) On this evidence it is reasonable to guess that every operator on $\mathbf{L}^2$ has a kernel, i.e., that every operator is an integral operator. This guess is wrong, hopelessly wrong. The trouble is not with wild operators, and it is not with wild measures; it arises already if the operator is the identity and if the measure is Lebesgue measure (in the line or in any interval). In fact, if μ is Lebesgue measure, then the identity is not an integral operator. The proof is accessible, but it reveals more about integrals than about operators; see, for instance, [66, p. 41].

What about the reverse problem? Under what conditions does a kernel induce an operator? Since the question includes the corresponding question for matrices, it is not reasonable to look for necessary and sufficient conditions. A somewhat special sufficient condition, which is nevertheless both natural and useful, is that the kernel be square integrable.

Suppose, to be quite precise, that X is a measure space with σ-finite measure μ, and suppose that K is a complex-valued measurable function on $X \times X$ such that $|K|^2$ is integrable with respect to the product measure $\mu \times \mu$. It follows that, for almost every x, the function $y \mapsto K(x, y)$ is in $\mathbf{L}^2(\mu)$, and hence that the product function $y \mapsto K(x, y)f(y)$ is integrable whenever $f \in \mathbf{L}^2(\mu)$. Since, moreover,

$$\int \left| \int K(x, y)f(y)d\mu(y) \right|^2 d\mu(x)$$
$$\leqq \int \left(\int |K(x, y)|^2 \, d\mu(y) \cdot \int |f(y)|^2 \, d\mu(y) \right) d\mu(x) = \|K\|^2 \cdot \|f\|^2$$

(where $\|K\|$ is the norm of K in $\mathbf{L}^2(\mu \times \mu)$), it follows that the equation

$$(Af)(x) = \int K(x, y)f(y)d\mu(y)$$

defines an operator (with kernel K) on $\mathbf{L}^2(\mu)$. The inequality implies also that

$$\|A\| \leqq \|K\|.$$

Integral operators with kernels of this type (i.e., kernels in $\mathbf{L}^2(\mu \times \mu)$) are called *Hilbert–Schmidt operators*. A good reference for their properties is [127].

The correspondence $K \mapsto A$ is a one-to-one linear mapping from $\mathbf{L}^2(\mu \times \mu)$ to operators on $\mathbf{L}^2(\mu)$. If A has a kernel K (in $\mathbf{L}^2(\mu \times \mu)$), then A^* has the kernel $\tilde{K}$ defined by

$$\tilde{K}(x, y) = (K(y, x))^*.$$

93

If A and B have kernels H and K (in $\mathbf{L}^2(\mu \times \mu)$), then AB has the kernel HK defined by

$$(HK)(x, y) = \int H(x, z)K(z, y)d\mu(z).$$

The proofs of all these algebraic assertions are straightforward computations with integrals.

On the analytic side, the situation is just as pleasant. If $\{K_n\}$ is a sequence of kernels in $\mathbf{L}^2(\mu \times \mu)$ such that $K_n \to K$ (in the norm of $\mathbf{L}^2(\mu \times \mu)$), and if the corresponding operators are A_n (for K_n) and A (for K), then $\|A_n - A\| \to 0$. The proof is immediate from the inequality between the norm of an integral operator and the norm of its kernel.

Problem 173. *Every Hilbert–Schmidt operator is compact.*

These considerations apply, in particular, when the space is the set of positive integers with the counting measure. It follows that if the entries of a matrix are square-summable, then it is bounded (in the sense that it defines an operator) and compact (in view of the assertion of Problem 173). It should also be remarked that the Schur test (Problem 45) for the boundedness of a matrix has a straightforward generalization to a theorem about kernels; see [21].

174. Compact versus Hilbert–Schmidt.

Problem 174. *Is every compact operator a Hilbert–Schmidt operator?*

175. Limits of operators of finite rank. Every example of a compact operator seen so far (diagonal operators, weighted shifts, integral operators) was proved to be compact by showing it to be a limit of operators of finite rank. That is no accident.

Problem 175. *Every compact operator is the limit (in the norm) of operators of finite rank.*

The generalization of the assertion to arbitrary Banach spaces was an unsolved problem for a long time. It is now known to be false; see [29] and [41].

176. Ideals of operators. An ideal of operators is *proper* if it does not contain every operator. An easy example of an ideal of operators on a Hilbert space is the set of all operators of finite rank on that space; if the space is infinite-dimensional, that ideal is proper. Another example is the set of all compact operators; again, if the space is infinite-dimensional, that ideal is proper. The second of these examples is closed; in the infinite-dimensional case the first one is not.

Problem 176. *If* **H** *is a separable Hilbert space, then the collection of compact operators is the only non-zero closed proper ideal of operators on* **H**.

Similar results hold for non-separable spaces, but the formulations and proofs are fussier and much less interesting.

177. Compactness on bases. An orthonormal sequence converges weakly to 0, and a compact operator is (w → s) continuous. It follows that if an operator A is compact, then A maps each orthonormal sequence onto a strong null sequence. To what extent is the converse true?

It is plain enough what the converse means, but there are some closely related questions that deserve a look. Could it be, for instance, that if an operator A maps *some* orthonormal sequence onto a strong null sequence, then A is compact? Certainly not—that's absurd. Reason: the orthonormal sequence in question could be just half of an orthonormal basis on the other half of which A is large. (Example: a projection with infinite rank and nullity.) Very well: try again. Could it be that (1) if A maps some orthonormal *basis* onto a strong null sequence, then A is compact? (To avoid the irrelevant distractions of large cardinal numbers, it is best to assume here that the underlying Hilbert space is separable, and that the orthonormal bases to be studied come presented as sequences.) It is conceivable that the answer to question (1) is yes.

The plain converse question originally asked is this: (2) if A maps *every* orthonormal basis onto a strong null sequence, then is A compact? It is conceivable that the answer to question (2) is no.

The implications between the possible answers to (1) and (2) are clear: if (1) is yes, then so is (2); if (2) is no, then so is (1); in the other two cases the answer to one question leaves the other one open. What are the facts?

Problem 177. *If an operator A (on a Hilbert space of dimension $\aleph_0$) maps an orthonormal basis onto a sequence that converges strongly to* 0, *is* A *compact? What if* A *maps every orthonormal basis onto a strong null sequence?*

178. Square root of a compact operator. It is easy to construct non-compact operators whose square is compact; in fact, it is easy to construct non-compact operators that are nilpotent of index 2. (Cf. Problem 96.) What about the normal case?

Problem 178. *Do there exist non-compact normal operators whose square is compact?*

179. Fredholm alternative. The principal spectral fact about a compact operator (normal or no) on a Hilbert space is that a non-zero number can get into the spectrum via the point spectrum only.

177

178

179

Problem 179. *If C is compact, then* spec $C - \{0\} \subset \Pi_0(C)$.

Equivalently: if C is compact, and if λ is a non-zero complex number, then either λ is an eigenvalue of C or $C - \lambda$ is invertible. In this form the statement is frequently called the *Fredholm alternative*. It has a facetious but substantially accurate formulation in terms of the equation $(C - \lambda)f = g$, in which g is regarded as given and f as unknown; according to that formulation, if the solution is unique, then it exists.

Corollary. *A compact operator whose point spectrum is empty is quasi-nilpotent.*

180   **180. Range of a compact operator.**

Problem 180. *Every (closed) subspace included in the range of a compact operator is finite-dimensional.*

Corollary. *Every non-zero eigenvalue of a compact operator has finite multiplicity.*

181   **181. Atkinson's theorem.** An operator A is called a *Fredholm operator* if (1) ran A is closed and both ker A and ran$^{\perp} A$ are finite-dimensional. (The last two conditions can be expressed by saying that the nullity and the co-rank of A are finite.) An operator A is *invertible modulo the ideal of operators of finite rank* if (2) there exists an operator B such that both $1 - AB$ and $1 - BA$ have finite rank. An operator A is *invertible modulo the ideal of compact operators* if (3) there exists an operator B such that both $1 - AB$ and $1 - BA$ are compact.

Problem 181. *An operator A is (1) a Fredholm operator if and only if it is (2) invertible modulo the ideal of operators of finite rank, or, alternatively, if and only if it is (3) invertible modulo the ideal of compact operators.*

The result is due to Atkinson [7].

182   **182. Weyl's theorem.** The process of adding a compact operator to a given one is sometimes known as *perturbation*. The accepted attitude toward perturbation is that compact operators are "small"; the addition of a compact operator cannot (or should not) make for radical changes.

Problem 182. *If the difference between two operators is compact, then their spectra are the same except for eigenvalues. More explicitly: if $A - B$ is compact, and if $\lambda \in$ spec $A - \Pi_0(A)$, then $\lambda \in$ spec B.*

Note that for $B = 0$ the statement follows from Problem 179.

96

183. Perturbed spectrum. The spectrum of an operator changes, of course, **183**
when a compact operator is added to it, but in some sense not very much.
Eigenvalues may come and go, but otherwise the spectrum remains invariant.
In another sense, however, the spectrum can be profoundly affected by the
addition of a compact operator.

> **Problem 183.** *There exists a unitary operator U and there exists a com-*
> *pact operator C such that the spectrum of U + C is the entire unit disc.*

184. Shift modulo compact operators. Weyl's theorem (Problem 182) implies **184**
that if U is the unilateral shift and if C is compact, then the spectrum of
$U + C$ includes the unit disc. (Here is a small curiosity. The reason the spectrum of $U + C$ includes the unit disc is that U has no eigenvalues. The
adjoint U^* has many eigenvalues, so that this reasoning does not apply to
it, but the conclusion does. Reason: the spectrum of $U^* + C$ is obtained
from the spectrum of $U + C^*$ by reflection through the real axis, and C^*
is just as compact as C.) More is true [137]: every point of the open unit
disc is an eigenvalue of $(U + C)^*$.

It follows from the preceding paragraph that $U + C$ can never be invertible (the spectrum cannot avoid 0), and it follows also that $U + C$ can
never be quasinilpotent (the spectrum cannot consist of 0 alone). Briefly:
if invertibility and quasinilpotence are regarded as good properties, then
not only is U bad, but it cannot be improved by a perturbation. Perhaps the
best property an operator can have (and U does not have) is normality;
can a perturbation improve U in this respect?

> **Problem 184.** *If U is the unilateral shift, does there exist a compact*
> *operator C such that U + C is normal?*

Freeman [44] has a result that is pertinent to this circle of ideas; he
proves that, for a large class of compact operators C, the perturbed shift
$U + C$ is similar to the unperturbed shift U.

185. Distance from shift to compact operators. An increasingly valuable part **185**
of the study of operators is a subject called non-commutative approximation theory. The following question is a small sample of it.

> **Problem 185.** *What is the distance from the unilateral shift to the set*
> *of all compact operators?*

CHAPTER 20

Examples of Compactness

186. Bounded Volterra kernels. Integral operators are generalized matrices. Experience with matrices shows that the more zeros they have, the easier they are to compute with; triangular matrices, in particular, are usually quite tractable. Which integral operators are the right generalizations of triangular matrices? For the answer it is convenient to specialize drastically the measure spaces considered; in what follows the only X will be the unit interval, and the only μ will be Lebesgue measure. (The theory can be treated somewhat more generally; see [115].)

A *Volterra kernel* is a kernel K in $\mathbf{L}^2(\mu \times \mu)$ such that $K(x, y) = 0$ when $x < y$. Equivalently: a Volterra kernel is a Hilbert–Schmidt kernel that is triangular in the sense that it vanishes above the diagonal ($x = y$) of the unit square. In view of this definition, the effect of the integral operator A (*Volterra operator*) induced by a Volterra kernel K can be described by the equation

$$(Af)(x) = \int_0^x K(x, y) f(y) dy.$$

If the diagonal terms of a finite triangular matrix vanish, then the matrix is nilpotent. Since the diagonal of the unit square has measure 0, and since from the point of view of Hilbert space sets of measure 0 are negligible, the condition of vanishing on the diagonal does not have an obvious continuous analogue. It turns out nevertheless that the zero values of a Volterra kernel above the diagonal win out over the non-zero values below.

Problem 186. *A Volterra operator with a bounded kernel is quasi-nilpotent.*

Caution: "bounded" here refers to the kernel, not to the operator; the assumption is that the kernel is bounded almost everywhere in the unit square.

187. Unbounded Volterra kernels. How important is the boundedness **187** assumption in Problem 186?

Problem 187. *Is every Volterra operator quasinilpotent?*

188. Volterra integration operator. The simplest non-trivial Volterra **188** operator is the one whose kernel is the characteristic function of the triangle $\{\langle x, y\rangle : 0 \leq y \leq x \leq 1\}$. Explicitly this is the Volterra operator V defined on $\mathbf{L}^2(0, 1)$ by

$$(Vf)(x) = \int_0^x f(y)dy.$$

In still other words, V is indefinite integration, with the constant of integration adjusted so that every function in the range of V vanishes at 0. (Note that every function in the range of V is continuous. Better: every vector in the range of V, considered as an equivalence class of functions modulo sets of measure 0, contains a unique continuous function.)

Since V^* is the integral operator whose kernel is the "conjugate transpose" of the kernel of V, so that the kernel of V^* is the characteristic function of the triangle $\{\langle x, y\rangle : 0 \leq x \leq y \leq 1\}$, it follows that $V + V^*$ is the integral operator whose kernel is equal to the constant function 1 almost everywhere. (The operators V^* and $V + V^*$ are of course not Volterra operators.) This is a pleasantly simple integral operator; a moment's reflection should serve to show that it is the projection whose range is the (one-dimensional) space of constants. It follows that Re V has rank 1; since $V = \text{Re } V + i \text{ Im } V$, it follows that V is a perturbation (by an operator of rank 1 at that) of a skew Hermitian operator.

The theory of Hilbert–Schmidt operators in general and Volterra operators in particular answers many questions about V. Thus, for instance, V is compact (because it is a Hilbert–Schmidt operator), and it is quasinilpotent (because it is a Volterra operator). There are many other natural questions about V; some are easy to answer and some are not. Here is an easy one: does V annihilate any non-zero vectors? (Equivalently: "does V have a non-trivial kernel?", but that way terminological confusion lies.) The answer is no. If $\int_0^x f(y)dy = 0$ for almost every x, then, by continuity, the equation holds for every x. Since the functions in the range of V are not only continuous but, in fact, differentiable almost everywhere, the equation can be differentiated; the result is that $f(x) = 0$ for almost every x. As for natural questions that are not so easily disposed of, here is a simple sample.

Problem 188. *What is the norm of V?*

189   **189. Skew-symmetric Volterra operator.** There is an operator V_0 on $L^2(-1, +1)$ (Lebesgue measure) that bears a faint formal resemblance to the operator V on $L^2(0, 1)$; by definition

$$(V_0 f)(x) = \int_{-x}^{+x} f(y)dy.$$

Note that V_0 is the integral operator induced by the kernel that is the characteristic function of the butterfly $\{\langle x, y \rangle : 0 \leqq |y| \leqq |x| \leqq 1\}$.

Problem 189. *Find the spectrum and the norm of the skew-symmetric Volterra operator V_0.*

190   **190. Norm 1, spectrum {1}.** Every finite matrix is unitarily equivalent to a triangular matrix. If a triangular matrix has only 1's on the main diagonal, then its norm is at least 1; the norm can be equal to 1 only in case the matrix is the identity. The conclusion is that on a finite-dimensional Hilbert space the identity is the only contraction with spectrum {1}. The reasoning that led to this conclusion was very finite-dimensional; can it be patched up to yield the same conclusion for infinite-dimensional spaces?

Problem 190. *Is there an operator A, other than 1, such that* spec A = $\{1\}$ *and* $\|A\| = 1$?

191   **191. Donoghue lattice.** One of the most important, most difficult, and most exasperating unsolved problems of operator theory is the problem of invariant subspaces. The question is simple to state: does every operator on an infinite-dimensional Hilbert space have a non-trivial invariant subspace? "Non-trivial" means different from both 0 and the whole space; "invariant" means that the operator maps it into itself. For finite-dimensional spaces there is, of course, no problem; as long as the complex field is used, the fundamental theorem of algebra implies the existence of eigenvectors.

According to a dictum of Pólya's, for each unanswered question there is an easier unanswered question, and the scholar's first task is to find the latter. Even that dictum is hard to apply here; many weakenings of the invariant subspace problem are either trivial or as difficult as the full-strength problem. If, for instance, in an attempt to get a positive result, "subspace" is replaced by "linear manifold" (not necessarily closed), then the answer is yes, and easy. (For an elegant discussion, see [124].) If, on the other hand, "Hilbert space" is replaced by "Banach space", the chances of finding a counterexample are greater, but, despite periodically circulating rumors, no construction has yet been verified.

Positive results are known for some special classes of operators. The cheapest way to get one is to invoke the spectral theorem and to conclude that normal operators always have non-trivial invariant subspaces. The earliest non-trivial result along these lines is the assertion that compact operators always have non-trivial invariant subspaces [6]. That result has been generalized [12, 58, 91, 93], but the generalizations are still closely tied to compactness. Non-compact results are few; here is a sample. If A is a contraction such that neither of the sequences $\{A^n\}$ and $\{A^{*n}\}$ tends strongly to 0, then A has a non-trivial invariant subspace [100]. A bird's eye view of the subject is in [71], a more extensive bibliography is in [40], and a detailed treatment in [111].

It is helpful to approach the subject from a different direction: instead of searching for counterexamples, study the structure of some non-counter-examples. One way to do this is to fix attention on a particular operator and to characterize all its invariant subspaces; the first significant step in this direction is the work of Beurling [13] (Problem 157).

Nothing along these lines is easy. The second operator whose invariant subspaces have received detailed study is the Volterra integration operator ([17, 37, 84, 122]). The results for it are easier to describe than for the shift, but harder to prove. If $(Vf)(x) = \int_0^x f(y)dy$ for f in $L^2(0, 1)$, and if, for each α in $[0, 1]$, M_α is the subspace of those functions that vanish almost everywhere on $[0, \alpha]$, then M_α is invariant under V; the principal result is that every invariant subspace of V is one of the M_α's. An elegant way of obtaining these results is to reduce the study of the Volterra integration operator (as far as invariant subspaces are concerned) to that of the uni-lateral shift; this was done in [123].

The collection of all subspaces invariant under some particular operator is a lattice (closed under the formation of intersections and spans). One way to state the result about V is to say that its lattice of invariant subspaces is anti-isomorphic to the closed unit interval. ("Anti-" because as α grows M_α shrinks.) The lattice of invariant subspaces of V^* is in an obvious way isomorphic to the closed unit interval.

Is there an operator whose lattice of invariant subspaces is isomorphic to the positive integers? The question must be formulated with a little more care: every invariant subspace lattice has a largest element. The exact formulation is easy: is there an operator for which there is a one-to-one and order-preserving correspondence $n \mapsto M_n$, $n = 0, 1, 2, 3, \cdots, \infty$, be-tween the indicated integers (including ∞) and all invariant subspaces? The answer is yes. The first such operator was discovered by Donoghue [37]; a wider class of them is described in [103].

Suppose that $\{\alpha_n\}$ is a monotone sequence ($\alpha_n \geqq \alpha_{n+1}$, $n = 0, 1, 2, \cdots$) of positive numbers ($\alpha_n > 0$) such that $\sum_{n=0}^{\infty} \alpha_n{}^2 < \infty$. The unilateral weighted shift with the weight sequence $\{\alpha_n\}$ will be called a *monotone l^2 shift*. The span of the basis vectors e_n, e_{n+1}, e_{n+2}, $\cdots$ is invariant under such a shift, $n = 0, 1, 2, \cdots$. The orthogonal complement, i.e., the span M_n of

$e_0, \cdots, e_{n-1}$, is invariant under the adjoint, $n = 1, 2, 3, \cdots$; the principal result is that every invariant subspace of that adjoint is one of these orthogonal complements.

Problem 191. *If A is the adjoint of a monotone l^2 shift, and if $\mathbf{M}$ is a non-trivial subspace invariant under A, then there exists an integer $n \, (= 1, 2, 3, \cdots)$ such that $\mathbf{M} = \mathbf{M}_n$.*

CHAPTER 21

Subnormal Operators

192. Putnam–Fuglede theorem. Some of the natural questions about   **192**
normal operators have the same answers for finite-dimensional spaces as for
infinite-dimensional ones, and the techniques used to prove the answers are
the same. Some questions, on the other hand, are properly infinite-dimensional, in the sense that for finite-dimensional spaces they are either mean-
ingless or trivial; questions about shifts, or, more generally, questions
about subnormal operators are likely to belong to this category (see Problem 195). Between these two extremes there are the questions for which
the answers are invariant under change of dimension, but the techniques
are not. Sometimes, to be sure, either the question or the answer must be
reformulated in order to bring the finite and the infinite into harmony. As
for the technique, experience shows that an infinite-dimensional proof can
usually be adapted to the finite-dimensional case; to say that the techniques
are different means that the natural finite-dimensional techniques are not
generalizable to infinite-dimensional spaces. It should be added, however,
that sometimes the finite and the infinite proofs are intrinsically different,
so that neither can be adapted to yield the result of the other; a case in
point is the statement that any two bases have the same cardinal number.
A familiar and typical example of a theorem whose statement is easily
generalizable from the finite to the infinite, but whose proof is not, is the
spectral theorem. A more striking example is the Fuglede commutativity
theorem. It is more striking because it was for many years an unsolved
problem. For finite-dimensional spaces the statement was known to be
true and trivial; for infinite-dimensional spaces it was unknown.

The Fuglede theorem (cf. Solution 146) can be formulated in several
ways. The algebraically simplest formulation is that if A is a normal
operator and if B is an operator that commutes with A, then B commutes

with A^* also. Equivalently: if A^* commutes with A, and A commutes with B, then A^* commutes with B. In the latter form the assertion is that in a certain special situation commutativity is transitive. (In general it is not.)

The operator A plays a double role in the Fuglede theorem; the modified assertion, obtained by splitting the two roles of A between two normal operators, is true and useful. Here is a precise formulation.

Problem 192. *If A_1 and A_2 are normal operators and if B is an operator such that $A_1 B = BA_2$, then $A_1^* B = BA_2^*$.*

Observe that the Fuglede theorem is trivial in case B is Hermitian (even if A is not necessarily normal); just take the adjoint of the assumed equation $AB = BA$. The Putnam generalization (i.e., Problem 192) is, however, not obvious even if B is Hermitian; the adjoint of $A_1 B = BA_2$ is, in that case, $BA_1^* = A_2^* B$, which is not what is wanted.

Corollary. *If two normal operators are similar, then they are unitarily equivalent.*

Is the product of two commutative normal operators normal? The answer is yes, and the proof is the same for spaces of all dimensions; the proof seems to need the Fuglede theorem. In this connection it should be mentioned that the product of not necessarily commutative normal operators is very reluctant to be normal. A pertinent positive result was obtained by Wiegmann [156]; it says that if **H** is a finite-dimensional Hilbert space and if A and B are normal operators on **H** such that AB is normal, then BA also is normal. Away from finite-dimensional spaces even this result becomes recalcitrant. It remains true for compact operators [157], but it is false in the general case [85].

193. Algebras of normal operators. The properties of an operator are intimately connected with how it enters various algebraic structures. Thus, for instance, it is trivial that if an operator belongs to a commutative algebra that is closed under the formation of adjoints, then that operator is normal. If, conversely, an algebra is closed under the formation of adjoints and consists of normal operators only, then that algebra is commutative. (Proof: if $A + iB$ and $C + iD$ are in the algebra, with A, B, C, and D Hermitian, then A, B, C, and D are in the algebra, because adjoints are formable, and therefore so are $A + iC$, $A + iD$, $B + iC$, and $B + iD$; the assumed normality implies that everything commutes.) Question: what if the condition on adjoints is dropped?

Problem 193. *Is an algebra of normal operators necessarily commutative?*

194. Spectral measure of the unit disc. One of the techniques that can be used to prove the Fuglede theorem is to characterize in terms of the geometry of

Hilbert space the spectral subspaces associated with a normal operator. That technique is useful in other contexts too.

A necessary and sufficient condition that a complex number have modulus less than or equal to 1 is that all its powers have the same property. This trivial observation extends to complex-valued functions:

$$\{x: |\varphi(x)| \leqq 1\} = \{x: |\varphi(x)|^n \leqq 1, n = 1, 2, 3, \cdots\}.$$

There is a close connection between complex-valued functions and normal operators. The operatorial analogue of the preceding numerical observations should be something like this: if A is a normal operator on a Hilbert space $\mathbf{H}$, then the set $\mathbf{E}$ of those vectors f in $\mathbf{H}$ for which $\|A^n f\| \leqq \|f\|$, $n = 1, 2, 3, \cdots$, should be, in some sense, the part of $\mathbf{H}$ on which A is below 1. This is true; the precise formulation is that $\mathbf{E}$ is a subspace of $\mathbf{H}$ and the projection on $\mathbf{E}$ is the value of the spectral measure associated with A on the closed unit disc in the complex plane. The same result can be formulated in a more elementary manner in the language of multiplication operators.

Problem 194. *If A is the multiplication operator induced by a bounded measurable function φ on a measure space, and if $D = \{z: |z| \leqq 1\}$, then a necessary and sufficient condition that an element f in $\mathbf{L}^2$ be such that $\chi_{\varphi^{-1}(D)} f = f$ is that $\|A^n f\| \leqq \|f\|$ for every positive integer n.*

Here, as usual, χ denotes the characteristic function of the set indicated by its subscript.

By translations and changes of scale the spectral subspaces associated with all discs can be characterized similarly; in particular, a necessary and sufficient condition that a vector f be invariant under multiplication by the characteristic function of $\{x: |\varphi(x)| \leqq \varepsilon\}$ $(\varepsilon > 0)$ is that $\|A^n f\| \leqq \varepsilon^n \|f\|$ for all n. One way this result can sometimes be put to good use is this: if, for some positive number ε, there are no f's in $\mathbf{L}^2$ (other than 0) such that $\|A^n f\| \leqq \varepsilon^n \|f\|$ for all n, then the subspace of f's that vanish on the complement of the set $\{x: |\varphi(x)| \leqq \varepsilon\}$ is 0, and therefore the set $\{x: |\varphi(x)| \leqq \varepsilon\}$ is (almost) empty. Conclusion: under these circumstances $|\varphi(x)| > \varepsilon$ almost everywhere, and consequently the operator A is invertible.

195. Subnormal operators. The theory of normal operators is so successful that much of the theory of non-normal operators is modeled after it. A natural way to extend a successful theory is to weaken some of its hypotheses slightly and hope that the results are weakened only slightly. One weakening of normality is quasinormality (see Problem 137). Subnormal operators constitute a considerably more useful and deeper generalization, which goes in an altogether different direction. An operator is *subnormal* if it has a normal extension. More precisely, an operator A on a Hilbert space $\mathbf{H}$ is subnormal if there exists a normal operator B on a Hilbert space $\mathbf{K}$ such that $\mathbf{H}$ is a

195

subspace of **K**, the subspace **H** is invariant under the operator B, and the restriction of B to **H** coincides with A.

Every normal operator is trivially subnormal. On finite-dimensional spaces every subnormal operator is normal, but that takes a little proving; cf. Solution 202 or Problem 203. A more interesting and typical example of a subnormal operator is the unilateral shift; the bilateral shift is a normal extension.

Problem 195. *Every quasinormal operator is subnormal.*

Normality implies quasinormality, but not conversely (witness the unilateral shift). The present assertion is that quasinormality implies subnormality, but, again, the converse is false. To get a counterexample, add a non-zero scalar to the unilateral shift. The result is just as subnormal as the unilateral shift, but a straightforward computation shows that if it were also quasinormal, then the unilateral shift would be normal.

196 **196. Quasinormal invariants.** The invariant subspace problem is easy for normal operators; for subnormal operators it was very hard and yielded, after many years, only to a subtle analytic approach [24]. In this respect, as in all others, quasinormal operators are between the two.

Problem 196. *Every quasinormal operator on a space of dimension greater than 1 has a non-trivial invariant subspace.*

197 **197. Minimal normal extensions.** A normal extension B (on **K**) of a subnormal operator A (on **H**) is *minimal* if there is no reducing subspace of B between **H** and **K**. In other words, B is minimal over A if whenever **M** reduces B and **H** $\subset$ **M**, it follows that **M** = **K**. What is the right article for minimal normal extensions: "a" or "the"?

Problem 197. *If B_1 and B_2 (on **K**$_1$ and **K**$_2$) are minimal normal extensions of the subnormal operator A on **H**, then there exists an isometry U from **K**$_1$ onto **K**$_2$ that carries B_1 onto B_2 (i.e., $UB_1 = B_2 U$) and is equal to the identity on **H**.*

In view of this result, it is permissible to speak of "the" minimal normal extension of a subnormal operator, and everyone does. Typical example: the minimal normal extension of the unilateral shift is the bilateral shift.

198 **198. Polynomials in the shift.** The explicit determination of the minimal normal extension of a subnormal operator can be a non-trivial problem; any special case where that extension is accessible is worth looking at.

Problem 198. *If U is the unilateral shift and p is a polynomial, then $p(U)$ is a subnormal operator; what is its minimal normal extension?*

106

199. Similarity of subnormal operators. For normal operators similarity **199** implies unitary equivalence (Problem 192). Subnormal operators are designed to imitate the properties of normal ones; is this one of the respects in which they succeed?

> **Problem 199.** *Are two similar subnormal operators necessarily unitarily equivalent?*

200. Spectral inclusion theorem. If an operator A is a restriction of an **200** operator B to an invariant subspace **H** of B, and if f is an eigenvector of A (i.e., $f \in \mathbf{H}$ and $Af = \lambda f$ for some scalar λ), then f is an eigenvector of B. Differently expressed: if $A \subset B$, then $\Pi_0(A) \subset \Pi_0(B)$, or, as an operator grows, its point spectrum grows. An equally easy verification shows that as an operator grows, its approximate point spectrum grows. In view of these very natural observations, it is tempting to conjecture that as an operator grows, its spectrum grows, and hence that, in particular, if A is subnormal and B is its minimal normal extension, then spec $A \subset$ spec B. The first non-trivial example of a subnormal operator shows that this conjecture is false: if A is the unilateral shift and B is the bilateral shift, then spec A is the unit disc, whereas spec B is only the perimeter of the unit disc. It turns out that this counterexample illustrates the general case better than do the plausibility arguments based on eigenvalues, exact or approximate.

> **Problem 200.** *If A is subnormal and if B is its minimal normal extension, then* spec $B \subset$ spec A.

Reference: [51].

201. Filling in holes. The spectral inclusion theorem (Problem 200) for **201** subnormal operators can be sharpened in an interesting and surprising manner. The result is that the spectrum of a subnormal operator is always obtained from the spectrum of its minimal normal extension by "filling in some of the holes". This informal expression can be given a precise technical meaning. A *hole* in a compact subset of the complex plane is a bounded component of its complement.

> **Problem 201.** *If A is subnormal, if B is its minimal normal extension, and if Δ is a hole of* spec B, *then Δ is either included in or disjoint from* spec A.

202. Extensions of finite co-dimension. **202**

> **Problem 202.** *Can a subnormal but non-normal operator on a Hilbert space* **H** *have a normal extension to a Hilbert space* **K** *when* dim($\mathbf{K} \cap \mathbf{H}^{\perp}$) *is finite?*

203

203. Hyponormal operators. If A (on **H**) is subnormal, with normal extension B (on **K**), what is the relation between A^* and B^*? The answer is best expressed in terms of the projection P from **K** onto **H**. If f and g are in **H**, then

$$(A^*f, g) = (f, Ag) = (f, Bg) = (B^*f, g) = (B^*f, Pg) = (PB^*f, g).$$

Since the operator PB^* on **K** leaves **H** invariant, its restriction to **H** is an operator on **H**, and, according to the preceding chain of equations, that restriction is equal to A^*. That is the answer:

$$A^*f = PB^*f$$

for every f in **H**.

This relation between A^* and B^* has a curious consequence. If $f \in$ **H**, then

$$\|A^*f\| = \|PB^*f\| \leq \|B^*f\| = \|Bf\| \text{ (by normality)} = \|Af\|.$$

The result ($\|A^*f\| \leq \|Af\|$) can be reformulated in another useful way; it is equivalent to the operator inequality

$$AA^* \leq A^*A.$$

Indeed: $\|A^*f\|^2 = (AA^*f, f)$ and $\|Af\|^2 = (A^*Af, f)$.

The curious inequality that subnormal operators always satisfy can also be obtained from an illuminating matrix calculation. Corresponding to the decomposition **K** = **H** $\oplus$ **H**$^\perp$, every operator on **K** can be expressed as an operator matrix, and, in particular, that is true for B. It is easy to express the relation ($A \subset B$) between A and B in terms of the matrix of B; a necessary and sufficient condition for it is that (1) the principal (northwest) entry is A, and (2) the one below it (southwest) is 0. The condition (2) says that **H** is invariant under B, and (1) says that the restriction of B to **H** is A. Thus

$$B = \begin{pmatrix} A & R \\ 0 & S \end{pmatrix},$$

so that

$$B^* = \begin{pmatrix} A^* & 0 \\ R^* & S^* \end{pmatrix}.$$

Since B is normal, it follows that the matrix

$$B^*B - BB^* = \begin{pmatrix} A^*A & A^*R \\ R^*A & R^*R + S^*S \end{pmatrix} - \begin{pmatrix} AA^* + RR^* & RS^* \\ SR^* & SS^* \end{pmatrix}$$

must vanish. This implies that

$$A^*A - AA^* = RR^*,$$

and hence that

$$A^*A - AA^* \geq 0.$$

108

There is a curious lack of symmetry here: why should A^*A play a role so significantly different from that of AA^*? A little meditation on the unilateral shift may help. If $A = U$, the unilateral shift, then A is subnormal, and $A^*A = 1$, whereas AA^* is a non-trivial projection; clearly $A^*A \geqq AA^*$. If $A = U^*$, then A is not subnormal. (Reason: if it were, then it would satisfy the inequality $A^*A \geqq AA^*$, i.e., $UU^* \geqq U^*U$, and then U would be normal.) If it were deemed absolutely essential, symmetry could be restored to the universe by the introduction of the dual concept of co-subnormality. (Proposed definition: the adjoint is subnormal.) If A is co-subnormal in this sense, then $AA^* \geqq A^*A$. An operator A such that $A^*A \geqq AA^*$ has been called *hyponormal*. (The dual kind might be called co-hyponormal. Note that "hypo" is in Greek what "sub" is in Latin. The nomenclature is not especially suggestive, but this is how it grew, and it seems to be here to stay.) The result of the preceding paragraphs is that every subnormal operator is hyponormal. The dull dual result is, of course, that every co-subnormal operator is co-hyponormal.

On a finite-dimensional space every hyponormal operator is normal. The most efficient proof of this assertion is a trace argument, as follows. Since $\text{tr}(AB)$ is always equal to $\text{tr}(BA)$, it follows that $\text{tr}(A^*A - AA^*)$ is always 0; if $A^*A \geqq AA^*$, then $A^*A - AA^*$ is a positive operator with trace 0, and therefore $A^*A - AA^* = 0$. What was thus proved is a generalization of the statement that on a finite-dimensional space every subnormal operator is normal (cf. Problem 195).

Problem 203. *Give an example of a hyponormal operator that is not subnormal.*

This is not easy. The techniques used are almost sufficient to yield an intrinsic characterization of subnormality ([49, 16]). "Intrinsic" means that the characterization is expressed in terms of the action of the operator on the vectors in its domain, and not in terms of the existence of something outside that domain. The characterization is of "finite character", in the sense that it depends on the behavior of the operator on all possible finite sets of vectors. With still more work of the same kind an elegant topological characterization of subnormality can be obtained; this was first done by Bishop [15]. Bishop's result is easy to state: the set of all subnormal operators is exactly the strong closure of the set of all normal operators. (See Problem 225.)

204. Normal and subnormal partial isometries.   **204**

Problem 204. *A partial isometry is normal if and only if it is the direct sum of a unitary operator and zero; it is subnormal if and only if it is the direct sum of an isometry and zero.*

In both cases, one or the other of the direct summands may be absent.

205

205. Norm powers and power norms. The set **T** of those operators A such that $\|A^n\| = \|A\|^n$ for every positive integer n has at the very least, a certain curiosity value. If $A \in \mathbf{T}$, then $\|A^n\|^{1/n} = \|A\|$, and therefore $r(A) = \|A\|$; if, conversely, $r(A) = \|A\|$, then $\|A^n\| \leq \|A\|^n = (r(A))^n = r(A^n) \leq \|A^n\|$, so that equality holds all the way through. Conclusion: $A \in \mathbf{T}$ if and only if $r(A) = \|A\|$.

The definition of **T** implies that every normal operator belongs to **T** (and so does the conclusion of the preceding paragraph). For two-by-two matrices an unpleasant computation proves a strong converse: if $\|A^2\| = \|A\|^2$, then A is normal. Since neither the assertion nor its proof have any merit, the latter is omitted. As soon as the dimension becomes greater than 2, the converse becomes false. If, for example.

$$A = \begin{pmatrix} 1 & 0 & 0 \\ 0 & 0 & 0 \\ 0 & 1 & 0 \end{pmatrix},$$

then $\|A^n\| = 1$ for all n, but A is certainly not normal.

The quickest (but not the most elementary) proof of the direct assertion (if A is normal, then $A \in \mathbf{T}$) is to refer to the spectral theorem. Since for subnormal and hyponormal operators that theorem is not available, a natural question remains unanswered. The answer turns out to be affirmative.

Problem 205. *If A is hyponormal, then $\|A^n\| = \|A\|^n$ for every positive integer n.*

Corollary. *The only hyponormal quasinilpotent operator is 0.*

206

206. Compact hyponormal operators. It follows from the discussion of hyponormal operators on finite-dimensional spaces (Problem 203) that a hyponormal operator of finite rank (on a possibly infinite-dimensional space) is always normal. What about limits of operators of finite rank?

Problem 206. *Every compact hyponormal operator is normal.*

Reference: [4, 10, 136].

207

207. Hyponormal, compact imaginary part. An operator A is compact if and only if both Re A and Im A are compact. The result of Problem 206 can therefore be stated in this form: if A is hyponormal and both Re A and Im A are compact, then A is normal. What if only one of Re A and Im A is compact?

Problem 207. *Does there exist a hyponormal operator that is not normal but has compact imaginary part?*

It should be kept in mind that even if A is hyponormal, A^* need not be. What is trivially true, however, is that iA is hyponormal at the same time as A is and $\operatorname{Im}(iA) = \operatorname{Re} A$.

208. Hyponormal idempotents. A normal operator with one good property usually has several good properties. A typical example of this kind of statement is that an idempotent normal operator must be Hermitian (and hence a projection). To what extent do the useful generalizations of normality behave normally?

> **Problem 208.** *Is every quasinormal idempotent a projection? What about subnormal idempotents? What about hyponormal ones?*

209. Powers of hyponormal operators. Every power of a normal operator is normal. This trivial observation has as an almost equally trivial consequence the statement that every power of a subnormal operator is subnormal. For hyponormal operators the facts are different.

> **Problem 209.** *Give an example of a hyponormal operator whose square is not hyponormal.*

This is not easy. It is, in fact, bound to be at least as difficult as the construction of a hyponormal operator that is not subnormal (Problem 203), since any solution of Problem 209 is automatically a solution of Problem 203. The converse is not true; the hyponormal operator used in Solution 203 has the property that all its powers are hyponormal also.

111

CHAPTER 22

Numerical Range

210   **210. Toeplitz–Hausdorff theorem.** In early studies of Hilbert space (by Hilbert, Hellinger, Toeplitz, and others) the objects of chief interest were quadratic forms. Nowadays they play a secondary role. First comes an operator A on a Hilbert space $\mathbf{H}$, and then, apparently as an afterthought, comes the numerical-valued function $f \mapsto (Af, f)$ on $\mathbf{H}$. This is not to say that the quadratic point of view is dead; it still suggests questions that are interesting with answers that can be useful.

Most quadratic questions about an operator are questions about its numerical range, sometimes called its field of values. The *numerical range* of an operator A is the set $W(A)$ of all complex numbers of the form (Af, f), where f varies over all vectors on the unit sphere. (Important: $\|f\| = 1$, not $\|f\| \leq 1$.) The numerical range of A is the range of the restriction to the unit sphere of the quadratic form associated with A. One reason for the emphasis on the image of the unit sphere is that the image of the unit ball, and also the entire range, are easily described in terms of it, but not vice versa. (The image of the unit ball is the union of all the closed segments that join the origin to points of the numerical range; the entire range is the union of all the closed rays from the origin through points of the numerical range.)

The determination of the numerical range of an operator is sometimes easy. Here are some sample results. If

$$A = \begin{pmatrix} 1 & 0 \\ 0 & 0 \end{pmatrix},$$

then $W(A)$ is the closed unit interval (easy); if

$$A = \begin{pmatrix} 0 & 0 \\ 1 & 0 \end{pmatrix},$$

then $W(A)$ is the closed disc with center 0 and radius $\frac{1}{2}$ (easy, but more interesting); if

$$A = \begin{pmatrix} 0 & 0 \\ 1 & 1 \end{pmatrix},$$

then $W(A)$ is the closed elliptical disc with foci at 0 and 1, minor axis 1 and major axis $\sqrt{2}$ (analytic geometry at its worst). There is a theorem that covers all these cases. If A is a two-by-two matrix with distinct eigenvalues α and β, and corresponding eigenvectors f and g, so normalized that $\|f\| = \|g\| = 1$, then $W(A)$ is a closed elliptical disc with foci at α and β; if $\gamma = |(f, g)|$ and $\delta = \sqrt{1 - \gamma^2}$, then the minor axis is $\gamma|\alpha - \beta|/\delta$ and the major axis is $|\alpha - \beta|/\delta$. If A has only one eigenvalue α, then $W(A)$ is the (circular) disc with center α and radius $\frac{1}{2}\|A - \alpha\|$.

A couple of three-dimensional examples will demonstrate that the two-dimensional case is not typical. If

$$A = \begin{pmatrix} 0 & 0 & \lambda \\ 1 & 0 & 0 \\ 0 & 1 & 0 \end{pmatrix},$$

where λ is a complex number of modulus 1, then $W(A)$ is the equilateral triangle (interior and boundary) whose vertices are the three cube roots of λ. (Cf. Problem 216.) If

$$A = \begin{pmatrix} 0 & 0 & 0 \\ 1 & 0 & 0 \\ 0 & 0 & 1 \end{pmatrix},$$

then $W(A)$ is the union of all the closed segments that join the point 1 to points of the closed disc with center 0 and radius $\frac{1}{2}$. (Cf. Problem 216.)

The higher the dimension, the stranger the numerical range can be. If A is the Volterra integration operator (see Problem 188), then $W(A)$ is the set lying between the curves

$$t \mapsto \frac{1 - \cos t}{t^2} \pm i \frac{t - \sin t}{t^2}, \qquad 0 \leq t \leq 2\pi$$

(where the value at 0 is taken to be the limit from the right).

The following assertion describes the most important common property of all these examples.

Problem 210. *The numerical range of an operator is always convex.*

The result is known as the *Toeplitz–Hausdorff theorem*. Consideration of real and imaginary parts shows that it is a special case ($n = 2$) of the following general assertion: if $A_1, \cdots, A_n$ are Hermitian operators, then the set of all n-tuples of the form $\langle (A_1 f, f), \cdots, (A_n f, f) \rangle$, where

113

$\|f\| = 1$, is a convex subset of n-dimensional real Euclidean space. True or false, the assertion seems to be a natural generalization of the Toeplitz-Hausdorff theorem; it is a pity that it is so very false. It is false for $n = 3$ in dimension 2; counterexamples are easy to come by.

The first paper on the subject was by Toeplitz [142], who proved that the boundary of $W(A)$ is a convex curve, but left open the possibility that it had interior holes. Hausdorff [69] proved that it did not. Donoghue [36] re-examined the facts and presented some pertinent computations. The result about the Volterra integration operator is due to A. Brown.

211 **211. Higher-dimensional numerical range.** The numerical range can be regarded as the one-dimensional case of a multi-dimensional concept. To see how that goes, recall the expression of a projection P of rank 1 in terms of a unit vector f in its range:

$$Pg = (g, f)f$$

for all g. If A is an arbitrary operator, then PAP is an operator of rank 1, and therefore a finite-dimensional concept such as trace makes sense for it. The trace of PAP can be computed by finding the (one-by-one) matrix of the restriction of PAP to the range of P, with respect to the (one-element) basis $\{f\}$; since $Pf = f$, the value of that trace is

$$(PAPf, f) = (APf, Pf) = (Af, f).$$

These remarks can be summarized as follows: $W(A)$ is equal to the set of all complex numbers of the form tr PAP, where P varies over all projections of rank 1. Replace 1 by an arbitrary positive integer k, and obtain the k-*numerical range* of A, in symbols $W_k(A)$: it is the set of all complex numbers of the form tr PAP, where P varies over all projections of rank k. The ordinary numerical range is the k-numerical range with $k = 1$.

Problem 211. *Is the k-numerical range of an operator always convex?*

212 **212. Closure of numerical range.**

Problem 212. *Give examples of operators whose numerical range is not closed.*

Observe that in the finite-dimensional case the numerical range of an operator is a continuous image of a compact set, and hence necessarily compact.

213 **213. Numerical range of a compact operator.** The numerical range of an operator isn't always closed, not even if the operator is compact. Solution 212 exhibited a compact operator A such that $W(A) = (0, 1]$; that is, 0 is in the closure of $W(A)$ but not in $W(A)$ itself. The number 0 plays a

special role with respect to the spectrum of a compact operator; does it play an equally special role with respect to the numerical range?

Problem 213. *Is there a compact operator A such that $0 \in W(A)$ but $W(A)$ is not closed?*

214. Spectrum and numerical range.   **214**

Problem 214. *The closure of the numerical range includes the spectrum.*

The trivial corollary that asserts that if $A = B + iC$, with B and C Hermitian, then spec $A \subset \overline{W(B)} + i\overline{W(C)}$ is the *Bendixson–Hirsch theorem*.

215. Quasinilpotence and numerical range. If A is a quasinilpotent operator,   **215**
then, by Problem 214, $0 \in \overline{W(A)}$. By Solution 212, the set $W(A)$ may fail to be closed, so that from $0 \in \overline{W(A)}$ it does not follow that $0 \in W(A)$. Is it true just the same?

Problem 215. *Give an example of a quasinilpotent operator A such that $0 \notin W(A)$.*

Observe that any such example is a solution of Problem 212.

216. Normality and numerical range. Can the closure of the numerical range   **216**
be very much larger than the spectrum? The answer is yes. A discouraging example is

$$\begin{pmatrix} 0 & 0 \\ 1 & 0 \end{pmatrix};$$

the spectrum is small ($\{0\}$), but the numerical range is large ($\{z : |z| \leq \frac{1}{2}\}$). Among normal operators such extreme examples do not exist; for them the closure of the numerical range is as small as the universal properties of spectra and numerical ranges permit.

To formulate the result precisely, it is necessary to introduce the concept of the *convex hull* of a set M, in symbols conv M. By definition, conv M is the smallest convex set that includes M; in other words, conv M is the intersection of all the convex sets that include M. It is a non-trivial fact of finite-dimensional Euclidean geometry that the convex hull of a compact set is closed. Perhaps the most useful formulation of this fact for the plane goes as follows: the convex hull of a compact set is the intersection of all the closed half planes that include it. A useful reference for all this is [144].

So much for making convex sets out of closed sets. The reverse process of making closed sets out of convex sets is much simpler to deal with; it is true and easy to prove that the closure of a convex set is convex.

Problem 216. *The closure of the numerical range of a normal operator is the convex hull of its spectrum.*

As an application consider the matrix

$$\begin{pmatrix} 0 & 0 & \lambda \\ 1 & 0 & 0 \\ 0 & 1 & 0 \end{pmatrix},$$

where $|\lambda| = 1$. Since this matrix is unitary, and therefore normal, the result just proved implies that its numerical range is the convex hull of its eigenvalues. The eigenvalues are easy to compute (they are the cube roots of λ), and this proves the assertion (made in passing in Problem 210) that the numerical range of this particular matrix is the triangle whose vertices are the cube roots of λ.

The general result includes the special assertion that the numerical range of every finite diagonal matrix is the convex hull of its diagonal entries. A different generalization of this special assertion is that the numerical range of a direct sum is the convex hull of the numerical ranges of its summands. The proof of the generalization is straightforward. For an example, consider the direct sum of

$$\begin{pmatrix} 0 & 0 \\ 1 & 0 \end{pmatrix}$$

and (1), and recapture the assertion (made in passing in Problem 210) about the domed-cone shape of the numerical range of

$$\begin{pmatrix} 0 & 0 & 0 \\ 1 & 0 & 0 \\ 0 & 0 & 1 \end{pmatrix}.$$

217   **217. Subnormality and numerical range.**

Problem 217. *Does the conclusion of Problem 216 remain true if in the hypothesis "normal" is replaced by "subnormal"?*

218   **218. Numerical radius.** The numerical range, like the spectrum, associates a set with each operator; it is a set-valued function of operators. There is a closely related numerical function w, called the *numerical radius*, defined by

$$w(A) = \sup\{|\lambda| : \lambda \in W(A)\}.$$

(Cf. the definition of spectral radius, Problem 88.) Some of the properties of the numerical radius lie near the surface; others are quite deep.

It is easy to prove that w is a norm. That is: $w(A) \geqq 0$, and $w(A) = 0$ if and only if $A = 0$; $w(\alpha A) = |\alpha| \cdot w(A)$ for each scalar α; and $w(A + B) \leqq$

$w(A) + w(B)$. This norm is equivalent to the ordinary operator norm, in the sense that each is bounded by a constant multiple of the other:

$$\tfrac{1}{2}\|A\| \leqq w(A) \leqq \|A\|.$$

(See [50, p. 33].) The norm w has many other pleasant properties; thus, for instance, $w(A^*) = w(A)$, $w(A^*A) = \|A\|^2$, and w is unitarily invariant, in the sense that $w(U^*AU) = w(A)$ whenever U is unitary.

Since spec $A \subset \overline{W(A)}$ (Problem 214), there is an easy inequality between the spectral radius and the numerical radius:

$$r(A) \leqq w(A).$$

The existence of quasinilpotent (or, for that matter, nilpotent) operators shows that nothing like the reverse inequality could be true.

Problem 218. (a) *If $w(1 - A) < 1$, then A is invertible.* (b) *If $w(A) = \|A\|$, then $r(A) = \|A\|$.*

219. Normaloid, convexoid, and spectraloid operators. If A is normal, then **219** $w(A) = \|A\|$. Wintner called an operator A with $w(A) = \|A\|$ *normaloid*. Another useful (but nameless) property of a normal operator A (Problem 216) is that $\overline{W(A)}$ is the convex hull of spec A. To have a temporary label for (not necessarily normal) operators with this property, call them *convexoid*. Still another (nameless) property of a normal operator A is that $r(A) = w(A)$; call an operator with this property *spectraloid*. It is a consequence of Problem 218 that every normaloid operator is spectraloid. It is also true that every convexoid operator is spectraloid. Indeed, since the closed disc with center 0 and radius $r(A)$ includes spec A and is convex, it follows that if A is convexoid, then that disc includes $W(A)$. This implies that $w(A) \leqq r(A)$, and hence that A is spectraloid.

Problem 219. *Discuss the implication relations between the properties of being convexoid and normaloid.*

220. Continuity of numerical range. In what sense is the numerical range a **220** continuous function of its argument? (Cf. Problems 102 and 103.) The best way to ask the question is in terms of the *Hausdorff metric* for compact subsets of the plane. To define that metric, write

$$M + (\varepsilon) = \{z + \alpha : z \in M, |\alpha| < \varepsilon\}$$

for each set M of complex numbers and each positive number ε. In this notation, if M and N are compact sets, the Hausdorff distance $d(M, N)$ between them is the infimum of all positive numbers ε such that both $M \subset N + (\varepsilon)$ and $N \subset M + (\varepsilon)$.

Since the Hausdorff metric is defined for compact sets, the appropriate function to discuss is $\overline{W}$, not W. As for the continuity question, it still has as many interpretations as there are topologies for operators. Is $\overline{W}$ weakly continuous? strongly? uniformly? And what about w? The only thing that is immediately obvious is that if $\overline{W}$ is continuous with respect to any topology, then so is w, and consequently, if w is discontinuous, then so is $\overline{W}$.

Problem 220. *Discuss the continuity of $\overline{W}$ and w in the weak, strong, and uniform operator topologies.*

221. Power inequality. The good properties of the numerical range and the numerical radius have to do with convexity and linearity; the relations between the numerical range and the multiplicative properties of operators are less smooth. Thus, for instance, w is certainly not multiplicative, i.e., $w(AB)$ is not always equal to $w(A)w(B)$. (Example with commutative normal operators: if

$$A = \begin{pmatrix} 1 & 0 \\ 0 & 0 \end{pmatrix} \quad \text{and } B = \begin{pmatrix} 0 & 0 \\ 0 & 1 \end{pmatrix},$$

then $w(A) = w(B) = 1$ and $w(AB) = 0$.) The next best thing would be for w to be submultiplicative ($w(AB) \leq w(A)w(B)$), but that is false too. (Example: if

$$A = \begin{pmatrix} 0 & 0 \\ 1 & 0 \end{pmatrix} \quad \text{and } B = \begin{pmatrix} 0 & 1 \\ 0 & 0 \end{pmatrix},$$

then $w(A) = w(B) = \frac{1}{2}$ and $w(AB) = 1$.) Since $w(AB) \leq \|AB\| \leq \|A\| \cdot \|B\|$, it follows that for normal operators w is submultiplicative (because if A and B are normal, then $\|A\| = w(A)$ and $\|B\| = w(B)$), and for operators in general $w(AB) \leq 4w(A)w(B)$ (because $\|A\| \leq 2w(A)$ and $\|B\| \leq 2w(B)$). The example used to show that w is not submultiplicative shows also that the constant 4 is best possible here.

Commutativity sometimes helps; here it does not. Examples of commutative operators A and B for which $w(AB) > w(A)w(B)$ are a little harder to come by, but they exist. Here is one:

$$A = \begin{pmatrix} 0 & 0 & 0 & 0 \\ 1 & 0 & 0 & 0 \\ 0 & 1 & 0 & 0 \\ 0 & 0 & 1 & 0 \end{pmatrix}$$

and $B = A^2$. It is easy to see that $w(A^2) = w(A^3) = \frac{1}{2}$. The value of $w(A)$ is slightly harder to compute, but it is not needed; the almost obvious relation $w(A) < 1$ will do. Indeed: $w(AB) = w(A^3) = \frac{1}{2} > w(A) \cdot \frac{1}{2} = w(A)w(B)$.

The only shred of multiplicative behavior that has not yet been ruled out is the *power inequality*

$$w(A^n) \leq (w(A))^n.$$

118

This turns out to be true, but remarkably tricky. Even for two-by-two matrices there is no simple computation that yields the result. If not the dimension but the exponent is specialized, if, say, $n = 2$, then relatively easy proofs exist, but even they require surprisingly delicate handling. The general case requires either brute force or ingenuity.

Problem 221. *If A is an operator such that $w(A) \leqq 1$, then $w(A^n) \leqq 1$ for every positive integer n.*

The statement is obviously a consequence of the power inequality. To show that it also implies the power inequality, reason as follows. If $w(A) = 0$, then $A = 0$, and everything is trivial. If $w(A) \neq 0$, then write $B = A/w(A)$, note that $w(B) \leqq 1$, use the statement of Problem 221 to infer that $w(B^n) \leqq 1$, and conclude that $w(A^n) \leqq (w(A))^n$.

Generalizations of the theorem are known. Here is a nice one: if p is a polynomial such that $p(0) = 0$ and $|p(z)| \leqq 1$ whenever $|z| \leqq 1$, and if A is an operator such that $w(A) \leqq 1$, then $w(p(A)) \leqq 1$. With a little care, polynomials can be replaced by analytic functions, and, with a lot of care, the unit disc (which enters by the emphasis on the inequality $|z| \leqq 1$) can be replaced by other compact convex sets.

The first proof of the power inequality is due to C. A. Berger; the first generalizations along the lines mentioned in the preceding paragraph were derived by J. G. Stampfli. The first published version, in a quite general form, appears in [86]. An interesting generalization along completely different lines appears in [101].

119

CHAPTER 23

Unitary Dilations

222. Unitary dilations. Suppose that $\mathbf{H}$ is a subspace of a Hilbert space $\mathbf{K}$, and let P be the (orthogonal) projection from $\mathbf{K}$ onto $\mathbf{H}$. Each operator B on $\mathbf{K}$ induces in a natural way an operator A on $\mathbf{H}$ defined for each f in $\mathbf{H}$ by

$$Af = PBf.$$

The relation between A and B can also be expressed by

$$AP = PBP.$$

Under these conditions the operator A is called the *compression* of B to $\mathbf{H}$ and B is called a *dilation* of A to $\mathbf{K}$. This geometric definition of compression and dilation is to be contrasted with the customary concepts of restriction and extension: if it happens that $\mathbf{H}$ is invariant under B, then it is not necessary to project Bf back into $\mathbf{H}$ (it is already there), and, in that case, A is the restriction of B to $\mathbf{H}$ and B is an extension of A to $\mathbf{K}$. Restriction-extension is a special case of compression-dilation, the special case in which the operator on the larger space leaves the smaller space invariant.

There are algebraic roads that lead to compressions and dilations, as well as geometric ones. One such road goes via quadratic forms. It makes sense to consider the quadratic form associated with B and to consider it for vectors of $\mathbf{H}$ only (i.e., to restrict it to $\mathbf{H}$). This restriction is a quadratic form on $\mathbf{H}$, and, therefore, it is induced by an operator on $\mathbf{H}$; that operator is the compression A. In other words, compression and dilation for operators are not only analogous to (and generalizations of) restriction and extension, but, in the framework of quadratic forms, they *are* restriction and extension: the quadratic form of A is the restriction of the quadratic form of B to $\mathbf{H}$, and the quadratic form of B is an extension of the quadratic form of A to $\mathbf{K}$.

Still another manifestation of compressions and dilations in Hilbert space theory is in connection with operator matrices. If $\mathbf{K}$ is decomposed into $\mathbf{H}$

120

and $\mathbf{H}^{\perp}$, and, correspondingly, operators on $\mathbf{K}$ are written in terms of matrices (whose entries are operators on $\mathbf{H}$ and $\mathbf{H}^{\perp}$ and linear transformations between $\mathbf{H}$ and $\mathbf{H}^{\perp}$), then a necessary and sufficient condition that B be a dilation of A is that the matrix of B have the form

$$\begin{pmatrix} A & X \\ Y & Z \end{pmatrix}.$$

Problem 222. (a) If $\|A\| \leq 1$, then A has a unitary dilation. (b) If $0 \leq A \leq 1$, then A has a dilation that is a projection.

Note that in both cases the assumptions are clearly necessary. If A has a dilation B that is a contraction, then $\|Af\| = \|PBf\| \leq \|Bf\| \leq \|f\|$ for all f in $\mathbf{H}$, and if A has a positive dilation B, then $(Af, f) = (Bf, f) \geq 0$ for all f in $\mathbf{H}$.

Corollary. *Every operator has a normal dilation.*

223. Images of subspaces. The range of an operator is not necessarily closed, **223** and, all the more, the image of a subspace under an operator is not necessarily closed. If the operator is very well behaved (bounded from below), all is well. Just how well does it have to behave?

Problem 223. *Is the image of a subspace under a projection closed?*

224. Weak closures and dilations. Dilations have easy but sometimes sur- **224** prising applications to the weak operator topology. Basic weak neighborhoods were defined (Problem 107) in terms of finite sets of vectors. An equivalent definition is in terms of projections of finite rank.

Assertion: a base for the weak operator topology is the collection of all sets of the form

$$\{A: \|F(A - A_0)F\| < \varepsilon\},$$

where F is a projection of finite rank and ε is a positive number. An efficient way to prove the assertion is to compare the pseudonorms

$$\|A\|_{f,g} = |(Af, g)|$$

defined by vectors with the ones

$$\|A\|_F = \|FAF\|$$

defined by projections of finite rank. Given vectors $f_1, \cdots, f_n, g_1, \cdots, g_n$, let F be the projection whose range is the span of all the f's and g's, and note that

$$\|A\|_{f_j, g_j} = |(FAFf_j, g_j)| \leq \|FAF\| \cdot M^2,$$

where $M = \max\{\|f_1\|, \cdots, \|f_n\|, \|g_1\|, \cdots, \|g_n\|\}$. In the reverse direction, given a projection F of finite rank, let $\{e_1, \cdots, e_n\}$ be an orthonormal basis for ran F and note that

$$\|A\|_F^2 \leq \sum_i \sum_j |(FAFe_j, e_i)|^2 \leq \sum_i \sum_j \|A\|_{e_j, e_i}^2.$$

121

It follows from the definition of the weak topology via projections that *if A_0 is an operator on an infinite-dimensional Hilbert space* **H** *and if* **S** *is a set of operators on* **H** *such that every compression of A_0 to a finite-dimensional subspace has a dilation in* **S**, *then A_0 itself belongs to the weak closure of* **S**. Indeed: the hypothesis means that for every projection F of finite rank there exists an operator S_0 in **S** such that $FA_0F = FS_0F$. The hypothesis implies (with room to spare) that for each ε the set

$$\{A: \|F(A - A_0)F\| < \varepsilon\}$$

meets **S**, and, consequently, that A_0 belongs to the weak closure of **S**.

Problem 224. *On an infinite-dimensional Hilbert space, what are the weak closures of the sets* **U** (*unitary operators*), **N** (*normal operators*), *and* **P** (*projections*)?

Sometimes it is good to know the weak *sequential* closures of these sets; for some pertinent information see [98].

225 **225. Strong closures and extensions.** The strong operator topology stands in the same relation to extension as the weak operator topology to dilation (see Problem 224). Assertion: a base for the strong operator topology is the collection of all sets of the form

$$\{A: \|(A - A_0)F\| < \varepsilon\},$$

where F is a projection of finite rank and ε is a positive number. An efficient way to prove the assertion is to compare the pseudonorms

$$\|A\|_f = \|Af\|$$

defined by vectors with the ones

$$\|A\|_F = \|AF\|$$

defined by projections of finite rank. Given vectors $f_1, \cdots, f_n$, let F be the projection whose range is the span of all the f's and note that

$$\|A\|_{f_j} = \|AFf_j\| \leq \|AF\| \cdot M,$$

where $M = \max\{\|f_1\|, \cdots, \|f_n\|\}$. In the reverse direction, given a projection F of finite rank, let $\{e_1, \cdots, e_n\}$ be an orthonormal basis for ran F, and note that, for all f,

$$\|AFf\| = \left\| A \sum_j (Ff, e_j)e_j \right\|$$

$$= \left\| \sum_j (f, e_j)Ae_j \right\| \leq \|f\| \cdot \sum_j \|Ae_j\|,$$

so that

$$\|A\|_F \leq \sum_j \|A\|_{e_j}.$$

One application of this approach to the strong topology is a simple proof that the set of nilpotent operators of index 2 is strongly dense; see Solution 111. It is sufficient to prove that every operator A of finite rank belongs to the strong closure of the set of nilpotents of index 2. Let $\mathbf{H}_0$ be a finite-dimensional subspace that includes ran A, let $\mathbf{H}_1$ be a subspace of the same dimension as $\mathbf{H}_0$ and orthogonal to it, and consider the matrix corresponding to A with respect to the decomposition $\mathbf{H} = \mathbf{H}_0 \oplus \mathbf{H}_1 \oplus (\mathbf{H}_0 \oplus \mathbf{H}_1)^{\perp}$:

$$\begin{pmatrix} A_0 & * & * \\ 0 & 0 & 0 \\ 0 & 0 & 0 \end{pmatrix}.$$

If B is the operator whose matrix with respect to the same decomposition is

$$\begin{pmatrix} A_0 & \dfrac{2\|A_0\|A_0}{\varepsilon} & 0 \\ \dfrac{-\varepsilon A_0}{2\|A_0\|} & -A_0 & 0 \\ 0 & 0 & 0 \end{pmatrix},$$

then B is nilpotent of index 2; if F is the projection

$$\begin{pmatrix} 1 & 0 & 0 \\ 0 & 0 & 0 \\ 0 & 0 & 0 \end{pmatrix},$$

then $\|(A - B)F\| < \varepsilon$. That is: B belongs to the strong neighborhood of A determined by ε and F.

The technique is based on the nilpotence of all matrices of the form

$$\begin{pmatrix} A & \dfrac{1}{\alpha}A \\ -\alpha A & -A \end{pmatrix};$$

it is related to the alternative proof of the weak density of the set of normal operators (Solution 224), based on the normality of all matrices of the form

$$\begin{pmatrix} A & A^* \\ A^* & A \end{pmatrix}.$$

Another application of the same technique, based on the idempotence of all matrices of the form

$$\begin{pmatrix} A & \dfrac{1}{\alpha}A \\ \alpha(1 - A) & 1 - A \end{pmatrix}$$

(with $\alpha = \varepsilon/2\|1 - A\|$) shows that the set of all idempotent operators is strongly dense.

The strong density of nilpotents and idempotents is part of the unbounded pathology of operator theory; in bounded sets it cannot happen. That is: the set of all nilpotent contractions of index 2 is strongly closed, and so is the set of all idempotent contractions. Proof: Solution 113.

It follows from the definition of the strong topology via projections that *if A_0 is an operator on an infinite-dimensional Hilbert space* **H** *and* **S** *is a set of operators on* **H** *such that every restriction of A_0 to a finite-dimensional subspace has an extension in* **S**, *then A_0 itself belongs to the strong closure of* **S**. (The restriction of A_0 to a subspace makes sense whether the subspace is invariant under A_0 or not; in any event the restriction is a bounded linear transformation from the subspace into **H**.) Indeed: the hypothesis means that for every projection F of finite rank there exists an operator S_0 in **S** such that $A_0 F = S_0 F$. The hypothesis implies, with room to spare, that for each ε the set

$$\{A: \|(A - A_0)F\| < \varepsilon\}$$

meets **S**, and, consequently, that A_0 belongs to the strong closure of **S**.

As an example, consider the set **S** of subnormal operators on **H**. If A_0 is subnormal, then, by definition, A_0 has a normal extension to a larger space **K**. The same is true, therefore, of every restriction of A_0 to a finite-dimensional subspace **M** of **H**. Since the pairs $\langle M, K \rangle$ are obviously isomorphic to the corresponding pairs $\langle M, H \rangle$, it follows that **S** is a subset of the strong closure of the set of normal operators on **H**. The fact is that **S** is strongly closed (Problem 203), but that's harder to prove.

Problem 225. *On an infinite-dimensional Hilbert space, what are the strong closures of the sets* **U** *(unitary operators) and* **P** *(projections)? What about the set of co-isometries?*

226. Strong limits of hyponormal operators.

Problem 226. *What is the strong closure of the set of hyponormal operators?*

227. Unitary power dilations. The least unitary looking contraction is 0, but even it has a unitary dilation. The construction of Solution 222 exhibits it as

$$\begin{pmatrix} 0 & 1 \\ 1 & 0 \end{pmatrix}.$$

The construction is canonical, in a sense, but it does not have many useful algebraic properties. It is not necessarily true, for instance, that the square of a dilation is a dilation of the square; indeed, the square of the dilation of 0 exhibited above is

$$\begin{pmatrix} 1 & 0 \\ 0 & 1 \end{pmatrix},$$

124

which is not a dilation of the square of 0. Is there a unitary dilation of 0 that is fair to squares? The answer is yes:

$$\begin{pmatrix} 0 & 0 & 1 \\ 1 & 0 & 0 \\ 0 & 1 & 0 \end{pmatrix}$$

is an example. The square of this dilation is

$$\begin{pmatrix} 0 & 1 & 0 \\ 0 & 0 & 1 \\ 1 & 0 & 0 \end{pmatrix},$$

which is a dilation of the square of 0. Unfortunately, however, this dilation is not perfect either; its cube is

$$\begin{pmatrix} 1 & 0 & 0 \\ 0 & 1 & 0 \\ 0 & 0 & 1 \end{pmatrix},$$

which is not a dilation of the cube of 0. The cube injustice can be remedied by passage to

$$\begin{pmatrix} 0 & 0 & 0 & 1 \\ 1 & 0 & 0 & 0 \\ 0 & 1 & 0 & 0 \\ 0 & 0 & 1 & 0 \end{pmatrix},$$

but then fourth powers fail. There is no end to inductive greed; the clearly suggested final demand is for a unitary dilation of 0 with the property that all its powers are dilations of 0. In matrix language the demand is for a unitary matrix with the property that one of its diagonal entries is 0 and that, moreover, the corresponding entry in all its powers is also 0. Brief meditation on the preceding finite examples, or just inspired guessing, might suggest the answer; the bilateral shift will work, with the $\langle 0, 0 \rangle$ entry playing the distinguished role. (Caution: the unilateral shift is not unitary.) The general definition suggested by the preceding considerations is this: an operator B is a *power dilation* (sometimes called a *strong dilation*) of an operator A if B^n is a dilation of A^n for $n = 1, 2, 3, \cdots$.

Problem 227. *Every contraction has a unitary power dilation.*

In all fairness to dilations, it should be mentioned that they all have at least one useful algebraic property: if B is a dilation of A, then B^* is a dilation of A^*. The quickest proof is via quadratic forms: if $(Af, f) = (Bf, f)$ for each f in the domain of A, then, for the same f's, $(A^*f, f) = (f, Af) = (Af, f)^* = (Bf, f)^* = (f, Bf) = (B^*f, f)$. One consequence of this is that if B is a power dilation of A, then B^* is a power dilation of A^*.

The power dilation theorem was first proved by Nagy [95]. The subject has received quite a lot of attention since then; good summaries of results are in [96] and [94]. An especially interesting aspect of the theory concerns minimal unitary power dilations. Their definition is similar to that of minimal normal extensions (Problem 197), and they too are uniquely determined by the given operator (to within unitary equivalence). The curious fact is that knowledge of the minimal unitary power dilation of an operator is not so helpful as one might think. Schreiber [128] proved that all strict contractions (see Problem 153) on separable Hilbert spaces have the same minimal unitary power dilation, namely a bilateral shift; Nagy [97] extended the result to non-separable spaces.

228

228. Ergodic theorem. If u is a complex number of modulus 1, then the averages

$$\frac{1}{n} \sum_{j=0}^{n-1} u^j$$

from a convergent sequence. This is an amusing and simple piece of classical analysis, whose generalizations are widely applicable. To prove the statement, consider separately the cases $u = 1$ and $u \neq 1$. If $u = 1$, then each average is equal to 1, and the limit is 1. If $u \neq 1$, then

$$\left| \frac{1}{n} \sum_{j=0}^{n-1} u^j \right| = \left| \frac{1 - u^n}{n(1 - u)} \right| \leq \frac{1}{n|1 - u|},$$

and the limit is 0.

The most plausible operatorial generalization of the result of the preceding paragraph is known as the *mean ergodic theorem* for unitary operators; it asserts that if U is a unitary operator on a Hilbert space, then the averages

$$\frac{1}{n} \sum_{j=0}^{n-1} U^j$$

form a strongly convergent sequence. A more informative statement of the ergodic theorem might go on to describe the limit; it is, in fact, the projection whose range is the subspace $\{f : Uf = f\}$, i.e., the subspace of fixed points of U.

It is less obvious that a similar ergodic theorem is true not only for unitary operators but for all contractions.

Problem 228. *If A is a contraction on a Hilbert space* **H**, *then*

$$\left\{ \frac{1}{n} \sum_{j=0}^{n-1} A^j \right\}$$

is a strongly convergent sequence of operators on **H**.

229. von Neumann's inequality. If F is a bounded complex-valued function **229** defined on a set M, write

$$\|F\|_M = \sup\{|F(\lambda)|: \lambda \in M\}.$$

If A is a normal operator with spectrum Λ, and if F is a bounded Borel measurable function on Λ, then $\|F(A)\| \leqq \|F\|_\Lambda$. (Equality does not hold in general; F may take a few large values that have no measure-theoretically detectable influence on $F(A)$.) It is not obvious how this inequality can be generalized to non-normal operators. There are two obstacles: in general, $F(A)$ does not make sense, and, when it does, the result can be false. There is an easy way around both obstacles: consider only such functions F for which $F(A)$ does make sense, and consider only such sets, in the role of Λ, for which the inequality does hold. A viable theory can be built on these special considerations.

If the only functions considered are polynomials, then they can be applied to every operator. If, however, the spectrum of the operator is too small, the inequality between norms will fail. If, for instance, A is quasinilpotent and $p(z) = z$, then $\|p(A)\| = \|A\|$ and $\|p\|_{\text{spec } A} = 0$; the inequality $\|p(A)\| \leqq \|p\|_{\text{spec } A}$ holds only if $A = 0$. The earliest positive result, which is still the most incisive and informative statement along these lines, is sometimes known as the von Neumann inequality. (Reference: [151], [70].)

Problem 229. *If* $\|A\| \leqq 1$ *and if D is the closed unit disc, then*

$$\|p(A)\| \leqq \|p\|_D$$

for every polynomial p.

The general context to which the theorem belongs is the theory of spectral sets. That theory is concerned with rational functions instead of just polynomials. Roughly speaking, a spectral set for an operator is a set such that the appropriate norm inequality holds for all rational functions on the set. Precisely, a *spectral set* for A is a set M such that spec $A \subset M$ and such that if F is a bounded rational function on M (i.e., a rational function with no poles in the closure of M), then $\|F(A)\| \leqq \|F\|_M$. (Note that the condition on the poles of the admissible F's implies that $F(A)$ makes sense for each such F.) It turns out that the theory loses no generality if the definition of spectral set demands that the set be closed, or even compact, and that is usually done. To demand the norm inequality for polynomials only does, however, seriously change the definition. A moderately sophisticated complex function argument (cf. [90]) can be used to show that the polynomial definition and the rational function definition are the same in case the set in question is sufficiently simple. (For this purpose a set is sufficiently simple if it is compact and its complement is connected.) In view of the last remark, the von Neumann inequality is frequently stated as follows: the closed unit disc is a spectral set for every contraction.

CHAPTER 24

Commutators

230. Commutators. A mathematical formulation of the famous Heisenberg uncertainty principle is that a certain pair of linear transformations P and Q satisfies, after suitable normalizations, the equation $PQ - QP = 1$. It is easy enough to produce a concrete example of this behavior; consider $\mathbf{L}^2(-\infty, +\infty)$ and let P and Q be the differentiation transformation and the position transformation, respectively (that is, $(Pf)(x) = f'(x)$ and $(Qf)(x) = xf(x)$). These are not bounded linear transformations, of course, their domains are far from being the whole space, and they misbehave in many other ways. Can this misbehavior be avoided?

To phrase the question precisely, define a *commutator* as an operator of the form $PQ - QP$, where P and Q are operators on a Hilbert space. More general uses of the word can be found in the literature (e.g., commutators on Banach spaces), and most of them do not conflict with the present definition; the main thing that it is intended to exclude is the unbounded case. The question of the preceding paragraph can be phrased this way: "Is 1 a commutator?" The answer is no.

Problem 230. *The only scalar commutator is* 0.

The finite-dimensional case is easy to settle. The reason is that in that case the concept of trace is available. Trace is linear, and the trace of a product of two factors is independent of their order. It follows that the trace of a commutator is always zero; the only scalar with trace 0 is 0 itself. That settles the negative statement. More is known: in fact a finite square matrix is a commutator if and only if it has trace 0 ([135], [2]).

For the general (not necessarily finite-dimensional) case, two beautiful proofs are known, quite different from one another; they are due to Wintner

128

[160] and Wielandt [158]. Both apply, with no change, to arbitrary complex normed algebras with unit. A *normed algebra* is a normed vector space that is at the same time an algebra such that

$$\|fg\| \leqq \|f\| \cdot \|g\|$$

for all f and g. A *unit* in a normed algebra is, of course, an element e such that $ef = fe = f$ for all f; it is customary to require, moreover, that $\|e\| = 1$. The algebraic character of the Wintner and Wielandt proofs can be used to get more information about commutators, as follows.

The identity is a projection; it is the unique projection with nullity 0. (Recall that the nullity of an operator is the dimension of its kernel.) What about a projection (on an infinite-dimensional Hilbert space) with nullity 1; can it be a commutator? Intuition cries out for a negative answer, and, for once, intuition is right [55]. Consider the normed algebra of all operators and in it the ideal of compact operators. The quotient algebra is a normed algebra. In that algebra the unit element is not a commutator (by Wintner and Wielandt); translated back to operators, this means that the identity cannot be equal to the sum of a commutator and a compact operator. Since a projection with nullity 1 is a very special example of such a sum, the proof is complete. The following statement summarizes what the proof proves.

Corollary. *The sum of a compact operator and a non-zero scalar is not a commutator.*

The corollary gives a sufficient condition that an operator be a non-commutator; the most surprising fact in this subject is that on separable spaces the condition is necessary also [22]. In other words: on a separable space every operator that is not the sum of a non-zero scalar and a compact operator is a commutator. The proof is not short.

231. Limits of commutators. Granted that the identity is not a commutator, **231** is it at least a limit of commutators? Do there, in other words, exist sequences $\{P_n\}$ and $\{Q_n\}$ of operators such that $\|1 - (P_n Q_n - Q_n P_n)\| \to 0$ as $n \to \infty$? The Brown–Pearcy characterization of commutators (see Problem 230) implies that the answer is yes. (See also Problem 235.) A more modest result is more easily accessible.

Problem 231. *If $\{P_n\}$ and $\{Q_n\}$ are bounded sequences of operators (i.e., if there exists a positive number α such that $\|P_n\| \leqq \alpha$ and $\|Q_n\| \leqq \alpha$ for all n), and if the sequence $\{P_n Q_n - Q_n P_n\}$ converges in the norm to an operator C, then $C \neq 1$.*

In other words: the identity cannot be the limit of commutators formed from bounded sequences. Reference: [20].

232

232. Kleinecke–Shirokov theorem. The result of Problem 230 says that if $C = PQ - QP$ and if C is a scalar, then $C = 0$. How does the proof use the assumption that C is a scalar? An examination of Wielandt's proof suggests at least part of the answer: it is important that C commutes with P. Commutators with this sort of commutativity property have received some attention; the original question $(PQ - QP = 1?)$ fits into the context of their theory. An easy way for $PQ - QP$ to commute with P is for it to be equal to P. Example:

$$P = \begin{pmatrix} 0 & 0 \\ 1 & 0 \end{pmatrix}, \qquad Q = \begin{pmatrix} 1 & 0 \\ 0 & 0 \end{pmatrix}.$$

If that happens, then an easy inductive argument proves that $P^nQ - QP^n = nP^n$, and this implies that

$$n\|P^n\| \leqq 2\|P^n\| \cdot \|Q\|$$

for every positive integer n. Since it is impossible that $n \leqq 2\|Q\|$ for all n, it follows that $P^n = 0$ for some n, i.e., that $P\ (= PQ - QP)$ is nilpotent.

The first general theorem of this sort is due to Jacobson [79], who proved, under suitable finiteness assumptions, that if $C = PQ - QP$ and C commutes with P, then C is nilpotent. This is a not unreasonable generalization of the theorem about scalars; after all the only nilpotent scalar is 0. In infinite-dimensional Hilbert spaces finiteness conditions are not likely to be satisfied. Kaplansky conjectured that if nilpotence is replaced by its appropriate generalization, quasinilpotence, then the Jacobson theorem will extend to operators, and he turned out to be right. The proof was discovered, independently, by Kleinecke [88] and Shirokov [134].

Problem 232. *If P and Q are operators, if $C = PQ - QP$, and if C commutes with P, then C is quasinilpotent.*

233

233. Distance from a commutator to the identity. By Wintner and Wielandt, commutators cannot be equal to 1; by Brown–Pearcy, commutators can come arbitrarily near to 1. Usually, however, a commutator is anxious to stay far from 1.

Problem 233. (a) *If $C = PQ - QP$ and if P is hyponormal (hence, in particular, if P is an isometry, or if P is normal), then $\|1 - C\| \geqq 1$.*
(b) *If C commutes with P, then $\|1 - C\| \geqq 1$.*

If the underlying Hilbert space is finite-dimensional, then it is an easy exercise in linear algebra to prove that $\|1 - C\| \geqq 1$ for all commutators C.

234

234. Operators with large kernels. As far as the construction of commutators is concerned, all the results of the preceding problems are negative; they all say that something is not a commutator.

130

To get a positive result, suppose that $\mathbf{H}$ is an infinite-dimensional Hilbert space and consider the infinite direct sum $\mathbf{H} \oplus \mathbf{H} \oplus \mathbf{H} \oplus \cdots$. Operators on this large space can be represented as infinite matrices whose entries are operators on $\mathbf{H}$. If, in particular, A is an arbitrary operator on $\mathbf{H}$ (it could even be the identity), then the matrix

$$P = \begin{pmatrix} 0 & A & 0 & 0 \\ 0 & 0 & A & 0 \\ 0 & 0 & 0 & A \\ 0 & 0 & 0 & 0 \\ & & & & \ddots \end{pmatrix}$$

defines an operator; if

$$Q = \begin{pmatrix} 0 & 0 & 0 & 0 \\ 1 & 0 & 0 & 0 \\ 0 & 1 & 0 & 0 \\ 0 & 0 & 1 & 0 \\ & & & & \ddots \end{pmatrix},$$

then it can be painlessly verified that

$$PQ - QP = \begin{pmatrix} A & 0 & 0 & 0 \\ 0 & 0 & 0 & 0 \\ 0 & 0 & 0 & 0 \\ 0 & 0 & 0 & 0 \\ & & & & \ddots \end{pmatrix}.$$

Since the direct sum of infinitely many copies of $\mathbf{H}$ is the direct sum of the first copy and the others, and since the direct sum of the others is isomorphic (unitarily equivalent) to $\mathbf{H}$, it follows that every two-by-two operator matrix of the form

$$\begin{pmatrix} A & 0 \\ 0 & 0 \end{pmatrix}$$

is a commutator [52, 53].

It is worth while reformulating the result without matrices. Call a subspace $\mathbf{M}$ of a Hilbert space $\mathbf{H}$ *large* if dim $\mathbf{M}$ = dim $\mathbf{H}$. (The idea has appeared before, even if the word has not; cf. Problem 142.) In this language, if $\mathbf{H}$ is infinite-dimensional, then $\mathbf{H}$ (regarded as one of the axes of the direct sum $\mathbf{H} \oplus \mathbf{H}$) is a large subspace of $\mathbf{H} \oplus \mathbf{H}$. If the matrix of an operator on $\mathbf{H} \oplus \mathbf{H}$ is

$$\begin{pmatrix} A & 0 \\ 0 & 0 \end{pmatrix},$$

then that operator has a large kernel, and, moreover, that kernel reduces A. If, conversely, an operator on an infinite-dimensional Hilbert space has a large reducing kernel, then that operator can be represented by a matrix of the form

$$\begin{pmatrix} A & 0 \\ 0 & 0 \end{pmatrix}.$$

(Represent the space as the direct sum of the kernel and its orthogonal complement. If the dimension of that orthogonal complement is too small, enlarge it by adjoining "half" the kernel.) In view of these remarks the matrix result of the preceding paragraph can be formulated as follows: every operator with a large reducing kernel is a commutator. This result can be improved [106].

Problem 234. *Every operator with a large kernel is a commutator.*

Corollary 1. *On an infinite-dimensional Hilbert space commutators are strongly dense.*

Corollary 2. *Every operator on an infinite-dimensional Hilbert space is the sum of two commutators.*

Corollary 2 shows that nothing like a trace can exist on the algebra of all operators on an infinite-dimensional Hilbert space. The reason is that a linear functional that deserves the name "trace" must vanish on all commutators, and hence, by Corollary 2, identically.

235. Direct sums as commutators.

Problem 235. *If an operator A on a separable Hilbert space is not a scalar, then the infinite direct sum $A \oplus A \oplus A \oplus \cdots$ is a commutator.*

Even though this result is far from a complete characterization of commutators, it answers many of the obvious questions about them. Thus, for instance, it is an immediate corollary that the spectrum of a commutator is quite arbitrary; more precisely, each non-empty compact subset of the plane (i.e., any set that can be a spectrum at all) is the spectrum of some commutator. Another immediate corollary is that the identity is the limit (in the norm) of commutators; compare Problems 231 and 233.

The techniques needed for the proof contain the germ (a very rudimentary germ, to be sure) of what is needed for the general characterization of commutators [22].

236. Positive self-commutators. The *self-commutator* of an operator A is the operator $A^*A - AA^*$. The theory of self-commutators has some interest. It is known that a finite square matrix is a self-commutator if and only if it is

Hermitian and has trace 0 ([140]). An obvious place where self-commutators could enter is in the theory of hyponormal operators; a necessary and sufficient condition that A be hyponormal is that the self-commutator of A be positive. That self-commutators can be non-trivially positive is a relatively rare phenomenon (which, by the way, is strictly infinite-dimensional). It is natural to ask just how positive a self-commutator can be, and the answer is not very.

Problem 236. *A positive self-commutator cannot be invertible.*

Reference: [108].

237. Projections as self-commutators. If a self-commutator $C = A^*A - AA^*$ is positive, then, by Problem 236, C is not invertible. The easiest way for C to be not invertible is to have a non-trivial kernel. Among the positive operators with non-trivial kernels, the most familar ones are the projections. Can C be a projection, and, if so, how?

The most obvious way for C to be a projection is for A to be normal; in that case $C = 0$. Whatever other ways there might be, they can always be combined with a normal operator (direct sum) to yield still another way, which, however, is only trivially different. The interesting question here concerns what may be called *abnormal* operators, i.e., operators that have no normal direct summands. Otherwise said, A is abnormal if no non-zero subspace of the kernel of $A^*A - AA^*$ reduces A.

It is not difficult to produce an example of an abnormal operator whose self-commutator is a projection: the unilateral shift will do. If A is a non-normal isometry (i.e., the direct sum of a unilateral shift of non-zero multiplicity and a unitary operator—see Problem 149), then $\|A\| = 1$ and $C = A^*A - AA^* = 1 - AA^*$ is the projection onto the kernel of A^*. What is interesting is that in the presence of the norm condition ($\|A\| = 1$) this is the only way to produce examples.

Problem 237. (a) *If A is an abnormal operator of norm 1, such that $A^*A - AA^*$ is a projection, then A is an isometry.* (b) *Does the statement remain true if the norm condition is not assumed?*

238. Multiplicative commutators. The word "commutator" occurs in two distinct mathematical contexts. In ring theory it means $PQ - QP$ (additive commutators); in group theory it means $PQP^{-1}Q^{-1}$ (multiplicative commutators). A little judicious guessing about trace versus determinant, and, more generally, about logarithm versus exponential, is likely to lead to the formulation of multiplicative analogues of the results about additive commutators. Some of those analogues are true. What about the analogue of the additive theorem according to which the only scalar that is an additive commutator is 0?

Problem 238. *If* **H** *is an infinite-dimensional Hilbert space, then a necessary and sufficient condition that a scalar α acting on* **H** *be a multiplicative commutator is that* $|\alpha| = 1$.

For finite-dimensional spaces determinants can be brought into play. The determinant of a multiplicative commutator is 1, and the only scalars whose determinants are 1 are the roots of unity of order equal to the dimension of the space. This proves that on an n-dimensional space a necessary condition for a scalar α to be a multiplicative commutator is that $\alpha^n = 1$; a modification of the argument that works for infinite-dimensional spaces shows that the condition is sufficient as well.

It turns out that the necessity proof is algebraic, just as in the additive theory, in the sense that it yields the same necessary condition for an arbitrary complex normed algebra with unit. From this, in turn, it follows, just as in the additive theory, that if a commutator is congruent to a scalar modulo the ideal of compact operators, then that scalar must have modulus 1.

239

239. Unitary multiplicative commutators. The positive assertion of Problem 238 can be greatly strengthened. One of the biggest steps toward the strengthened theory is the following assertion.

Problem 239. *On an infinite-dimensional Hilbert space every unitary operator is a multiplicative commutator.*

240

240. Commutator subgroup. The *commutator subgroup* of a group is the smallest subgroup that contains all elements of the form $PQP^{-1}Q^{-1}$; in other words, it is the subgroup generated by all commutators (multiplicative ones, of course). The set of all invertible operators on a Hilbert space is a multiplicative group; in analogy with standard finite-dimensional terminology, it may be called the *full linear group* of the space.

Problem 240. *What is the commutator subgroup of the full linear group of an infinite-dimensional Hilbert space?*

CHAPTER 25

Toeplitz Operators

241. Laurent operators and matrices. Multiplications are the prototypes of normal operators, and most of the obvious questions about them (e.g., those about numerical range, norm, and spectrum) have obvious answers. (This is not to say that every question about them has been answered.) Multiplications are, moreover, not too sensitive to a change of space; aside from the slightly fussy combinatorics of atoms, and aside from the pathology of the uncountable, what happens on the unit interval or the unit circle is typical of what can happen anywhere.

If φ is a bounded measurable function on the unit circle, then the multiplication induced by φ on L^2 (with respect to normalized Lebesgue measure μ) is sometimes called the *Laurent operator* induced by φ, in symbols L_φ. The matrix of L_φ with respect to the familiar standard orthonormal basis in L^2 ($e_n(z) = z^n$, $n = 0, \pm 1, \pm 2, \cdots$) has a simple form, elegantly related to φ. To describe the relation, define a *Laurent matrix* as a (bilaterally) infinite matrix $\langle \lambda_{ij} \rangle$ such that

$$\lambda_{i+1, j+1} = \lambda_{ij}$$

for all i and j ($= 0, \pm 1, \pm 2, \cdots$). In words: a Laurent matrix is one all of whose diagonals (parallel to the main diagonal) are constants.

Problem 241. *A necessary and sufficient condition that an operator on L^2 be a Laurent operator L_φ is that its matrix $\langle \lambda_{ij} \rangle$ with respect to the basis $\{e_n : n = 0, \pm 1, \pm 2, \cdots\}$ be a Laurent matrix; if that condition is satisfied, then $\lambda_{ij} = \alpha_{i-j}$, where $\varphi = \sum_n \alpha_n e_n$ is the Fourier expansion of φ.*

242. Toeplitz operators and matrices. Laurent operators (multiplications) are distinguished operators on L^2 (of the unit circle), and H^2 is a distinguished

135

subspace of L^2; something interesting is bound to happen if Laurent operators are compressed to H^2. The description of what happens is called the theory of Toeplitz operators. Explicitly: if P is the projection from L^2 onto H^2, and if φ is a bounded measurable function, then the *Toeplitz operator* T_φ induced by φ is defined by

$$T_\varphi f = P(\varphi \cdot f)$$

for all f in H^2. The simplest non-trivial example of a Laurent operator is the bilateral shift $W (= L_{e_1})$; correspondingly, the simplest non-trivial example of a Toeplitz operator is the unilateral shift $U (= T_{e_1})$.

There is a natural basis in L^2; the matrix of a Laurent operator with respect to that basis has an especially simple form. The corresponding statements are true about H^2 and Toeplitz operators. To state them, define a *Toeplitz matrix* as a (unilaterally) infinite matrix $\langle \lambda_{ij} \rangle$ such that

$$\lambda_{i+1, j+1} = \lambda_{ij}$$

for all i and j ($= 0, 1, 2, \cdots$). In words: a Toeplitz matrix is one all of whose diagonals (parallel to the main diagonal) are constants. The structural differences between the Laurent theory and the Toeplitz theory are profound, but the difference between the two kinds of matrices is superficial and easy to describe; for Laurent matrices both indices go both ways from 0, but for Toeplitz matrices they go forward only.

Problem 242. *A necessary and sufficient condition that an operator on H^2 be a Toeplitz operator T_φ is that its matrix $\langle \lambda_{ij} \rangle$ with respect to the basis $\{e_n : n = 0, 1, 2, \cdots\}$ be a Toeplitz matrix; if that condition is satisfied, then $\lambda_{ij} = \alpha_{i-j}$, where $\varphi = \sum_n \alpha_n e_n$ is the Fourier expansion of φ.*

The necessity of the condition should not be surprising: in terms of an undefined but self-explanatory phrase, it is just that the compressed operator has the compressed matrix.

The unilateral shift U does for Toeplitz operators what the bilateral shift W does for Laurent operators—but does it differently.

Corollary 1. *A necessary and sufficient condition that an operator A on H^2 be a Toeplitz operator is that $U^*AU = A$.*

Since W is unitary, there is no difference between $W^*AW = A$ and $AW = WA$. The corresponding equations for U say quite different things. The first, $U^*AU = A$, characterizes Toeplitz operators. The second, $AU = UA$, characterizes analytic Toeplitz operators (see Problem 147). The Toeplitz operator T_φ induced by φ is called *analytic* in case φ is analytic (see Problem 33), i.e., in case φ is not only in L^∞ but in H^∞. (To justify the definition, note that the statement of Problem 242 implies that the correspondence $\varphi \mapsto T_\varphi$ is one-to-one.) Observe that an analytic Toeplitz operator

136

is subnormal; it is not only a compression but a restriction of the corresponding Laurent operator.

Corollary 2. *The only compact Toeplitz operator is 0.*

243. Toeplitz products. The algebraic structure of the set of all Laurent **243** operators holds no surprises: everything is true and everything is easy. The mapping $\varphi \mapsto L_\varphi$ from bounded measurable functions to operators is an algebraic homomorphism (it preserves unit, linear operations, multiplication, and conjugation), and an isometry (supremum norm to operator norm); the spectrum of L_φ is the essential range of φ. Since the Laurent operators constitute the commutant of W (Problem 146), and since the product $W^{-1}AW$ is weakly continuous in its middle factor, it follows that the set of all Laurent operators is weakly (and hence strongly) closed.

Some of the corresponding Toeplitz statements are true and easy, but some are hard, or false, or unknown. The easiest statements concern unit, linear operations, and conjugation: since both the mappings $\varphi \mapsto L_\varphi$ and $L_\varphi \mapsto (PL_\varphi)|\mathbf{H}^2$ ($=$ the restriction of PL_φ to $\mathbf{H}^2 = T_\varphi$) preserve the algebraic structures named, the same is true of their composite, which is $\varphi \mapsto T_\varphi$. (The preservation of adjunction is true for compressions in general; see Problem 227.) The argument that proved that the set of all Laurent operators is weakly closed works for Toeplitz operators too; just replace $W^{-1}AW$ by U^*AU (cf. Corollary 1 of Problem 242).

It is a trivial consequence of the preceding paragraph that a Toeplitz operator T_φ is Hermitian if and only if φ is real; indeed $T_\varphi = T_\varphi^*$ if and only if $\varphi = \varphi^*$. It is also true that T_φ is positive if and only if φ is positive. Indeed, since $(T_\varphi f, f) = (L_\varphi f, f)$ whenever $f \in \mathbf{H}^2$, it follows that T_φ is positive if and only if $(L_\varphi f, f) \geqq 0$ for all f in $\mathbf{H}^2$. The latter condition is equivalent to this one: $(W^n L_\varphi f, W^n f) \geqq 0$ whenever $f \in \mathbf{H}^2$ (and n is an arbitrary integer). Since W commutes with L_φ, the condition can also be expressed in this form: $(L_\varphi W^n f, W^n f) \geqq 0$ whenever $f \in \mathbf{H}^2$. Since the set of all $W^n f$'s, with f in $\mathbf{H}^2$, is dense in $\mathbf{L}^2$, the condition is equivalent to $L_\varphi \geqq 0$, and hence to $\varphi \geqq 0$.

The easiest statements about the multiplicative properties of Toeplitz operators are negative: the set of all Toeplitz operators is certainly not commutative and certainly not closed under multiplication. A counterexample for both assertions is given by the unilateral shift and its adjoint. Both U and U^* are Toeplitz operators, but the product U^*U (which is equal to the Toeplitz operator 1) is not the same as the product UU^* (which is not a Toeplitz operator). One way to prove that UU^* is not a Toeplitz operator is to use Corollary 1 of Problem 242: since $U^*(UU^*)U = (U^*U)(U^*U) = 1$ ($\neq UU^*$), everything is settled. Alternatively, this negative result could have been obtained via Problem 242 by a direct look at the matrix of UU^*.

When is the product of two Toeplitz operators a Toeplitz operator? The answer is: rarely. Reference: [19].

Problem 243. *A necessary and sufficient condition that the product $T_\varphi T_\psi$ of two Toeplitz operators be a Toeplitz operator is that either φ^* or ψ be analytic; if the condition is satisfied, then $T_\varphi T_\psi = T_{\varphi\psi}$.*

The Toeplitz operator T_φ induced by φ is called *co-analytic* in case φ is co-analytic (see Problem 33). In this language, Problem 243 says that the product of two Toeplitz operators is a Toeplitz operator if and only if the first factor is co-analytic or the second one is analytic.

Corollary. *A necessary and sufficient condition that the product of two Toeplitz operators be zero is that at least one factor be zero.*

Concisely: among the Toeplitz operators there are no zero-divisors.

244. Compact Toeplitz products. The Toeplitz mapping $\varphi \mapsto T_\varphi$ from functions to operators is not multiplicative; is it at least multiplicative modulo compact operators?

Problem 244. *If φ and ψ are in $\mathbf{L}^\infty$, does it follow that $T_\varphi T_\psi - T_{\varphi\psi}$ is compact?*

245. Spectral inclusion theorem for Toeplitz operators. Questions about the norms and the spectra of Toeplitz operators are considerably more difficult than those for Laurent operators. As for the norm of T_φ, for instance, all that is obvious at first glance is that $\|T_\varphi\| \leqq \|L_\varphi\|\ (= \|\varphi\|_\infty)$; that much is obvious because T is a compression of L. About the spectrum of T nothing is obvious, but there is a relatively easy inequality ([68]) that answers some of the natural questions.

Problem 245. *If L and T are the Laurent and the Toeplitz operators induced by a bounded measurable function, then $\Pi(L) \subset \Pi(T)$.*

This is a spectral inclusion theorem, formally similar to Problem 200; here, too, the "larger" operator has the smaller spectrum. The result raises a hope that it is necessary to nip in the bud. If T_φ is bounded from below, so that $0 \notin \Pi(T_\varphi)$, then, by Problem 245, $0 \notin \Pi(L_\varphi)$. This is equivalent to L_φ being bounded from below and hence to φ being bounded away from 0. If the converse were true, then the spectral structure of T_φ would be much more easily predictable from φ than in fact it is; unfortunately the converse is false. If, indeed, $\varphi = e_{-1}$, then φ is bounded away from 0, but $T_\varphi e_0 = P e_{-1} = 0$, so that T_φ has a non-trivial kernel.

Although the spectral behavior of Toeplitz operators is relatively bad, Problem 245 can be used to show that in some respects Toeplitz operators behave as if they were normal. Here are some samples.

Corollary 1. *If φ is a bounded measurable function, then $r(T_\varphi) = \|T_\varphi\| = \|\varphi\|_\infty$.*

Corollary 1 says, among other things, that the correspondence $\varphi \mapsto T_\varphi$ is norm-preserving; this recaptures the result (cf. Problem 242) that that correspondence is one-to-one.

Corollary 2. *There are no quasinilpotent Toeplitz operators other than 0.*

Corollary 3. *Every Toeplitz operator with a real spectrum is Hermitian.*

Corollary 4. *The closure of the numerical range of a Toeplitz operator is the convex hull of its spectrum.*

246. Continuous Toeplitz products. The multiplicative properties of T_φ are **246** bad both absolutely and relatively (relative to compact operators, that is). The example in Solution 244 used discontinuous functions; is that what made it work?

Problem 246. *If φ and ψ are continuous, does it follow that $T_\varphi T_\psi - T_{\varphi\psi}$ is compact?*

247. Analytic Toeplitz operators. The easiest Toeplitz operators are the **247** analytic ones, but even for them much more care is needed than for multiplications. The operative word is "analytic". Recall that associated with each φ in $\mathbf{H}^\infty$ there is a function $\tilde\varphi$ analytic in the open unit disc D (see Problem 35). The spectral behavior of T_φ is influenced by the complex analytic behavior of $\tilde\varphi$ rather than by the merely set-theoretic behavior of φ. Reference: [159].

Problem 247. *If $\varphi \in \mathbf{H}^\infty$, then the spectrum of T_φ is the closure of the image of the open unit disc D under the corresponding element $\tilde\varphi$ of $\tilde{\mathbf{H}}^\infty$; in other words spec $T_\varphi = \overline{\tilde\varphi(D)}$.*

Here is still another way to express the result. If $\varphi \in \mathbf{L}^\infty$, then the spectrum of L_φ is the essential range of φ; if $\varphi \in \mathbf{H}^\infty$, then the spectrum of T_φ is what may be called the essential range of $\tilde\varphi$.

248. Eigenvalues of Hermitian Toeplitz operators. Can an analytic Toeplitz **248** operator have an eigenvalue? Except in the trivial case of scalar operators, the answer is no. The reason is that if φ is analytic and $\varphi \cdot f = \lambda f$ for some f in $\mathbf{H}^2$, then the F. and M. Riesz theorem (Problem 158) implies that either $\varphi = \lambda$ or $f = 0$. Roughly speaking, the reason is that an analytic function cannot take a constant value on a set of positive measure without being a constant. For Hermitian Toeplitz operators this reasoning does not apply: there is nothing to stop a non-constant real-valued function from being constant on a set of positive measure.

Problem 248. *Given a real-valued function φ in $\mathbf{L}^\infty$, determine the point spectrum of the Hermitian Toeplitz operator T_φ.*

249   **249. Zero-divisors.** Can operators that are not small have small products? There are many ways to interpret the question. A trivial way is to interpret "small" to mean "zero", and in that case the answer is trivially yes: there are operators that are zero-divisors. It is relevant to recall, however, that with this interpretation the answer is no for Toeplitz operators.

Another way to interpret "small" is as "compact", but the answer for operators in general doesn't change; it is still yes. Indeed: there are non-compact zero-divisors. One way to make the question more challenging might seem to be to insist that the factors be "good" operators, e.g., Hermitian operators with kernel 0. The answer, however, is still the same, and still easy. Example: if $A = 1$ (on an infinite-dimensional space) and $B = \mathrm{diag}(1, \frac{1}{2}, \frac{1}{3}, \cdots)$ (on the same space), then $A \oplus B$ and $B \oplus A$ are non-compact Hermitian operators with kernel 0, whose product is compact.

Very well, to rule out this sort of construction, rule out eigenvalues. Do there exist two Hermitian operators with no eigenvalues whose product is compact? Now at last the answer is not quite so near the surface, but it turns out that it is still yes. Unless, however, the approach is right, the construction can be quite laborious.

At least one question pertinent to the present context still remains: what about Toeplitz operators?

> **Problem 249.** *Do there exist non-zero Toeplitz operators whose product is compact? Equivalently: are there Toeplitz zero-divisors modulo the compact operators?*

250   **250. Spectrum of a Hermitian Toeplitz operator.**

> **Problem 250.** *Given a real-valued function φ in $\mathbf{L}^\infty$, determine the spectrum of the Hermitian Toeplitz operator T_φ.*

For more recent and more general studies of the spectra of Toeplitz operators, see Widom [154, 155].

HINTS

Chapter 1. Vectors

Problem 1. Polarize.

Problem 2. Consider a perturbation of the given form by a small multiple of a strictly positive one.

Problem 3. Use uniqueness: if $f = \sum_j \alpha_j e_j$, then $\xi(f) = \sum_j \alpha_j \xi(e_j)$.

Problem 4. Use inner products to reduce the problem to the strict convexity of the unit disc.

Problem 5. Consider characteristic functions in $L^2(0, 1)$. An alternative hint, for those who know about spectral measures, is to contemplate spectral measures.

Problem 6. If f and g are normalized crinkled arcs, define φ on $[0, 1]$ so that $\|g(t)\| = \|f(\varphi(t))\|$.

Problem 7. A countably infinite set has an uncountable collection of infinite subsets such that the intersection of two distinct ones among them is always finite.

Problem 8. Consider the differences $e_n - e_{n-1}$ formed from an orthonormal basis $\{\cdots, e_{-1}, e_0, e_1, \cdots\}$.

Problem 9. Omit an infinite subset by omitting one element at a time.

Problem 10. Determine the orthogonal complement of the span.

Problem 11. Use Solution 10.

Problem 12. $|(e_j - f_j, e_i)| = |(e_i - f_i, f_j)|$.

Chapter 2. Spaces

Problem 13. Prove that $\mathbf{M} + \mathbf{N}$ is complete. There is no loss of generality in assuming that $\dim \mathbf{M} = 1$.

Problem 14. In an infinite-dimensional space there always exist two subspaces whose vector sum is different from their span.

Problem 15. If $g + h \in \mathbf{L}$, with g in $\mathbf{M}$ and h in $\mathbf{N}$, then $g \in \mathbf{L} \cap \mathbf{M}$. For the converse: use Solution 14.

Problem 16. How many basis elements can an open ball of diameter $\sqrt{2}$ contain?

Problem 17. Given a countable basis, use rational coefficients. Given a countable dense set, approximate each element of a basis close enough to exclude all other basis elements.

Problem 18. Fit infinitely many balls of the same radius inside any given ball of positive radius.

Chaper 3. Weak Topology

Problem 19. Consider orthonormal sets. Caution: is weak closure the same as weak sequential closure?

Problem 20. Expand $\| f_n - f \|^2$.

Problem 21. Use the definition of weak convergence, $(f_n, g) \to (f, g)$, with $g = f$.

Problem 22. The span of a weakly dense set is the whole space.

Problem 23. Consider the set of all complex-valued functions ξ on $\mathbf{H}$ such that $|\xi(f)| \leq \| f \|$ for all f, endowed with the product topology, and show that the linear functionals of norm less than or equal to 1 form a closed subset.

Problem 24. Given a countable dense set, define all possible basic weak neighborhoods of each of its elements, using finite subsets of itself for the vector parameters and reciprocals of positive integers for the numerical parameters of the neighborhoods; show that the resulting collection of neighborhoods is a base for the weak topology. Alternatively, given an orthonormal basis $\{e_1, e_2, e_3, \cdots\}$, define a metric by

$$d(f, g) = \sum_j \frac{1}{2^j} |(f - g, e_j)|.$$

Problem 25. Given a vector in the open unit ball, add suitable multiples of orthonormal vectors to convert it to a unit vector.

Problem 26. If the unit ball is weakly metrizable, then it is weakly separable.

Problem 27. If the conclusion is false, then construct, inductively, an orthonormal sequence such that the inner product of each term with a suitable element of the given weakly bounded set is very large; then form a suitable (infinite) linear combination of the terms of that orthonormal sequence.

Problem 28. Construct a sequence that has a weak cluster point but whose norms tend to ∞.

Problem 29. Consider partial sums and use the principle of uniform boundedness.

Problem 30. (a) Given an unbounded linear functional ξ, use a Hamel basis to construct a Cauchy net $\{g_J\}$ such that $(f, g_J) \to \xi(f)$ for each f. (b) If $\{g_n\}$ is a weak Cauchy sequence, then $\xi(f) = \lim_n (f, g_n)$ defines a bounded linear functional.

Chapter 4. Analytic Functions

Problem 31. The value of an analytic function at the center of a disc is equal to its average over the disc. This implies that evaluation at a point of D is a bounded linear functional on $\mathbf{A}^2(D)$, and hence that Cauchy sequences in the norm are Cauchy sequences in the sense of uniform convergence on compact sets.

Problem 32. What is the connection between the concepts of convergence appropriate to power series and Fourier series?

Problem 33. Is conjugation continuous?

Problem 34. Is the Fourier series of a product the same as the formal product of the Fourier series?

Problem 35. A necessary and sufficient condition that

$$\sum_{n=0}^{\infty} |\alpha_n|^2 < \infty$$

is that the numbers

$$\sum_{n=0}^{\infty} |\alpha_n|^2 r^{2n} \qquad (0 < r < 1)$$

be bounded. Use continuity of the partial sums at $r = 1$.

Problem 36. Start with a well-behaved functional Hilbert space and adjoin a point to its domain.

Problem 37. To evaluate the Bergman and the Szegő kernels, use the general expression of a kernel function as a series.

Problem 38. Examine the finite-dimensional case to see whether or not the isometry of conjugation implies any restriction on the kernel function.

Problem 39. Use the kernel function of $\tilde{\mathbf{H}}^2$.

Problem 40. Approximate f, in the norm, by the values of $\tilde{f}$ on expanding concentric circles.

Problem 41. Use the maximum modulus principle and Fejér's theorem about the Cesàro convergence of Fourier series.

Problem 42. Assume that one factor is bounded and use Problem 34.

Problem 43. Given the Fourier expansion of an element of $\mathbf{H}^2$, first find the Fourier expansion of its real part, and then try to invert the process.

Chapter 5. Infinite Matrices

Problem 44. Treat the case of dimension $\aleph_0$ only. Construct the desired orthonormal set inductively; ensure that it is a basis by choosing every other element of it so that the span of that element and its predecessors includes the successive terms of a prescribed basis.

Problem 45. Write $\sum_j \alpha_{ij}\xi_j$ as

$$\sum_i (\sqrt{\alpha_{ij}}\,\sqrt{q_j})\left(\frac{\sqrt{\alpha_{ij}}\,\xi_j}{\sqrt{q_j}}\right),$$

and apply the Schwarz inequality.

Problem 46. Apply Problem 45 with $p_i = q_i = 1/\sqrt{i + \frac{1}{2}}$.

Problem 47. Look at the sum of the squares of the matrix entries. Note that the operator has rank 1.

Problem 48. Is it a Gramian?

Problem 49. Consider the Gramian matrix $\langle (f_j, f_i) \rangle$. Use the principle of uniform boundedness.

Chapter 6. Boundedness and Invertibility

Problem 50. For (a) and (b), extend an orthonormal basis to a Hamel basis; for (c) imitate Solution 27; for (d) use a matrix with a large but finite first row.

Problem 51. Apply the principle of uniform boundedness for linear functionals twice.

Problem 52. Prove that A^* is bounded from below by proving that the inverse image under A^* of the unit sphere in **H** is bounded.

Problem 53. (a) Use the x-axis and the graph of an operator; diminish the graph by restricting the operator to a subspace. (b) Form infinite direct sums.

Problem 54. Consider the graph of a linear transformation that maps a Hamel basis of a separable Hilbert space onto an orthonormal basis of a non-separable one.

Problem 55. Use Problem 54.

Problem 56. Write the given linear transformation as a matrix (with respect to orthonormal bases of **H** and **K**); if $\aleph_0 \leq \dim \mathbf{K} < \dim \mathbf{H}$, then there must be a row consisting of nothing but 0's.

Problem 57. Use Problem 56.

Problem 58. Apply Problem 52 to the mapping that projects the graph onto the domain.

Problem 59. The equation $Af = BXf$ uniquely determines an Xf in $\ker^\perp B$; to prove boundedness, use the closed graph theorem.

Problem 60. For a counterexample, look at unbounded diagonal matrices. For a proof, apply either the closed graph theorem or the principle of uniform boundedness.

Chapter 7. Multiplication Operators

Problem 61. $|\alpha_j| = \|Ae_j\|$ and

$$\sum_j |\alpha_j \xi_j|^2 \leq \left(\sup_j |\alpha_j|\right)^2 \cdot \sum_j |\xi_j|^2.$$

Problem 62. If $|\alpha_n| \geq n$, then the sequence $\{1/\alpha_n\}$ belongs to l^2.

Problem 63. The inverse operator must send e_n onto $(1/\alpha_n)e_n$.

Problem 64. If $\varepsilon > 0$ and if f is the characteristic function of a set of positive finite measure on which $|\varphi(x)| > \|\varphi\|_\infty - \varepsilon$, then

$$\|Af\| \geq (\|\varphi\|_\infty - \varepsilon) \cdot \|f\|.$$

Problem 65. If $\|A\| = 1$, then $\|\varphi^n \cdot f\| \leq \|f\|$ for every positive integer n and for every f in $\mathbf{L}^2$; this implies that $|\varphi(x)| \leq 1$ whenever $f(x) \neq 0$.

Problem 66. Imitate the discrete case (Solution 62), or prove that a multiplication is necessarily closed and apply the closed graph theorem.

Problem 67. Imitate the discrete case (Solution 63).

Problem 68. For the boundedness of the multiplication, use the closed graph theorem. For the boundedness of the multiplier, assume that if $x \in X$, then there exists an f in $\mathbf{H}$ such that $f(x) \neq 0$; imitate the "slick" proof in Solution 65.

Problem 69. Consider the set of all those absolutely continuous functions on $[0, 1]$ whose derivatives belong to $\mathbf{L}^2$.

Chapter 8. Operator Matrices

Problem 70. Necessity: if an operator commutes with each entry of the matrix, then it commutes with each entry of the inverse. Sufficiency: use Cramer's rule.

Problem 71. Multiply on the right by

$$\begin{pmatrix} 1 & 0 \\ T & 1 \end{pmatrix},$$

148

with T chosen so as to annihilate the lower left entry of the product. Look for counterexamples formed out of the operator on l^2 defined by

$$\langle \xi_0, \xi_1, \xi_2, \cdots \rangle \mapsto \langle 0, \xi_0, \xi_1, \xi_2, \cdots \rangle,$$

and its adjoint.

Problem 72. If a finite-dimensional subspace is invariant under an invertible operator, then it is invariant under the inverse.

Chapter 9. Properties of Spectra

Problem 73. The kernel of an operator is the orthogonal complement of the range of its adjoint.

Problem 74. To prove $\Pi_0(p(A)) \subset p(\Pi_0(A))$, given α in $\Pi_0(p(A))$, factor $p(\lambda) - \alpha$. Use the same technique for Π, and, for Γ, apply the result with A^* in place of A.

Problem 75. For Π: if $\| f_n \| = 1$, then the numbers $\| P^{-1} f_n \|$ are bounded from below by $1/\| P \|$. For Γ: the range of $P^{-1}AP$ is included in the image under P^{-1} of the range of A.

Problem 76. Pretend that it is legitimate to expand $(1 - AB)^{-1}$ into a geometric series.

Problem 77. Prove that the complement is open.

Problem 78. Suppose that $\lambda_n \notin \operatorname{spec} A$, $\lambda \in \operatorname{spec} A$, and $\lambda_n \to \lambda$. If $f \neq 0$ and $f \perp \operatorname{ran}(A - \lambda)$, then

$$\frac{(A - \lambda)(A - \lambda_n)^{-1} f}{\| (A - \lambda_n)^{-1} f \|} \to 0.$$

Chapter 10. Examples of Spectra

Problem 79. If A is normal, then $\Pi_0(A) = (\Pi_0(A^*))^*$.

Problem 80. Use Problem 79.

Problem 81. If $\varphi \cdot f = \lambda f$ almost everywhere, then $\varphi = \lambda$ whenever $f \neq 0$.

149

Problem 82. Verify that $U^*\langle \xi_0, \xi_1, \xi_2, \cdots \rangle = \langle \xi_1, \xi_2, \xi_3, \cdots \rangle$. Compute that $\Pi_0(U)$ is empty and $\Pi_0(U^*)$ is the open unit disc. If $|\lambda| < 1$, then $U - \lambda$ is bounded from below.

Problem 83. If all the vectors in a convex subset of a Hilbert space are eigenvectors of an operator A, then they all belong to the same eigenvalue of A.

Problem 84. Represent W as a multiplication.

Problem 85. Use a spanning set of eigenvectors of A^* for the domain; for each f in that domain, define the multiplier as the conjugate of the corresponding eigenvalue.

Chapter 11. Spectral Radius

Problem 86. If λ_0 is not in the spectrum of A and if $|\lambda - \lambda_0|$ is sufficiently small, then

$$\rho_A(\lambda) = (A - \lambda_0)^{-1} \sum_{n=0}^{\infty} ((A - \lambda_0)^{-1}(\lambda - \lambda_0))^n.$$

Problem 87. Apply Liouville's theorem on bounded entire functions to the resolvent.

Problem 88. Write

$$\tau(\lambda) = \left(A - \frac{1}{\lambda}\right)^{-1}.$$

Use the analyticity of the resolvent to conclude that τ is analytic for $|\lambda| < 1/r(A)$, and then use the principle of uniform boundedness.

Problem 89. Look for a diagonal operator D such that $AD = DB$.

Problem 90. If $A = S^{-1}BS$, then the matrix of S must be lower triangular; find the matrix entries in row $n + 1$, column n, $n = 0, 1, 2, \cdots$.

Problem 91. For the norm: S is an isometry, and therefore $\|P\| = \|SP\|$. For the spectral radius: use Problem 88.

Problem 92. Look at weighted shifts.

Problem 93. Imitate the coordinate technique used for the unweighted unilateral shift.

Problem 94. Consider a sequence dense in an interval and let the weights contain arbitrarily long blocks of each term of the sequence.

Problem 95. If $f = \langle \xi_0, \xi_1, \xi_2, \cdots \rangle \in l^2(p)$, write

$$Uf = \langle \sqrt{p_0}\xi_0, \sqrt{p_1}\xi_1, \sqrt{p_2}\xi_2, \cdots \rangle,$$

and prove that U is an isometry from $l^2(p)$ onto l^2 that transforms the shift on $l^2(p)$ onto a weighted shift on l^2.

Problem 96. Try unilateral weighted shifts; apply Solution 91.

Problem 97. If $f(A) = 0$, factor out the largest possible power of z from $f(z)$.

Problem 98. Try unilateral weighted shifts with infinitely many zero weights; apply Solution 91.

Chapter 12. Norm Topology

Problem 99. Think of projections on $\mathbf{L}^2(0, 1)$.

Problem 100. If A_0 is invertible, then $1 - AA_0^{-1} = (A_0 - A)A_0^{-1}$; use the geometric series trick to prove that A is invertible and to obtain a bound on $\|A^{-1}\|$.

Problem 101. Add a small scalar.

Problem 102. Find the spectral radius of both A_k and A_k^{-1}.

Problem 103. The distance from $A - \lambda$ to the set of singular operators is positive on the complement of spec A. Alternatively, the norm of the resolvent is bounded on the complement of Λ_0; the reciprocal of a bound is a suitable ε. For a sequential proof, use the fact that the set of singular operators is closed.

Problem 104. Approximate a weighted unilateral shift with positive spectral radius by weighted shifts with enough zero weights to make them nilpotent.

Problem 105. What does it mean that $\lambda \notin \liminf_n$ spec A_n? Recall that, for normal operators, spectral radius is equal to norm.

Problem 106. In the finite-dimensional case spectrum is continuous. In the infinite-dimensional case, modify Solution 104.

Chapter 13. Operator Topologies

Problem 107. For the first part, assume that $|(A_n f, g)| < \varepsilon$ for all unit vectors g, and replace g by $A_n f/\|A_n f\|$. For the second part, imitate the first part.

Problem 108. For a counterexample with respect to the strong topology, consider the projections onto a decreasing sequence of subspaces.

Problem 109. Imitate Solution 21.

Problem 110. For a counterexample with respect to the strong topology, consider the powers of the adjoint of the unilateral shift.

Problem 111. The set of all nilpotent operators of index 2 is strongly dense.

Problem 112. Use nets.

Problem 113. (a) Use the principle of uniform boundedness. (b) Look at powers of the unilateral shift.

Problem 114. Consider square roots of $\begin{pmatrix} 1 & 0 \\ 0 & 1 \end{pmatrix}$ with one axis nailed down and the other slipping out far away.

Problem 115. Use the slip-away technique of Solution 114.

Chapter 14. Strong Operator Topology

Problem 116. Use Problem 20.

Problem 117. Let U be the unilateral shift; use suitable multiples of powers of U^* for the A's, and, similarly, use suitable multiples of powers of U for the B's.

Problem 118. Consider the adjoint of the unilateral shift.

Problem 119. Consider an increasing sequence of projections; consider powers of the adjoint of the unilateral shift.

Problem 120. If $\{A_n\}$ is increasing and converges to A weakly, then the positive square root of $A - A_n$ converges to 0 strongly. For a counterexample with respect to the uniform topology, consider sequences of projections.

Problem 121. The B_n's form a bounded increasing sequence.

Problem 122. Study the sequence of powers of EFE.

Chapter 15. Partial Isometries

Problem 123. If N is a neighborhood of $F(\lambda)$, then $F^{-1}(N)$ is a neighborhood of λ. If $\lambda \notin F(\text{spec } A)$, then some neighborhood of λ is disjoint from $F(\text{spec } A)$.

Problem 124. Compare $\|Af\|^2$ with $\|\sqrt{A}f\|^2$.

Problem 125. Represent A as multiplication operator by a function φ, say; if p is a non-zero polynomial such that $p(A)$ is diagonal, then consider, for each eigenvalue λ of $p(A)$, the set $\varphi^{-1}\{z: p(z) = \lambda\}$.

Problem 126. Use the Weierstrass polynomial approximation theorem in the plane.

Problem 127. If U is a partial isometry with initial space $\mathbf{M}$, evaluate (U^*Uf, f) when $f \in \mathbf{M}$ and when $f \perp \mathbf{M}$; if U^*U is a projection with range $\mathbf{M}$, do the same thing.

Problem 128. The only troublesome part is to find a co-isometry U and a non-reducing subspace $\mathbf{M}$ such that $U\mathbf{M} = \mathbf{M}$; for this let U be the adjoint of the unilateral shift and let $\mathbf{M}$ be the (one-dimensional) subspace of eigenvectors belonging to a non-zero eigenvalue.

Problem 129. For closure: A is a partial isometry if and only if $A = AA^*A$. For connectedness: if U is a partial isometry, if V is an isometry, and if $\|U - V\| < 1$, then U is an isometry.

Problem 130. For rank: the restriction of U to the initial space of V is one-to-one. For nullity: if $f \in \ker V$ and $f \perp \ker U$, then

$$\|Uf - Vf\| = \|f\|.$$

Problem 131. Find a unitary operator that matches up initial spaces, and another that matches up final spaces, and find continuous curves that join each of them to the identity.

Problem 132. If U is a unitary operator matrix of size 2 that transforms $M(A)$ onto $M(B)$, then it transforms $M(A)M(A)^*$ onto $M(B)M(B)^*$.

Problem 133. If a compact subset Λ of the closed unit disc contains 0, find a contraction A with spectrum Λ, and extend A to a partial isometry.

Chapter 16. Polar Decomposition

Problem 134. Put $P^2 = A^*A$, and define U by $UPf = Af$ on ran P and by $Uf = 0$ on ker P.

Problem 135. Every partial isometry has a maximal enlargement.

Problem 136. To prove that maximal partial isometries are extreme points, use Problem 4. To prove the converse, show that every contraction is the average of two maximal partial isometries; use Problem 135.

Problem 137. If UP commutes with P^2, then it commutes with P, so that $UP - PU$ annihilates ran P.

Problem 138. Look for 2-dimensional counterexamples to three of the four possible questions; use the polar decomposition of A to prove the fourth.

Problem 139. The question and the answer are mildly interesting but the method is not: if A is a weighted shift, just compute $(A^*A)A$ and $A(A^*A)$.

Problem 140. For the positive result, apply Problem 135. For the negative one: a left-invertible operator that is not right-invertible cannot be the limit of right-invertible operators.

Problem 141. Consider polar decompositions UP and join both U and P to 1.

Chapter 17. Unilateral Shift

Problem 142. Assume that **H** is separable, and argue that it is enough to prove the existence of two orthogonal reducing subspaces of infinite dimension. Prove it by the consideration of spectral measures.

Problem 143. Apply Problem 142, and factor the given unitary operator into two operators, one of which shifts the resulting two-way sequence of subspaces forward and the other backward.

Problem 144. (a) If a normal operator has a one-sided inverse, then it is invertible. (b) Since 1 is an approximate eigenvalue of the unilateral shift, the same is true of the real part. (c) There is no invertible operator within 1 of the unilateral shift.

Problem 145. If $V^2 = U^*$, then dim ker $V \leq 1$ and V maps the underlying Hilbert space onto itself.

Problem 146. If W commutes with an operator A, and if ψ is a bounded measurable function on the circle, then, by the Fuglede commutativity theorem, $\psi(W)$ commutes with A. Put $Ae_0 = \varphi$, prove that $A\psi = \varphi \cdot \psi$, and use the technique of Solution 65.

Problem 147. Begin as for Solution 146; use Solution 65; imitate Solution 66.

Problem 148. Every function in $\mathbf{H}^\infty$ is the limit almost everywhere of a bounded sequence of polynomials; cf. Solution 41.

Problem 149. If V is an isometry on $\mathbf{H}$, and if $\mathbf{N}$ is the orthogonal complement of the range of V, then $\bigcap_{n=0}^\infty V^n \mathbf{H} = \bigcap_{n=0}^\infty (V^n \mathbf{N})^\perp$.

Problem 150. Use Problem 149, and recall that -1 belongs to the spectrum of the unilateral shift.

Problem 151. If U has multiplicity m and V is a square root of U^*, then, by Sylvester's law of nullity, $m \leq 2$ null V. If m is finite, then the reverse inequality holds.

Problem 152. If $\|A\| \leq 1$ and $A^n \to 0$ strongly, write $T = \sqrt{1 - A^*A}$ and assign to each vector f the sequence

$$\langle Tf, TAf, TA^2f, \cdots \rangle.$$

Problem 153. If $r(A) < 1$, then $\sum_{n=0}^\infty \|A^n\|^2 z^n$ converges at $z = 1$, and consequently an equivalent norm is defined by $\|f\|_0^2 = \sum_{n=0}^\infty \|A^n f\|^2$.

Problem 154. If $A = S^{-1}CS$, then $\|A^n\| \leq \|S^{-1}\| \cdot \|C^n\| \cdot \|S\|$.

Problem 155. Write $\mathbf{N} = \mathbf{M} \cap (U\mathbf{M})^\perp$ and apply the results of Solution 149. To prove dim $\mathbf{N} = 1$, assume the existence of two orthogonal unit vectors f and g in $\mathbf{N}$ and use Parseval's equation to compute $\|f\|^2 + \|g\|^2$. It is helpful to regard U as the restriction of the bilateral shift.

Problem 156. Prove that $\mathbf{M}_k^\perp(\lambda)$ is invariant under U^*.

Problem 157. Use Problem 155 to express **M** in terms of a wandering subspace **N**, and examine the Fourier expansion of a unit vector in **N**.

Problem 158. Given f in H^2, let **M** be the least subspace of H^2 that contains f and is invariant under U, and apply Problem 157 to **M**.

Problem 159. Necessity: consider a Hermitian operator that commutes with A (and hence with A^* and with A^*A), and examine its matrix. Sufficiency: assume $\{\alpha_n\}$ periodic of period p; let M_j be the span of the e_j's with $n \equiv j \pmod p$; observe that each vector has a unique representation in the form $f_0 + \cdots + f_p$, with f_j in M_j; for each measurable subset E of the circle, consider the set of all those f's for which $f_j(z) = 0$ for all j and for all z in the complement of E.

Chapter 18. Cyclic Vectors

Problem 160. For the simple shift, consider a vector $\langle \xi_0, \xi_1, \xi_2, \cdots \rangle$ such that

$$\lim_k \frac{1}{|\xi_k|^2} \sum_{n=1}^{\infty} |\xi_{n+k}|^2 = 0.$$

For shifts of higher multiplicity, form vectors whose components are subsequences of this sequence $\{\xi_n\}$.

Problem 161. Consider shifts of multiplicity greater than 1.

Problem 162. There exists an orthonormal sequence $\{f_n\}$ such that $\|(1 - U^*)f_n\| \to 0$ and hence there exist projections P_n of rank 2 such that $(1 - P_n)(1 - U) \to 1 - U$.

Problem 163. For each f and g in l^2, consider $\langle g^*, -f^* \rangle$, where the stars indicate coordinatewise complex conjugation.

Problem 164. Approximate $A^{*n}f$ by $p(A)f$, with p a polynomial.

Problem 165. If **K** is the cyclic subspace spanned by a function that is never 0, and if φ is a bounded measurable function, then $\varphi\mathbf{K} \subset \mathbf{K}$; use Fejér's theorem.

Problem 166. If f is a cyclic vector of A, and if $0 < \|\alpha A\| < 1$, then $(1 - \alpha A)^p f$ is a cyclic vector of A for each positive integer p.

Problem 167. (a) Consider the first basis vector. (b) Gram–Schmidt.

Problem 168. Consider $2U^*$ (where U is the unilateral shift acting on l^2). Construct a vector f by stringing together suitable multiples of the initial segments of the vectors in a countable dense set, separated by suitably long sequences of zeroes.

Chapter 19. Properties of Compactness

Problem 169. Use nets. In the discussion of $(w \to s)$ continuity recall that a basic weak neighborhood depends on a finite set of vectors, and consider the orthogonal complement of their span.

Problem 170. To prove self-adjointness, use the polar decomposition.

Problem 171. Approximate by diagonal operators of finite rank.

Problem 172. If the restriction of a compact operator to an invariant subspace is invertible, then the subspace is finite-dimensional. Infer, via the spectral theorem, that the part of the spectrum of a normal compact operator that lies outside a closed disc with center at the origin consists of a finite number of eigenvalues with finite multiplicities.

Problem 173. Approximate by simple functions.

Problem 174. If A is a Hilbert–Schmidt operator, then the sum of the eigenvalues of A^*A is finite.

Problem 175. Use the polar decomposition and Problem 172.

Problem 176. Every operator of rank 1 belongs to every non-zero ideal. Every non-compact Hermitian operator is bounded from below on some infinite-dimensional invariant subspace; its restriction to such a subspace is invertible.

Problem 177. Consider the direct sum of a sequence of projections of rank 1. Consider the collection of all operators that map orthonormal sequences to strong null sequences.

Problem 178. Use the spectral theorem.

Problem 179. If C is compact, then $\Pi(C) - \{0\} \subset \Pi_0(C)$, and $\Pi_0(C)$ is countable.

Problem 180. Assume, with no loss, that ker $A = 0$. In that case, if **M** is a subspace included in ran A, the restriction of A to the inverse image of **M** is invertible.

Problem 181. From (1) to (2): the restriction of A to $\ker^\perp A$ is invertible. From (3) to (1): if $1 - BA$ is compact, apply Solution 179 to $1 - BA$.

Problem 182. Assume $\lambda = 0$; note that if B is invertible, then $A = B(1 + B^{-1}(A - B))$.

Problem 183. Perturb the bilateral shift by an operator of rank 1.

Problem 184. If C is compact and $U + C$ is normal, then the spectrum of $U + C$ is large; but the spectrum of $(U + C)^*(U + C)$ is small.

Problem 185. Given the shift U and a compact operator C, estimate the spectral radius of $U - C$.

Chapter 20. Examples of Compactness

Problem 186. If A is a Volterra operator with kernel bounded by c, then A^n is a Volterra operator with kernel bounded by $c^n/(n - 1)!$.

Problem 187. Can a Volterra operator have a non-zero eigenvalue?

Problem 188. Express V^*V as an integral operator. By differentiation convert the equation $V^*Vf = \lambda f$ into a differential equation, and solve it.

Problem 189. Identify $L^2(-1, +1)$ with $L^2(0, 1) \oplus L^2(0, 1)$, and determine the two-by-two operator matrix corresponding to such an identification. Caution: there is more than one interesting way of making the identification.

Problem 190. Put $A = (1 + V)^{-1}$, where V is the Volterra integration operator.

Problem 191. Reduce to the case where **M** contains a vector f with infinitely many non-zero Fourier coefficients; in that case prove that there exist scalars λ_n such that $\lambda_n A^n f \to e_0$, so that **M** contains e_0; use induction to conclude that **M** contains e_k for every positive integer k.

Chapter 21. Subnormal Operators

Problem 192. Apply Fuglede's theorem to two-by-two operator matrices made out of A_1, A_2, and B.

Problem 193. Given A and B, consider $A + B$ and $A + iB$.

Problem 194. If $\|A^n f\| \leqq \|f\|$ for all n, and if

$$M_r = \{x : |\varphi(x)| \geqq r > 1\},$$

then $\|f\|^2 \geqq \int_{M_r} r^{2n} |f|^2 \, d\mu$.

Problem 195. Show that $\ker A$ reduces A and throw it away. Once $\ker A = 0$, consider the polar decomposition of A, extend the isometric factor to a unitary two-by-two matrix, extend the positive factor to a positive two-by-two matrix, and do all this so that the two extensions commute.

Problem 196. If A commutes with a Hermitian B, then it commutes with the spectral measure of B.

Problem 197. The desired isometry U must be such that if $\{f_1, \cdots, f_n\}$ is a finite subset of **H**, then $U(\sum_j B_1^{*j} f_j) = \sum_j B_2^{*j} f_j$.

Problem 198. Let W be the bilateral shift, and consider the powers of the adjoint of the operator $p(W)$.

Problem 199. Consider the measure space consisting of the unit circle together with its center, with measure defined so as to be normalized Lebesgue measure in the circle and a unit mass at the center. Form a subnormal operator by restricting a suitable multiplication on L^2 to the closure of the set of all polynomials.

Problem 200. It is sufficient to prove that if A is invertible, then so is B. Use Problem 194.

Problem 201. Both $\Delta - \text{spec } A$ and $\Delta \cap \text{spec } A$ are open. Use Problem 78.

Problem 202. Every finite-dimensional subspace invariant under a normal operator B reduces B.

Problem 203. If A (on **H**) is subnormal, and if $f_0, \cdots, f_n$ are vectors in **H**, then the matrix $\langle (A^i f_j, A^j f_i) \rangle$ is positive definite. A weighted shift with weights $\{\alpha_0, \alpha_1, \alpha_2, \cdots\}$ is hyponormal if and only if $|\alpha_n|^2 \leqq |\alpha_{n+1}|^2$ for all n.

Problem 204. Use Problem 149.

Problem 205. If A is hyponormal, then

$$\|A^n f\|^2 \leqq \|A^{n+1}\| \cdot \|A^{n-1}\| \cdot \|f\|^2$$

for every vector f.

Problem 206. If A is hyponormal, then the span of the eigenvectors of A reduces A. If A is compact also, then consider the restriction of A to the orthogonal complement of that span, and apply Problem 179 and Problem 205.

Problem 207. Consider the eigenvectors of the imaginary part.

Problem 208. Given a hyponormal idempotent P, decompose the space into ran P and ran$^\perp P$.

Problem 209. Try a linear combination of the unilateral shift and its adjoint.

Chapter 22. Numerical Range

Problem 210. A set in the plane is convex if and only if its intersection with every straight line is connected. Note incidentally that the general Toeplitz–Hausdorff theorem is equivalent to its 2-dimensional special case.

Problem 211. If $\mathbf{M}$ and $\mathbf{N}$ are k-dimensional Hilbert spaces and if T is a linear transformation from $\mathbf{M}$ to $\mathbf{N}$, then there exist orthonormal bases $\{f_1, \cdots, f_k\}$ for $\mathbf{M}$, and $\{g_1, \cdots, g_k\}$ for $\mathbf{N}$, and there exist positive scalars $\alpha_1, \cdots, \alpha_k$ such that $Tf_i = \alpha_i g_i$, $i = 1, \cdots, k$. If P and Q are projections of rank k, apply this statement to the restriction of QP to the range of P, and apply the Toeplitz–Hausdorff theorem k times.

Problem 212. Try a diagonal operator. Try the unilateral shift.

Problem 213. The quadratic form associated with a compact operator is weakly continuous on the unit ball.

Problem 214. The closure of the numerical range includes both the compression spectrum (the complex conjugate of the point spectrum of the adjoint) and the approximate point spectrum.

Problem 215. Let V be the Volterra integration operator and consider $1 - (1 + V)^{-1}$.

Problem 216. Use the spectral theorem; reduce the thing to be proved to the statement that if the values of a function are in the right half plane, then so is the value of its integral with respect to a positive measure.

Problem 217. Use Problems 200, 214, and 216.

Problem 218. (a) Prove the contrapositive. (b) If $\|A\| = 1$ and $(Af_n, f_n) \to 1$, then $Af_n - f_n \to 0$.

Problem 219. Write

$$M = \begin{pmatrix} 0 & 0 \\ 1 & 0 \end{pmatrix},$$

let N be a normal operator whose spectrum is the closed disc with center 0 and radius $\frac{1}{2}$, and consider

$$\begin{pmatrix} M & 0 \\ 0 & N \end{pmatrix} \quad \text{and} \quad \begin{pmatrix} M & 0 \\ 0 & 1 \end{pmatrix}.$$

Problem 220. If $\|A - B\| < \varepsilon$ and $\|f\| = 1$, then $(Af, f) \in W(B) + (\varepsilon)$. Let U be the unilateral shift and consider U^{*n}, $n = 1, 2, 3, \cdots$.

Problem 221. A necessary and sufficient condition that $w(A) \leq 1$ is that $\mathrm{Re}(1 - zA)^{-1} \geq 0$ for every z in the open unit disc. Write down the partial fraction expansion of $1/(1 - z^n)$ and replace z by zA.

Chapter 23. Unitary Dilations

Problem 222. (a) Suppose that the given Hilbert space is one-dimensional real Euclidean space and the dilation space is a plane. Examine the meaning of the assertion in this case, use analytic geometry to prove it, and let the resulting formulas suggest the solution in the general case. (b) Imitate (a).

Problem 223. Find an operator A, $0 \leq A \leq 1$, with non-closed range, and consider

$$\begin{pmatrix} A & \sqrt{A(1-A)} \\ \sqrt{A(1-A)} & 1-A \end{pmatrix}.$$

Problem 224. For **U** use Problem 222(a), and for **P** Problem 222(b), together, in both cases, with the projection characterization of the weak topology. Are the suggested closures indeed closed?

161

Problem 225. For **U** use Problem 149; for **P** note that every idempotent contraction is a projection; for co-isometries use Problem 152.

Problem 226. If an operator B is such that to every vector g there corresponds a hyponormal operator that agrees with B on both g and B^*g, then B itself is hyponormal.

Problem 227. Look for a bilaterally infinite matrix that does the job; use the techniques and results of Solution 222.

Problem 228. Use the spectral theorem to prove the assertion for unitary operators, and then use the existence of unitary power dilations to infer it for all contractions.

Problem 229. Find a unitary power dilation of A.

Chapter 24. Commutators

Problem 230. Wintner: assume that P is invertible and examine the spectral implications of $PQ = QP + 1$. Wielandt: assume $PQ - QP = 1$, evaluate $P^nQ - QP^n$, and use that evaluation to estimate its norm.

Problem 231. Consider the Banach space of all bounded sequences of vectors, modulo null sequences, and observe that each bounded sequence of operators induces an operator on that space.

Problem 232. Fix P and consider $\Delta Q = PQ - QP$ as a function of Q; determine $\Delta^n Q^n$.

Problem 233. (a) Generalize the formula for the "derivative" of a power to the non-commutative case, and imitate Wielandt's proof. (b) Use the Kleinecke–Shirokov theorem.

Problem 234. Represent the space as an infinite direct sum in such a way that all summands after the first are in the kernel. Examine the corresponding matrix representation of the given operator, and try to represent it as $PQ - QP$, where P is the pertinent unilateral shift.

Problem 235. Find an invertible operator T such that $A + T^{-1}AT$ has a non-zero kernel; apply Problem 234 to the direct sum of $A + T^{-1}AT$ with itself countably many times. Prove and use the lemma that if $B + C$ is a commutator, then so is $B \oplus C$.

Problem 236. If $C = A^*A - AA^* \geqq 0$, choose λ in $\Pi(A)$, find $\{f_n\}$ so that $\|f_n\| = 1$ and $(A - \lambda)f_n \to 0$, and prove that $Cf_n \to 0$.

Problem 237. (a) Prove that (1) A is quasinormal, (2) $\ker(1 - A^*A)$ reduces A, and (3) $\ker^\perp(1 - A^*A) \subset \ker(A^*A - AA^*)$. (b) Consider a weighted bilateral shift, with all the weights equal to either 1 or $\sqrt{2}$.

Problem 238. For sufficiency, try a (bilateral) diagonal operator and a bilateral shift; for necessity, adapt the Wintner argument from the additive theory.

Problem 239. Use Problem 142, and then try a diagonal operator matrix and a bilateral shift, in an operator matrix imitation of the technique that worked in Problem 238.

Problem 240. Use Problem 142, together with a multiplicative adaptation of the introduction to Problem 234, to prove that every invertible normal operator is the product of two commutators.

Chapter 25. Toeplitz Operators

Problem 241. For necessity: compute. For sufficiency: use Problem 146.

Problem 242. For necessity: compute. For sufficiency: write $A_n f = W^{*n}APW^n f$ for all f in $\mathbf{L}^2$, $n = 0, 1, 2, \cdots$, and prove that the sequence $\{A_n\}$ is weakly convergent.

Problem 243. If $\langle \gamma_{ij} \rangle$ is the matrix of $T_\varphi T_\psi$, then

$$\gamma_{i+1, j+1} = \gamma_{ij} + \alpha_{i+1}\beta_{-j-1},$$

where

$$\varphi = \sum_i \alpha_i e_i \quad \text{and} \quad \psi = \sum_j \beta_j e_j.$$

Problem 244. Consider the characteristic functions of a set and its complement.

Problem 245. Prove that $W^{*n}TPW^n \to L$ strongly, and use that to prove that if $0 \in \Pi(L)$, then $0 \in \Pi(T)$.

Problem 246. What if φ is a trigonometric polynomial?

Problem 247. Let K be the kernel function of $\tilde{\mathbf{H}}^2$, and, for a fixed y in D and a fixed $\tilde{f}$ in $\tilde{\mathbf{H}}^2$, write $\tilde{g}(z) = (\tilde{\varphi}(z) - \tilde{\varphi}(y))\tilde{f}(z)$. Since $\tilde{g}(y) = 0$, it follows that $\tilde{g} \perp K_y$ and hence that $\tilde{\varphi}(y)$ is in the (compression) spectrum of T_φ.

Problem 248. If φ is real and $T_\varphi f = 0$, then $\varphi \cdot f^* \cdot f$ is real and belongs to $\mathbf{H}^1$.

Problem 249. Modify Solution 244 so as to be able to apply Solution 246.

Problem 250. If φ is real and T_φ is invertible, then $\varphi \cdot f^* \in \mathbf{H}^2$, and this implies that sgn φ is constant.

SOLUTIONS

CHAPTER 1

Vectors

Solution 1. *The limit of a sequence of quadratic forms is a quadratic form.* **1**

PROOF. Associated with each function φ of two variables there is a function φ^- of one variable, defined by $\varphi^-(f) = \varphi(f, f)$; associated with each function ψ of one variable there is a function ψ^+ of two variables, defined by

$$\psi^+(f, g) = \psi(\tfrac{1}{2}(f + g)) - \psi(\tfrac{1}{2}(f - g))$$
$$+ i\psi(\tfrac{1}{2}(f + ig)) - i\psi(\tfrac{1}{2}(f - ig)).$$

If φ is a sesquilinear form, then $\varphi = \varphi^{-+}$; if ψ is a quadratic form, then $\psi = \psi^{+-}$. If $\{\psi_n\}$ is a sequence of quadratic forms and if $\psi_n \to \psi$ (that is, $\psi_n(f) \to \psi(f)$ for each vector f), then $\psi_n^+ \to \psi^+$ and $\psi_n^{+-} \to \psi^{+-}$. Since each ψ_n is a quadratic form, it follows that each ψ_n^+ is a sesquilinear form and hence that ψ^+ is one too. Since, moreover, $\psi_n = \psi_n^{+-}$, it follows that $\psi = \psi^{+-}$, and hence that ψ is a quadratic form.

The index set for sequences (i.e., the set of natural numbers) has nothing to do with the facts here; the proof is just as valid for ordered sequences of arbitrary length, and, more generally, for nets of arbitrary structure.

Solution 2. Yes, the Schwarz inequality is true for not necessarily strictly **2** positive forms, and one way to prove it is to reduce the general case to the strictly positive case. Indeed, given φ, let φ_+ be an arbitrary strictly positive, symmetric, sesquilinear form on the same space, and write, for each positive number ε,

$$\varphi_\varepsilon = \varphi + \varepsilon\varphi_+.$$

The form φ_ε is strictly positive; apply the Schwarz inequality to it and let ε tend to 0. As for finding a strictly positive form φ_+ (on every real or complex vector space): just use Hamel bases. If $\{e_j\}$ is one, write

$$\varphi_+\left(\sum_j \alpha_j e_j, \sum_j \beta_j e_j\right) = \sum_j \alpha_j \beta_j^*.$$

The sums are formally infinite but only finitely non-zero.

3 **Solution 3.** To motivate the approach, assume for a moment that it is already known that $\xi(f) = (f, g)$ for some g. Choose an arbitrary but fixed ortho-normal basis $\{e_j\}$ and expand g accordingly: $g = \sum_j \beta_j e_j$. Since

$$\beta_j = (g, e_j) = (e_j, g)^* = \xi(e_j)^*,$$

the vector g could be captured by writing

$$g = \sum_j \xi(e_j)^* e_j.$$

If the existence of the Riesz representation is known, this reasoning proves uniqueness and exhibits the coordinates of the representing vector. The main problem, from the point of view of the present approach to the existence proof, is to prove the convergence of the series $\sum_j \beta_j e_j$, where $\beta_j = \xi(e_j)^*$.

For each finite set J of indices, write $g_J = \sum_{j \in J} \beta_j e_j$. Then

$$\xi(g_J) = \sum_{j \in J} |\beta_j|^2,$$

and therefore

$$\sum_{j \in J} |\beta_j|^2 \leq \|\xi\| \cdot \|g_J\| = \|\xi\| \cdot \sqrt{\sum_{j \in J} |\beta_j|^2}.$$

This implies that

$$\sqrt{\sum_{j \in J} |\beta_j|^2} \leq \|\xi\|,$$

and hence that

$$\sum_j |\beta_j|^2 < \infty.$$

This result justifies writing $g = \sum_j \beta_j e_j$. If $f = \sum_j \alpha_j e_j$, then

$$\xi(f) = \sum_j \alpha_j \xi(e_j) = \sum_j \alpha_j \beta_j^* = (f, g),$$

and the proof is complete.

4 **Solution 4.** The boundary points of the closed unit ball are the vectors on the unit sphere (that is, the unit vectors, the vectors f with $\|f\| = 1$). The thing to prove therefore is that if $f = tg + (1 - t)h$, where $0 \leq t \leq 1$, $\|f\| = 1$, $\|g\| \leq 1$, and $\|h\| \leq 1$, then $f = g = h$. Begin by observing that

$$1 = (f, f) = (f, tg + (1 - t)h) = t(f, g) + (1 - t)(f, h).$$

168

Since $|(f, g)| \leqq 1$ and $|(f, h)| \leqq 1$, it follows that $(f, g) = (f, h) = 1$; this step uses the strict convexity of the closed unit disc. The result says that the Schwarz inequality degenerates, both for f and g and for f and h, and this implies that both g and h are multiples of f. Write $g = \alpha f$ and $h = \beta f$. Since $1 = (f, g) = (f, \alpha f) = \alpha^*$, and, similarly, $1 = \beta^*$, the proof is complete.

Solution 5. Since every infinite-dimensional Hilbert space has a subspace   **5** isomorphic to $\mathbf{L}^2(0, 1)$, it is sufficient to describe the construction for that special space. The description is easy. If $0 \leq t \leq 1$, let $f(t)$ be the characteristic function of the interval $[0, t]$; in other words, $(f(t))(s) = 1$ or 0 according as $0 \leqq s \leqq t$ or $t < s \leqq 1$. If $0 \leqq a \leqq b \leqq 1$, then

$$\| f(b) - f(a) \|^2 = \int |(f(b))(s) - (f(a))(s)|^2 \, ds$$

$$= \int_a^b ds = b - a;$$

this implies that f is continuous. The verifications of simplicity and of the orthogonality conditions are obvious.

As for the existence of tangents: it is easy to see that the difference quotients do not tend to a limit at any point. Indeed,

$$\left\| \frac{f(t + h) - f(t)}{h} \right\|^2 = \left| \frac{h}{h^2} \right| = \left| \frac{1}{h} \right|,$$

which shows quite explicitly that f is not differentiable anywhere.

Infinite-dimensionality was explicitly used in the particular proof given above, but that does not imply that it is unavoidable. Is it? An examination of the finite-dimensional situation is quite instructive.

Constructions similar to the one given above are familiar in the theory of spectral measures (cf. [50, p. 58]). If E is the spectral measure on the Borel sets of $[0, 1]$ such that $E(M)$ is, for each Borel set M, multiplication by the characteristic function of M, and if e is the function constantly equal to 1, then the curve f above is given by

$$f(t) = E([0, t])e.$$

This remark shows how to construct many examples of suddenly turning continuous curves: use different spectral measures and apply them to different vectors. It is not absolutely necessary to consider only continuous spectral measures whose support is the entire interval, but it is wise; those assumptions guarantee that every non-zero vector will work in the role of e.

Solution 6. It is convenient to begin with a few small auxiliary statements   **6** true about every normalized crinkled arc f.

(1)   If $0 \leqq s \leqq t \leqq 1$, then $(f(s), f(t)) = \| f(s) \|^2$. Proof: $(f(s), f(t)) = (f(s) - f(0), f(t) - f(s) + f(s))$.

(2) If $0 \leqq s \leqq t \leqq 1$, then $\| f(s) - f(t) \|^2 = \| f(t) \|^2 - \| f(s) \|^2$. Proof: immediate from (1).

(3) The mapping $t \mapsto \| f(t) \|$ is strictly monotone and continuous. Proof: immediate from (2).

(4) The non-zero values of f are linearly independent. Proof: if $0 < t_1 < t_2 < \cdots < t_n \leqq 1$, then the vectors $f(t_1)$, $f(t_2) - f(t_1)$, $\cdots$, $f(t_n) - f(t_{n-1})$ are pairwise orthogonal, hence linearly independent; the vectors $f(t_1), \cdots, f(t_n)$ span the same n-dimensional space, and hence they too must be linearly independent.

The ground is now prepared for the uniqueness proof. Suppose that f and g are normalized crinkled arcs. Define the mapping φ from [0, 1] to [0, 1] to be the following composition: first the mapping $t \mapsto \| g(t) \|$, and then the inverse of the mapping $t \mapsto \| f(t) \|$. The result is a reparametrization φ of [0, 1] such that $\| g(t) \| = \| f(\varphi(t)) \|$ for all t.

Now define a mapping U, first on the range of f only, by $U f(\varphi(t)) = g(t)$; note that, by the definition of φ, the mapping U is isometric. Since the non-zero elements of ran f are linearly independent, U can be extended to the linear (not necessarily closed) span of ran f so as to become linear. Since both ran f and ran g span $\mathbf{H}$, it follows that both the domain and the range of the extended U are dense in $\mathbf{H}$.

The next thing to prove is that the extended U is still isometric. The first step in this direction is to consider two numbers s and t, $0 \leqq s < t \leqq 1$, and observe that

$$\| U(f(\varphi(t)) - f(\varphi(s))) \|^2 = \| g(t) - g(s) \|^2$$
$$= \| g(t) \|^2 - \| g(s) \|^2$$
$$= \| f(\varphi(t)) \|^2 - \| f(\varphi(s)) \|^2$$
$$= \| f(\varphi(t)) - f(\varphi(s)) \|^2 ;$$

in other words, U is isometric on differences such as $f(\varphi(t)) - f(\varphi(s))$. The second and last step is to observe that each vector in the linear span of ran f is a finite linear combination of orthogonal differences such as $f(\varphi(t)) - f(\varphi(s))$, and its image under U is the same linear combination of the corresponding orthogonal differences $g(t) - g(s)$. Since the square norm is additive for orthogonal summands, it follows that U, as defined so far, is isometric.

The rest is trivial: extend by continuity and get a unitary operator U on $\mathbf{H}$ such that $g(t) = U f(\varphi(t))$.

7   **Solution 7.** *If the orthogonal dimension of a Hilbert space is infinite, then its linear dimension is greater than or equal to $2^{\aleph_0}$.*

(Recall that if either the linear dimension or the orthogonal dimension of a Hilbert space is finite, then so is the other, and the two are equal.)

PROOF. The main tool is the following curious piece of set theory, which has several applications: there exists a collection $\{J_t\}$, of cardinal number $2^{\aleph_0}$, consisting of infinite sets of positive integers, such that $J_s \cap J_t$ is finite whenever $s \neq t$. Here is a quick outline of a possible construction. Since there is a one-to-one correspondence between the positive integers and the rational numbers, it is sufficient to prove the existence of sets of rational numbers with the stated property. For each *real* number t, let J_t be an infinite set of rational numbers that has t as its only cluster point.

[An alternative construction of uncountably many "almost disjoint" subsets of a countable set is based on an elegant geometric observation of J. R. Buddenhagen [25]. Consider the (countable) set L of all those points in the plane whose coordinates are integers (i.e., the lattice points in the plane). For each positive real number t, let L_t be the part of L in the open band between the line of slope t through the origin and the parallel line at distance 2 above it. Each L_t is infinite; since, however, $L_s \cap L_t$ is bounded whenever $s \neq t$, it follows that the intersection of two distinct L_t's is always finite.]

Suppose now that $\{e_1, e_2, e_3, \cdots\}$ is a countably infinite orthonormal set in a Hilbert space $\mathbf{H}$, and let $f = \sum_n \xi_n e_n$ (Fourier expansion) be an arbitrary vector such that $\xi_n \neq 0$ for all n. If $\{J_t\}$ is a collection of sets of positive integers of the kind described above, write $f_t = \sum_{n \in J_t} \xi_n e_n$. Assertion: the collection $\{f_t\}$ of vectors is linearly independent. Suppose, indeed, that a finite linear combination of the f's vanishes, say $\sum_{i=1}^k \alpha_i f_{t_i} = 0$. Since, for each $i \neq 1$, the set J_{t_1} contains infinitely many integers that do not belong to J_{t_i}, it follows that J_{t_1} contains at least one integer, say n, that does not belong to any J_{t_i} ($i \neq 1$). It follows that $\alpha_1 \xi_n = 0$, and hence, since $\xi_n \neq 0$, that $\alpha_1 = 0$. The same argument proves, of course, that $\alpha_i = 0$ for each $i = 1, \cdots, k$.

This result is the main reason why the concept of linear dimension is of no interest in Hilbert space theory. In a Hilbert space context "dimension" always means "orthogonal dimension".

There are shorter solutions of the problem, but the preceding argument has the virtue of being elementary in a sense in which they are not. Thus, for instance, every infinite-dimensional Hilbert space may be assumed to include $L^2(0, 1)$, and the vectors $f(t)$, $0 < t \leq 1$, exhibited in Solution 4, constitute a linearly independent set with cardinal number $2^{\aleph_0}$. Alternatively, every infinite-dimensional Hilbert space may be assumed to include l^2, and the vectors

$$g(t) = \langle 1, t, t^2, \cdots \rangle, \qquad 0 < t < 1,$$

constitute a linearly independent set with cardinal number $2^{\aleph_0}$.

The problem as originally stated has at least one solution that is both short and elementary. Assertion: if $\{f_1, f_2, f_3, \cdots\}$ is a linearly independent sequence, then there exists a vector that is not a (finite) linear combination of the f's. Proof: the Gram-Schmidt orthogonalization process implies the existence of an orthonormal sequence $\{e_1, e_2, e_3, \cdots\}$ such that $\bigvee\{e_1, \cdots, e_n\} = \bigvee\{f_1, \cdots, f_n\}$ for $n = 1, 2, 3, \cdots$; if the Fourier expansion of a vector g with respect to the e's has infinitely many non-zero coefficients, then g does not belong to $\bigvee\{f_1, \cdots, f_n\}$ for any n.

An even shorter (but less elementary) solution consists of two words: Baire category. In more detail: if $\{f_1, f_2, f_3, \cdots\}$ is an arbitrary sequence, then $\mathbf{H}_n = \bigvee\{f_1, \cdots, f_n\}$ is, for each n, a proper subspace and, therefore, a nowhere dense set; it follows that the $\mathbf{H}_n$'s cannot fill up a complete metric space.

8   **Solution 8.** Let $\{e_n: n = 0, \pm 1, \pm 2, \cdots\}$ be an orthonormal basis, and let $\mathbf{T}$ be the set of all vectors of the form $f_n = e_n - e_{n-1}, n = 0, \pm 1, \pm 2, \cdots$.

How can a vector g be orthogonal to $\mathbf{T} - \{f_k\}$? If that happens, then $(g, e_n) = (g, e_{n-1})$ whenever $n \neq k$. Put $n = k + 1, k + 2, \cdots$ to get

$$(g, e_k) = (g, e_{k+1}) = (g, e_{k+2}) = \cdots;$$

put $n = k - 1, k - 2, \cdots$ to get

$$(g, e_{k-1}) = (g, e_{k-2}) = (g, e_{k-3}) = \cdots.$$

Since infinitely many Fourier coefficients can be equal only if they vanish, it follows that $g = 0$. This proves that $\mathbf{T} - \{f_k\}$ is always total (and implies, in particular, that $\mathbf{T}$ itself is total).

If, on the other hand, $j < k$, then $\mathbf{T} - \{f_j, f_k\}$ is not total. The argument of the preceding paragraph still applies, to be sure, but the result this time is that if $g \perp \mathbf{T} - \{f_j, f_k\}$, then

$$(g, e_k) = (g, e_{k+1}) = (g, e_{k+2}) = \cdots,$$

$$(g, e_{j-1}) = (g, e_{j-2}) = (g, e_{j-3}) = \cdots,$$

and

$$(g, e_j) = \cdots = (g, e_{k-1}).$$

These conditions imply that $(g, e_n) = 0$ when $n \geq k$ and when $n < j$; when $j \leq n < k$, the conditions amount to the equality of a finite number of Fourier coefficients, which is harmless. If, for instance, $g = e_j + \cdots + e_{k-1}$, then all the conditions are satisfied, and that implies that $\mathbf{T} - \{f_j, f_k\}$ is not total.

The proof is due to L. J. Wallen. An alternative proof (with different merits) goes as follows. Let $\{e_n: n = 1, 2, 3, \cdots\}$ be an orthonormal basis, let f be a vector in whose Fourier expansion,

$$f = \sum_n \alpha_n e_n,$$

every coefficient is different from 0, and let $\mathbf{T}$ be the set $\{f, e_1, e_2, e_3, \cdots\}$. Since the span of the set $\mathbf{T} - \{e_k\}$ contains the vector $f - \sum_{n \neq k} \alpha_n e_n$, it follows that $\mathbf{T} - \{e_k\}$ is total for each k. If, on the other hand, $j < k$, then the vector

$$g = \alpha_k{}^* e_j - \alpha_j{}^* e_k$$

is orthogonal to $\mathbf{T} - \{e_j, e_k\}$. Reason:

$$(f, g) = (\alpha_j e_j + \alpha_k e_k, \alpha_k{}^* e_j - \alpha_j{}^* e_k) = \alpha_j \alpha_k - \alpha_k \alpha_j = 0.$$

Conclusion: $\mathbf{T} - \{e_j, e_k\}$ is not total.

Solution 9. If the given set is $\{f_1, f_2, \cdots\}$ (there is no loss of generality in the assumption of countability), then omit f_1 and approximate it to within $\frac{1}{2}$ by a vector in $\bigvee \{f_2, \cdots, f_{n_1}\}$; then omit f_1, f_{n_1+1} and approximate them to within $\frac{1}{3}$ by vectors in $\bigvee \{f_{n_1+2}, \cdots, f_{n_2}\}$; then omit $f_1, f_{n_1+1}, f_{n_2+1}$ and approximate them to within $\frac{1}{4}$ by vectors in $\bigvee \{f_{n_2+2}, \cdots, f_{n_3}\}$; etc. The result is an infinite set $\{f_1, f_{n_1+1}, f_{n_2+1}, \cdots\}$ that is included in the span of its complement in $\{f_1, f_2, f_3, \cdots\}$. **9**

Solution 10. *If $0 < |\alpha| < 1$ and* **10**

$$f_k = \langle 1, \alpha^k, \alpha^{2k}, \alpha^{3k}, \cdots \rangle \qquad \text{for } k = 1, 2, 3, \cdots,$$

then the f_k's span l^2.

PROOF. Perhaps the quickest approach is to look for a vector f orthogonal to all the f_k's. If $f = \langle \xi_0, \xi_1, \xi_2, \cdots \rangle$, then

$$0 = (f, f_k) = \sum_{n=0}^{\infty} \xi_n \alpha^{*nk}.$$

In other words, the power series

$$\sum_{n=0}^{\infty} \xi_n z^n$$

vanishes for $z = \alpha^{*k}$ ($k = 1, 2, 3, \cdots$), and consequently it vanishes identically. Conclusion: $\xi_n = 0$ for all n, and therefore $f = 0$.

The phrasing of the problem is deceptive. The solution has nothing to do with the arithmetic structure of the powers α^k; the same method applies if the powers α^k are replaced by arbitrary numbers α_k (and, correspondingly, α^{nk} is replaced by $\alpha_k{}^n$), provided only that the numbers α_k cluster somewhere in the interior of the unit disc. (Note that if $\sum_{n=0}^{\infty} |\xi_n|^2 < \infty$, then the power series $\sum_{n=0}^{\infty} \xi_n z^n$ has radius of convergence greater than or equal to 1.)

The result is a shallow generalization of the well known facts about Vandermonde matrices, and the proof suggested above is adaptable to the finite-dimensional case. If $l_m{}^2$ is the m-dimensional Hilbert space of all sequences $\langle \xi_0, \cdots, \xi_{m-1} \rangle$ of length m ($= 1, 2, 3, \cdots$), and if the vectors f_k ($k = 1, \cdots, m$) are defined by $f_k = \langle 1, \alpha_k, \cdots, \alpha_k{}^{m-1} \rangle$ (where $0 \leq |\alpha_k| < 1$ and the α_k's are distinct), then the span of $\{f_1, \cdots, f_m\}$ is $l_m{}^2$. Indeed, if $f = \langle \xi_0, \cdots, \xi_{m-1} \rangle$ is orthogonal to each f_k, then $\sum_{n=0}^{m-1} \xi_n \alpha_k{}^{*n} = 0$, i.e., the polynomial $\sum_{n=0}^{m-1} \xi_n z^n$ of degree $m - 1$ (at most) vanishes at m distinct points, and hence identically.

11   **Solution 11.** The example described in Problem 10 works, and so do many of the generalizations mentioned in Solution 10. For instance, let $\langle \alpha_1, \alpha_2, \cdots \rangle$ be a sequence of distinct complex numbers in the unit disc, with $\alpha_k \to 0$. If $f_k = \langle 1, \alpha_k, \alpha_k^2, \alpha_k^3, \cdots \rangle$, then the set $\{f_k : k = 1, 2, 3, \cdots\}$ is total in l^2; since every infinite subsequence of the α_k's has the same properties as the original sequence, it follows that every infinite subset of the f_k's is total also.

A set such that every infinite subset of it is total deserves to be called *totally total*, or, abbreviated, t-total.

A final comment along these lines has to do with the concept of dimension. One way to define dimension was mentioned in Problem 7: prove that all bases have the same cardinal number, and then define dimension as the common cardinal number of all bases. A more direct way is to define the dimension of **H** as the minimal cardinal number of a total set in **H**. Caution: this does not say that dimension is the cardinal number of a minimal total set; the order of the words is important. The collection of cardinal numbers under consideration is well ordered, and hence, in particular, has a least element; the collection of total sets is only partially ordered. It is common mathematical experience that a proof of the existence of minimal sets of a certain kind is likely to prove a stronger statement, namely that every one of the sets of that kind includes a minimal one. For total sets in Hilbert space the mere existence of minimal ones is obvious (orthonormal bases), but the stronger statement is false. To say that the stronger statement is false is to say that there exists a total set such that no total subset of it is minimal, and the existence of such a set is obvious from Solution 11. (For further comments on dimension see Problems 54 and 55.)

Does every separable Banach space have a minimal total set? The question seems to be open. For sufficiently unpleasant topological vector spaces the answer is no; see [47, p. 214].

12   **Solution 12.** It seems that the more you know the harder this is. The following ingenious proof was offered by a student who at the time was completely innocent of the techniques of Hilbert space [143].

The basic observation is that

$$|(e_j - f_j, e_i)| = |(e_i - f_i, f_j)|$$

for all i and j. This is so trivial that it is hard to see how it could have any usable consequences. It does, however, yield

$$\sum_j \|e_j - f_j\|^2 = \sum_j \sum_i |(e_j - f_j, e_i)|^2 \quad \text{(by Parseval)}$$

$$= \sum_i \sum_j |(e_i - f_i, f_j)|^2$$

$$\leq \sum_i \|e_i - f_i\|^2 \quad \text{(by Bessel)}.$$

174

The assumed finiteness condition implies that the Bessel inequality must have been an equality for each i:

$$\sum_j |(e_i - f_i, f_j)|^2 = \|e_i - f_i\|^2.$$

(The elementary lemma that is being used is that if $0 \le a_i \le b_i$ for each i and $\sum_i a_i = \sum_i b_i < \infty$, then $a_i = b_i$ for each i.) It follows that

$$\sum_j (e_i - f_i, f_j)f_j = e_i - f_i,$$

so that

$$\sum_j (e_i, f_j)f_j = e_i$$

for each i. Consequence: each e_i belongs to the span of the f_j's, and therefore, indeed, the f_j's span $\mathbf{H}$.

There is an alternative proof, which is a shade less elementary, but whose technique is more often usable. Begin by choosing a positive integer n so that

$$\sum_{j>n} \|e_j - f_j\|^2 < 1,$$

and then define a linear transformation A, first on the linear combinations of the e_j's only, by writing

$$Ae_j = e_j \quad \text{if } j \le n,$$

and

$$Ae_j = f_j \quad \text{if } j > n.$$

If $f = \sum_j \xi_j e_j$ (finite sum), then

$$\|f - Af\|^2 = \left\| \sum_{j>n} \xi_j(e_j - f_j) \right\|^2$$

$$\le \sum_{j>n} |\xi_j|^2 \cdot \sum_{j>n} \|e_j - f_j\|^2$$

$$\le \|f\|^2 \cdot \sum_{j>n} \|e_j - f_j\|^2.$$

It follows that $1 - A$ is bounded (as far as it is defined) by

$$\sum_{j>n} \|e_j - f_j\|^2,$$

which is strictly less than 1. This implies ([50, p. 52]) that A has a (unique) extension to an invertible operator on $\mathbf{H}$ (which may as well be denoted by A again). The invertibility of A implies that the vectors $e_1, \cdots, e_n, f_{n+1}, f_{n+2}, \cdots$ (the images under A of $e_1, \cdots, e_n, e_{n+1}, e_{n+2}, \cdots$) span $\mathbf{H}$. It follows that if $\mathbf{M}$ is the span of $f_{n+1}, f_{n+2}, \cdots$, then dim $\mathbf{M}^\perp = n$. Conclusion: the vectors $f_1, \cdots, f_{n+1}, f_{n+2}, \cdots$ span $\mathbf{H}$.

175

CHAPTER 2

Spaces

13   **Solution 13.** It is sufficient to prove that if dim $\mathbf{M} = 1$, then $\mathbf{M} + \mathbf{N}$ is closed; the general case is obtained by induction on the dimension. Suppose, therefore, that $\mathbf{M}$ is spanned by a single vector f_0, so that $\mathbf{M} + \mathbf{N}$ consists of all the vectors of the form $\alpha f_0 + g$, where α is a scalar and $g \in \mathbf{N}$. If $f_0 \in \mathbf{N}$, then $\mathbf{M} + \mathbf{N} = \mathbf{N}$; in this case there is nothing to prove. If $f_0 \notin \mathbf{N}$, let g_0 be the projection of f_0 in $\mathbf{N}$; that is, g_0 is the unique vector in $\mathbf{N}$ for which $f_0 - g_0 \perp \mathbf{N}$.

Observe now that if g is a vector in $\mathbf{N}$, then

$$\|\alpha f_0 + g\|^2 = \|\alpha(f_0 - g_0) + (\alpha g_0 + g)\|^2$$
$$\geqq |\alpha|^2 \cdot \|f_0 - g_0\|^2$$

(since $f_0 - g_0 \perp \alpha g_0 + g$), or

$$|\alpha| \leqq \frac{\|\alpha f_0 + g\|}{\|f_0 - g_0\|},$$

and therefore

$$\|g\| = \|(\alpha f_0 + g) - \alpha f_0\|$$
$$\leqq \|\alpha f_0 + g\| + \frac{\|\alpha f_0 + g\|}{\|f_0 - g_0\|} \cdot \|f_0\|.$$

These inequalities imply that $\mathbf{M} + \mathbf{N}$ (the set of all $\alpha f_0 + g$'s) is closed. Indeed, if $\alpha_n f_0 + g_n \to h$, so that $\{\alpha_n f_0 + g_n\}$ is a Cauchy sequence, then the inequalities imply that both $\{\alpha_n\}$ and $\{g_n\}$ are Cauchy sequences. It follows that $\alpha_n \to \alpha$ and $g_n \to g$, say, with g in $\mathbf{N}$ of course, and consequently $h = \lim_n (\alpha_n f_0 + g_n) = \alpha f_0 + g$.

Solution 14. *The lattice of subspaces of a Hilbert space* **H** *is modular if* **14** *and only if* dim **H** $<$ $\aleph_0$ (*i.e.,* **H** *is finite-dimensional*); *it is distributive if and only if* dim **H** $\leq$ 1.

PROOF. If **H** is infinite-dimensional, then it has subspaces **M** and **N** such that **M** $\cap$ **N** $= 0$ and **M** + **N** $\neq$ **M** $\vee$ **N** (cf. Problem 52). Given **M** and **N**, find a vector f_0 in **M** $\vee$ **N** that does not belong to **M** + **N**, and let **L** be the span of **N** and f_0. By Problem 13, **L** is equal to the vector sum of **N** and the one-dimensional space spanned by f_0, i.e., every vector in **L** is of the form $\alpha f_0 + g$, where α is a scalar and g is in **N**.

Both **L** and **M** $\vee$ **N** contain f_0, and, therefore, so does their intersection. On the other hand, **L** $\cap$ **M** $= 0$. Reason: if $\alpha f_0 + g \in$ **M** (with g in **N**), then $\alpha f_0 \in$ **M** + **N**; this implies that $\alpha = 0$ and hence that $g = 0$. Conclusion: (**L** $\cap$ **M**) $\vee$ **N** = **N**, which does not contain f_0.

The preceding argument is the only part of the proof in which infinite-dimensionality plays any role. All the remaining parts depend on easy finite-dimensional geometry only. They should be supplied by the reader, who is urged to be sure he can do so before he abandons the subject.

Solution 15. The inclusion $\supset$ in the modular law is true for all lattices; just **15** observe that both **L** $\cap$ **M** and **N** are included in both **L** and **M** $\vee$ **N**.

Suppose now that **M** $\vee$ **N** = **M** + **N**. If $f \in$ **L** $\cap$ (**M** $\vee$ **N**), then (by assumption) $f = g + h$, with g in **M** and h in **N**. Since $-h \in$ **N** $\subset$ **L** and $f \in$ **L**, it follows that $f - h \in$ **L**, so that $g \in$ **L**, and therefore $g \in$ **L** $\cap$ **M**. Conclusion: $f \in$ (**L** $\cap$ **M**) + **N**. (This is in essence the usual proof that the normal subgroups of a group constitute a modular lattice.)

The reverse implication (if **M** + **N** $\neq$ **M** $\vee$ **N** for some **M** and **N**, then the modular law fails) is exactly what the proof of Solution 14 proves.

Solution 16. In a Hilbert space of dimension n ($< \aleph_0$) the (closed) unit ball is **16** a closed and bounded subset of $2n$-dimensional real Euclidean space, and therefore the closed unit ball is compact. It follows, since translations and changes of scale are homeomorphisms, that every closed ball is compact; since the open balls constitute a base for the topology, it follows that the space is locally compact.

Suppose, conversely, that **H** is a locally compact Hilbert space. The argument in the preceding paragraph reverses to this extent: the assumption of local compactness implies that each closed ball is compact, and, in particular, so is the closed unit ball. To infer finite-dimensionality, recall that the distance between two orthogonal unit vectors is $\sqrt{2}$, so that each open ball of diameter $\sqrt{2}$ (or less) can contain at most one element of each orthonormal basis. The collection of all open balls of diameter $\sqrt{2}$ is an open cover of the closed unit ball; the compactness of the latter implies that every orthonormal basis is finite, and hence that **H** is finite-dimensional.

17   **Solution 17.** If dim $\mathbf{H} \leqq \aleph_0$, then $\mathbf{H}$ has a countable orthonormal basis. Since every vector in $\mathbf{H}$ is the limit of finite linear combinations of basis vectors, it follows that every vector in $\mathbf{H}$ is the limit of such linear combinations with coefficients whose real and imaginary parts are rational. The set of all such rational linear combinations is countable, and consequently $\mathbf{H}$ is separable.

Suppose, conversely, that $\{f_1, f_2, f_3, \cdots\}$ is a countable set dense in $\mathbf{H}$. If $\{g_j\}$ is an orthonormal basis for $\mathbf{H}$, then for each index j there exists an index n_j such that $\|f_{n_j} - g_j\| < \sqrt{2}/2$. Since two open balls of radius $\sqrt{2}/2$ whose centers are distinct g_j's are disjoint, the mapping $j \mapsto n_j$ is one-to-one; this implies that the cardinal number of the set of indices j is not greater than $\aleph_0$.

The Gram–Schmidt process yields an alternative approach to the converse. Since that process is frequently described for linearly independent sequences only, begin by discarding from the sequence $\{f_n\}$ all terms that are linear combinations of earlier ones. Once that is done, apply Gram–Schmidt to orthonormalize. The resulting orthonormal set is surely countable; since its span is the same as that of the original sequence $\{f_n\}$, it is a basis.

18   **Solution 18.** Since a measure is, by definition, invariant under translation, there is no loss of generality in considering balls with center at 0 only. If $\mathbf{E}$ is such a ball, with radius r (>0), and if $\{e_1, e_2, e_3, \cdots\}$ is an infinite orthonormal set in the space, consider the open balls $\mathbf{E}_n$ with center at $(r/2)e_n$ and radius $r/4$; that is, $\mathbf{E}_n = \{f: \|f - (r/2)e_n\| < r/4\}$. If $f \in \mathbf{E}_n$, then

$$\|f\| \leqq \left\| f - \frac{r}{2} e_n \right\| + \left\| \frac{r}{2} e_n \right\| < r,$$

so that $\mathbf{E}_n \subset \mathbf{E}$. If $f \in \mathbf{E}_n$ and $g \in \mathbf{E}_m$, then

$$\left\| \frac{r}{2} e_n - \frac{r}{2} e_m \right\| \leqq \left\| \frac{r}{2} e_n - f \right\| + \|f - g\| + \left\| g - \frac{r}{2} e_m \right\|.$$

This implies that if $n \neq m$, then

$$\|f - g\| \geqq \frac{r\sqrt{2}}{2} - \frac{r}{4} - \frac{r}{4} > 0,$$

and hence that if $n \neq m$, then $\mathbf{E}_n$ and $\mathbf{E}_m$ are disjoint. Since, by invariance, all the $\mathbf{E}_n$'s have the same measure, it follows that $\mathbf{E}$ includes infinitely many disjoint Borel sets of the same positive measure, and hence that the measure of $\mathbf{E}$ must be infinite.

CHAPTER 3

Weak Topology

Solution 19. If **S** is a weakly closed set in **H** and if $\{f_n\}$ is a sequence of vectors in **S** with $f_n \to f$ (strong), then

$$|(f_n, g) - (f, g)| \leqq \|f_n - f\| \cdot \|g\| \to 0,$$

so that $f_n \to f$ (weak), and therefore $f \in \mathbf{S}$. This proves that weakly closed sets are strongly closed; in fact, the proof shows that the strong closure of each set is included in its weak closure. The falsity of the converse (i.e., that a strongly closed set need not be weakly closed) can be deduced from the curious observation that if $\{e_1, e_2, e_3, \cdots\}$ is an orthonormal sequence, then $e_n \to 0$ (weak). Reason: for each vector f, the inner products (f, e_n) are the Fourier coefficients of f, and, therefore, they are the terms of an absolutely square-convergent series. It follows that the set of all e_n's is not closed in the weak topology; in the strong topology it is discrete and therefore closed. Another way of settling the converse is to exhibit a strongly open set that is not weakly open; one such set is the open unit ball. To prove what needs proof, observe that in an infinite-dimensional space weakly open sets are unbounded.

It remains to prove that subspaces are weakly closed. If $\{f_n\}$ is a sequence in a subspace **M**, and if $f_n \to f$ (weak), then, by definition, $(f_n, g) \to (f, g)$ for every g. Since each f_n is orthogonal to $\mathbf{M}^\perp$, it follows that $f \perp \mathbf{M}^\perp$ and hence that $f \in \mathbf{M}$. This argument shows that **M** contains the limits of all weakly convergent sequences in **M**, but that does not yet justify the conclusion that **M** is weakly closed. At this point in this book the weak topology is not known to be metrizable; sequential closure may not be the same as closure. The remedy, however, is easy; just observe that the sequential argument works without the change of a single symbol if the word "sequence" is replaced by "net", and net closure is always the same as closure.

With a different proof a stronger theorem can be proved; it turns out that the crucial concept is convexity. To be precise, the theorem is that every strongly closed convex set is weakly closed. Reference: [39, p. 422].

20   **Solution 20.** The proof depends on a familiar trivial computation:

$$\| f_n - f \|^2 = (f_n - f, f_n - f) = \| f_n \|^2 - (f, f_n) - (f_n, f) + \| f \|^2.$$

Since $f_n \to f$ (weak), the terms with minus signs tend to $\| f \|^2$, and, by assumption, so does the first term. Conclusion: $\| f_n - f \|^2 \to 0$, as asserted.

21   **Solution 21.** If $f_n \to f$ (weak), then, for each g, $|(f, g) - (f_n, g)|$ is small when n is large; it follows, in particular, that $|(f, f)| \leqq |(f_n, f)| + \varepsilon$ for each $\varepsilon > 0$ when n is sufficiently large. Consequence: $\| f \|^2 \leqq \| f_n \| \cdot \| f \| + \varepsilon$ for large n, and therefore $\| f \|^2 \leqq (\liminf_n \| f_n \|) \cdot \| f \| + \varepsilon$. The proof is completed by letting ε tend to 0.

22   **Solution 22.** *Every weakly separable Hilbert space is separable.*

PROOF. The span of a countable set is always a (strongly) separable subspace; it is therefore sufficient to prove that if a set S is weakly dense in a Hilbert space H, then the span of S is equal to H. Looked at from the right point of view this is obvious. The span of S is, by definition, a (strongly closed) subspace, and hence, by Problem 19, it is weakly closed; being at the same time weakly dense in H, it must be equal to H.

Caution: it is not only more elegant but it is also safer to argue without sequences. It is not a priori obvious that if f is in the weak closure of S, then f is the limit of a sequence in S.

There is a one-sentence proof that the span of a weakly dense set is the whole space, as follows. Since a vector orthogonal to a set is orthogonal to the weak closure of the set, it follows that the only vector orthogonal to a weakly dense set is 0.

23   **Solution 23.** Given the Hilbert space H, for each f in H let D_f be the closed disc $\{z : |z| \leqq \| f \|\}$ in the complex plane, and let D be the Cartesian product of all the D_f's, with the customary product topology. For each g in the unit ball, the mapping $f \mapsto (f, g)$ is a point, say $\delta(g)$, in D. The mapping δ thus defined is a homeomorphism from the unit ball (with the weak topology) into D (with the product topology). Indeed, if $\delta(g_1) = \delta(g_2)$, that is, if $(f, g_1) = (f, g_2)$ for all f, then clearly $g_1 = g_2$, so that δ is one-to-one. As for continuity:

$$g_j \to g \text{ (weak)} \quad \text{if and only if } (f, g_j) \to (f, g)$$

for each f in H, and that, in turn, happens if and only if $\delta(g_j) \to \delta(g)$ in D. The Riesz theorem on the representation of linear functionals on H implies that the range of δ consists exactly of those elements ξ of $\mathbf{D}$ (complex-valued functions on $\mathbf{H}$) that are in fact linear functionals of norm less than or equal to 1 on $\mathbf{H}$.

The argument so far succeeded in constructing a homeomorphism δ from the unit ball into the compact Hausdorff space $\mathbf{D}$, and it succeeded in identifying the range of δ. The remainder of the argument will show that that range is closed (and therefore compact) in $\mathbf{D}$; as soon as that is done, the weak compactness of the unit ball will follow.

The property of being a linear functional is a property of "finite character". That is: ξ is a linear functional if and only if it satisfies equations (infinitely many of them) each of which involves only a finite number of elements of $\mathbf{H}$; this implies that the set of all linear functionals is closed in $\mathbf{D}$. In more detail, consider fixed pairs of scalars α_1 and α_2 and vectors f_1 and f_2, and form the subset $\mathbf{E}(\alpha_1, \alpha_2, f_1, f_2)$ of $\mathbf{D}$ defined by

$$\mathbf{E}(\alpha_1, \alpha_2, f_1, f_2) = \{\xi \in \mathbf{D} \colon \xi(\alpha_1 f_1 + \alpha_2 f_2) = \alpha_1 \xi(f_1) + \alpha_2 \xi(f_2)\}.$$

The assertion about properties of finite character amounts to this: the set of all linear functionals in $\mathbf{D}$ (the range of δ) is the intersection of all the sets of the form $\mathbf{E}(\alpha_1, \alpha_2, f_1, f_2)$. Since the definition of product topology implies that each of the functions $\xi \mapsto \xi(f_1)$, $\xi \mapsto \xi(f_2)$, and $\xi \mapsto \xi(\alpha_1 f_1 + \alpha_2 f_2)$ is continuous on $\mathbf{D}$, it follows that each set $\mathbf{E}(\alpha_1, \alpha_2, f_1, f_2)$ is closed, and hence that the range of δ is compact.

The proof above differs from the proof of a more general Tychonoff-Alaoglu theorem (the unit ball of the conjugate space of a Banach space is weak * compact) in notation only.

Solution 24. *In a separable Hilbert space the weak topology of the unit ball is metrizable.*   **24**

PROOF 1. Since the unit ball $\mathbf{H}_1$ of $\mathbf{H}$ is weakly compact (Problem 23), it is sufficient to prove the existence of a countable base for the weak topology of $\mathbf{H}_1$. For this purpose, let $\{h_j \colon j = 1, 2, 3, \cdots\}$ be a countable set dense in the space, and consider the basic weak neighborhoods (in $\mathbf{H}_1$) defined by

$$U(p, q, r) = \left\{ f \in \mathbf{H}_1 \colon |(f - h_p, h_j)| < \frac{1}{q}, \; j = 1, \cdots, r \right\},$$

where $p, q, r = 1, 2, 3, \cdots$. To prove: if $f_0 \in \mathbf{H}_1$, k is a positive integer, $g_1, \cdots, g_k$ are arbitrary vectors, and ε is a positive number, and if

$$\mathbf{U} = \{f \in \mathbf{H}_1 \colon |(f - f_0, g_i)| < \varepsilon, i = 1, \cdots, k\},$$

then there exist integers p, q, and r such that

$$f_0 \in \mathbf{U}(p, q, r) \subset \mathbf{U}.$$

The proof is based on the usual inequality device:

$$|(f - f_0, g_i)| \leq |(f - h_p, h_j)| + |(h_p - f_0, h_j)| + |(f - f_0, g_i - h_j)|$$
$$\leq |(f - h_p, h_j)| + \|h_p - f_0\| \cdot \|h_j\| + \|f - f_0\| \cdot \|g_i - h_j\|.$$

Argue as follows: for each i $(=1, \cdots, k)$ choose j_i so that $\|g_i - h_{j_i}\|$ is small, and choose p so that $\|h_p - f_0\|$ is very small. Specifically: choose q so that $1/q < \varepsilon/3$, choose j_i so that $\|g_i - h_{j_i}\| < 1/2q$, choose r so that $j_i \leq r$ for $i = 1, \cdots, k$, and, finally, choose p so that $\|h_p - f_0\| < 1/qm$, where $m = \max\{\|h_j\| : j = 1, \cdots, r\}$. If $j = 1, \cdots, r$, then

$$|(f_0 - h_p, h_j)| \leq \|f_0 - h_p\| \cdot \|h_j\| < \frac{1}{qm} \cdot m = \frac{1}{q},$$

so that $f_0 \in U(p, q, r)$. If $f \in U(p, q, r)$ and $i = 1, \cdots, k$, then

$$|(f - h_p, h_{j_i})| < \frac{1}{q} < \frac{\varepsilon}{3},$$

$$\|h_p - f_0\| \cdot \|h_{j_i}\| < \frac{1}{qm} \cdot m < \frac{\varepsilon}{3},$$

and

$$\|f - f_0\| \cdot \|g_i - h_{j_i}\| < 2 \cdot \frac{1}{2q} < \frac{\varepsilon}{3},$$

(recall that $\|f\| \leq 1$ and $\|f_0\| \leq 1$). It follows that $f \in U$, and the proof is complete.

PROOF 2. There is an alternative procedure that sheds some light on the problem and has the merit, if merit it be, that it exhibits a concrete metric for the weak topology of $\mathbf{H}_1$. Let $\{e_1, e_2, e_3, \cdots\}$ be an orthonormal basis for $\mathbf{H}$. (There is no loss of generality in assuming that the basis is infinite; in the finite-dimensional case all these topological questions become trivial.) For each vector f write

$$|f| = \sum_j \frac{1}{2^j} |(f, e_j)|;$$

since $|(f, e_j)| \leq \|f\|$, the series converges and defines a norm. If $d(f, g) = |f - g|$ whenever f and g are in $\mathbf{H}_1$, then d is a metric for $\mathbf{H}_1$. To show that d metrizes the weak topology of $\mathbf{H}_1$, it is sufficient to prove that $f_n \to 0$ (weak) if and only if $|f_n| \to 0$. (Caution: the metric d is defined for all $\mathbf{H}$ but its relation to the weak topology of $\mathbf{H}$ is not the same as its relation to the weak topology of $\mathbf{H}_1$. The uniform boundedness of the elements of $\mathbf{H}_1$ is what is needed in the argument below.)

Assume that $f_n \to 0$ (weak), so that, in particular, $(f_n, e_j) \to 0$ as $n \to \infty$ for each j. The tail of the series for $|f_n|$ is uniformly small for all n (in fact, the tail of the series for $|f|$ is uniformly small for all f in $\mathbf{H}_1$). In the present case the assumed weak convergence implies that each particular partial sum of the series for $|f_n|$ becomes small as n becomes large, and it follows that $|f_n| \to 0$.

Assume that $|f_n| \to 0$. Since the sum of the series for $|f_n|$ dominates each term, it follows that $(f_n, e_j) \to 0$ as $n \to \infty$ for each j. This implies that if g is a finite linear combination of e_j's, then $(f_n, g) \to 0$. Such linear combinations are dense. If $h \in \mathbf{H}$, then

$$|(f_n, h)| \leqq |(f_n, h - g)| + |(f_n, g)|.$$

Choose g so as to make $\|h - g\|$ small (and therefore $|(f_n, h - g)|$ will be just as small), and then choose n so as to make $|(f_n, g)|$ small. (This is a standard argument that is sometimes isolated as a lemma: a bounded sequence that satisfies the condition for weak convergence on a dense set is weakly convergent.) Conclusion: $f_n \to 0$ (weak).

Solution 25. *In an infinite-dimensional Hilbert space the weak sequential* **25**
closure of the unit sphere is the unit ball.

PROOF. In view of Solution 24, the weak closure of the unit sphere is the same as its weak sequential closure. What is to be proved, therefore, is that if $\|f\| < 1$, then f is the weak limit of a sequence of vectors f_n with $\|f_n\| = 1$. (Recall that the unit ball is weakly closed.) To do that, let $\{e_n\}$ be an orthonormal sequence of vectors orthogonal to f and write

$$f_n = f + (\sqrt{1 - \|f\|^2})e_n, n = 1, 2, 3, \cdots.$$

Since $e_n \to 0$ weakly, it follows that $f_n \to f$ weakly.

The answer to the second question is no; the result is an easy corollary of what was just proved. If $\mathbf{S}_r$ and $\mathbf{H}_r$ are the sphere and the ball of radius r (> 0) and center 0, then what is already known is that the weak sequential closure of $\mathbf{S}_r$ always includes $\mathbf{H}_r$, and hence, all the more, that the weak sequential closure of $\mathbf{S}_r$ includes $\mathbf{H}_s$ whenever $r \geqq s$. Consequence: the exterior of the unit ball is weakly dense (because it is the union of the $\mathbf{S}_r$'s with $r > 1$), but, obviously, the intersection of that exterior with the unit ball is empty, and is therefore not dense in the unit ball.

Solution 26. *If the weak topology of the unit ball in a Hilbert space* $\mathbf{H}$ *is* **26**
metrizable, then $\mathbf{H}$ *is separable.*

PROOF. If $\mathbf{H}_1$, the unit ball, is weakly metrizable, then it is weakly separable (since it is weakly compact). Let $\{f_n : n = 1, 2, 3, \cdots\}$ be a countable set weakly dense in $\mathbf{H}_1$. The set of all vectors of the form mf_n, $m, n = 1, 2, 3, \cdots$, is weakly dense in $\mathbf{H}$. (Reason: for fixed m, the mf_n's are weakly dense in $m\mathbf{H}_1$, and $\bigcup_m m\mathbf{H}_1 = \mathbf{H}$.) The proof is completed by recalling (Solution 22) that weakly separable Hilbert spaces are separable.

27 **Solution 27.** Suppose that $\mathbf{T}$ is a weakly bounded set in $\mathbf{H}$ and that, specifically, $|(f, g)| \leqq \alpha(f)$ for all g in $\mathbf{T}$. If $\mathbf{H}$ is finite-dimensional, the proof is easy. Indeed, if $\{e_1, \cdots, e_n\}$ is an orthonormal basis for $\mathbf{H}$, then

$$|(f, g)| = \left| \left(\sum_{i=1}^{n} (f, e_i) e_i, g \right) \right| = \left| \sum_{i=1}^{n} (f, e_i)(e_i, g) \right|$$

$$\leqq \sqrt{\sum_{i=1}^{n} |(f, e_i)|^2} \cdot \sqrt{\sum_{i=1}^{n} |(e_i, g)|^2}$$

$$\leqq \|f\| \cdot \sqrt{\sum_{i=1}^{n} (\alpha(e_i))^2},$$

and all is well.

Assume now that $\mathbf{H}$ is infinite-dimensional, and assume that the conclusion is false. A consequence of this assumption is the existence of an element g_1 of $\mathbf{T}$ and a unit vector e_1 such that $|(e_1, g_1)| \geqq 1$. Are the linear functionals induced by $\mathbf{T}$ (i.e., the mappings $f \mapsto (f, g)$ for g in $\mathbf{T}$) bounded on the orthogonal complement of the at most two-dimensional space spanned by e_1 and g_1? If so, then they are bounded on $\mathbf{H}$, contrary to the present assumption. A consequence of this argument is the existence of an element g_2 of $\mathbf{T}$ and a unit vector e_2 orthogonal to e_1 and g_1, such that

$$|(e_2, g_2)| \geqq 2(\alpha(e_1) + 2).$$

Continue in the same vein. Argue, as before, that the linear functionals induced by $\mathbf{T}$ cannot be bounded on the orthogonal complement of the at most four-dimensional space spanned by e_1, e_2, g_1, g_2; arrive, as before, to the existence of an element g_3 of $\mathbf{T}$ and a unit vector e_3 orthogonal to e_1, e_2 and g_1, g_2, and such that

$$|(e_3, g_3)| \geqq 3(\alpha(e_1) + \tfrac{1}{2}\alpha(e_2) + 3).$$

Induction yields, after n steps, an element g_{n+1} of $\mathbf{T}$ and a unit vector e_{n+1} orthogonal to $e_1, \cdots, e_n$ and $g_1, \cdots, g_n$, such that

$$|(e_{n+1}, g_{n+1})| \geqq (n + 1)\left(\sum_{i=1}^{n} \frac{1}{i} \alpha(e_i) + n + 1 \right).$$

Now put $f = \sum_{i=1}^{\infty} (1/i) e_i$. Since

$$|(f, g_{n+1})| = \left| \sum_{i=1}^{n} \frac{1}{i} (e_i, g_{n+1}) + \frac{1}{n+1} (e_{n+1}, g_{n+1}) \right|$$

$$\geqq - \sum_{i=1}^{n} \frac{1}{i} \alpha(e_i) + \frac{1}{n+1} (n + 1) \cdot \left(\sum_{i=1}^{n} \frac{1}{i} \alpha(e_i) + n + 1 \right)$$

$$= n + 1,$$

it follows that if $\mathbf{T}$ is not bounded, then it cannot be weakly bounded either.

This proof is due to D. E. Sarason. Special cases of it occur in [148, footnote 32] and [138, p. 59]; almost the general case is in [1, p. 45]. The production of non-category proofs (for Banach spaces) became a cottage industry for a while; a couple of elegant examples are [76] and [72].

Solution 28. Let $\{e_1, e_2, e_3, \cdots\}$ be an infinite orthonormal set in **H** and let **E**   **28**
be the set of all vectors of the form $\sqrt{n} \cdot e_n$, $n = 1, 2, 3, \cdots$. Assertion: the
origin belongs to the weak closure of **E**. Suppose indeed that

$$\{f : |(f, g_i)| < \varepsilon, i = 1, \cdots, k\}$$

is a basic weak neighborhood of 0. Since $\sum_{n=1}^{\infty} |(g_i, e_n)|^2 < \infty$ for each i,
it follows that $\sum_{n=1}^{\infty} (\sum_{i=1}^{k} |(g_i, e_n)|)^2 < \infty$. (The sum of a finite number of
square-summable sequences is square-summable.) It follows that there is at
least one value of n for which $\sum_{i=1}^{k} |(g_i, e_n)| < \varepsilon/\sqrt{n}$; (otherwise square
both sides and contemplate the harmonic series). If n is chosen so that
this inequality is satisfied, then, in particular, $|(g_i, e_n)| < \varepsilon/\sqrt{n}$ for each
i, and therefore $|(\sqrt{n} \cdot e_n, g_i)| < \varepsilon$ for each $i (=1, \ldots, k)$.

The weak non-metrizability of **H** can be established by proving that no
sequence in **E** converges weakly to 0. Since no infinite subset of **E** is bounded,
the desired result is an immediate consequence of the principle of uniform
boundedness.

The first construction of this kind is due to von Neumann [148, p. 380].
The one above is simpler; it was discovered by A. L. Shields.

It is sometimes good to remember that every cluster point of a set is the
limit of a net whose terms are in the set. The construction of such a net is a
standard technique of general topology; in the case at hand it can be described
as follows. The underlying directed set is the set **D** of all basic weak neighborhoods of 0 (ordered by reverse inclusion). If $D \in \mathbf{D}$, choose (!) an element
f_D of **E** in **D**. (The first paragraph above shows that such an element
always exists; in other words, for each D there is a positive integer n such
that $\sqrt{n} \, e_n \in D$.) The net $D \mapsto f_D$ converges weakly to 0. Proof: if D_0 is a
(basic) weak neighborhood of 0, then $D_0 \in \mathbf{D}$ and $f_D \in D_0$ whenever
$D \subset D_0$.

An alternative elegant proof was suggested by G. T. Adams. It proceeds by
contradiction, and does not construct something as explicitly as the preceding
argument, but it involves no computation. It goes as follows.

The weak closure of the unit sphere (i.e., of the set $\mathbf{S}_1 = \{f : \|f\| = 1\}$)
contains 0; see Problem 20. Obvious extension: for each positive integer n,
the weak closure of the sphere $\mathbf{S}_n = \{f : \|f\| = n\}$ contains 0. If there were a
metric d for the weak topology, then, for each n, there would exist a vector f_n
in $\mathbf{S}_n$ such that $d(f_n, 0) < 1/n$. This leads to a contradiction: the unbounded
sequence $\{f_n\}$ is weakly convergent (to 0). Conclusion: there is no such
metric.

Solution 29. Write $g_k = \langle \beta_1{}^*, \cdots, \beta_k{}^*, 0, 0, 0, \cdots \rangle$, so that clearly $g_k \in l^2$,   **29**
$k = 1, 2, 3, \cdots$. If $f = \langle \alpha_1, \alpha_2, \alpha_3, \cdots \rangle$ is in l^2, then $(f, g_k) = \sum_{j=1}^{k} \alpha_j \beta_j \to$
$\sum_{j=1}^{\infty} \alpha_j \beta_j$. It follows that, for each f in l^2, the sequence $\langle (f, g_k) \rangle$ is bounded,
i.e., that the sequence $\{g_k\}$ of vectors in l^2 is weakly bounded. Conclusion (from
the principle of uniform boundedness): there exists a positive constant β such
that $\|g_k\|^2 \leq \beta$ for all k, and, therefore, $\sum_{j=1}^{\infty} |\beta_j|^2 \leq \beta$.

The method generalizes to many measure spaces, including all σ-finite ones. Suppose that X is a measure space with σ-finite measure μ, and suppose that g is a measurable function on X with the property that its product with every function in $\mathbf{L}^2(\mu)$ belongs to $\mathbf{L}^1(\mu)$; the conclusion is that g belongs to $\mathbf{L}^2(\mu)$.

Let $\{E_k\}$ be an increasing sequence of sets of finite measure such that $\bigcup_k E_k = X$ and such that g is bounded on each E_k. (Here is where σ-finiteness comes in.) Write $g_k = \chi_{E_k} g^*$ (where χ_{E_k} is the characteristic function of E_k), $k = 1, 2, 3, \cdots$. The rest of the proof is the obvious modification of the preceding discrete proof; just replace sums by integrals.

For those who know about the closed graph theorem, it provides an alternative approach; apply it to the linear transformation $f \mapsto fg^*$ from $\mathbf{L}^2$ into $\mathbf{L}^1$. For a discussion of an almost, but not quite, sufficiently general version of the closed graph theorem, see Problem 58.

30   **Solution 30.** (a) The idea is that a sufficiently "large" Cauchy net can turn out to be anxious to converge to an "unbounded vector", i.e., to something not in the space. To make this precise, let ξ be an unbounded linear functional, fixed throughout what follows; on an infinite-dimensional Hilbert space such things always exist. (Use a Hamel basis to make one.) Then let $\{e_j\}$ be a Hamel basis, and, corresponding to each finite subset J of the index set, let $\mathbf{M}_J$ be the (finite-dimensional) subspace spanned by the e_j's with j in J. Consider the linear functional ξ_J that is equal to ξ on $\mathbf{M}_J$ and equal to 0 on $\mathbf{M}_J^\perp$. Since the ξ_J's are bounded (finite-dimensionality), there exists a net $J \mapsto g_J$ of vectors such that $\xi_J(f) = (f, g_J)$ for each f and for each J. (The finite sets J are ordered by inclusion, of course.) Given f_0, let J_0 be a finite set such that $f_0 \in \mathbf{M}_{J_0}$. If both J and K include J_0, then $(f, g_J) - (f, g_K) = 0$; it follows that $\{g_J\}$ is a weak Cauchy net. This Cauchy net cannot possibly converge weakly to anything. Suppose indeed that $g_J \to g$ weakly, so that $\xi_J(f_0) \to (f_0, g)$ for each fixed f_0. As soon as J_0 is so large that $f_0 \in \mathbf{M}_{J_0}$, then $\xi_{J_0}(f_0) = \xi(f_0)$; it follows that $\xi(f_0) = (f_0, g)$ for each f_0. Since ξ is unbounded, that is impossible.

(b) *Every Hilbert space is sequentially weakly complete.*

PROOF. If $\{g_n\}$ is a weak Cauchy sequence in $\mathbf{H}$, then $\{(f, g_n)\}$ is a Cauchy sequence, and therefore bounded, for each f in $\mathbf{H}$, so that $\{g_n\}$ is weakly bounded. It follows from the principle of uniform boundedness that $\{g_n\}$ is bounded. Since $\lim_n (f, g_n) = \xi(f)$ exists for each f in $\mathbf{H}$, and since the boundedness of $\{g_n\}$ implies that the linear functional ξ is bounded, it follows that there exists a vector g in $\mathbf{H}$ such that $\lim_n (f, g_n) = (f, g)$ for all f. This means that $g_n \to g$ (weak), so that $\{g_n\}$ does indeed have a weak limit.

CHAPTER 4

Analytic Functions

Solution 31. *For each region* D, *the inner-product space* $\mathbf{A}^2(D)$ *is complete.*

31

PROOF. It is convenient to present the proof in three steps.

(1) If D is open disc with center λ and radius r, and if $f \in \mathbf{A}^2(D)$, then

$$f(\lambda) = \frac{1}{\pi r^2} \int_D f(z)d\mu(z).$$

There is no loss of generality in restricting attention to the unit disc D_1 in the role of D, $D_1 = \{z : |z| < 1\}$; the general case reduces to this special case by an appropriate translation and change of scale. Suppose, accordingly, that $f \in \mathbf{A}^2$ $(= \mathbf{A}^2(D_1))$ with Taylor series $\sum_{n=0}^{\infty} \alpha_n z^n$, and let D_r be the disc $\{z : |z| < r\}$, $0 < r < 1$. In each D_r, $0 < r < 1$, the Taylor series of f converges uniformly, and, consequently, it is term-by-term integrable. This implies that

$$\int_{D_r} f(z)d\mu(z) = \sum_{n=0}^{\infty} \alpha_n \int_{D_r} z^n \, d\mu(z)$$

$$= \alpha_0 \cdot \pi r^2.$$

Since $|f|$ is integrable over D_1, it follows that $\int_{D_r} f \, d\mu \to \int_{D_1} f \, d\mu$ as $r \to 1$; since $\alpha_0 = f(0)$, the proof of (1) is complete.

Return now to the case of a general region D.

(2) If $v_\lambda(f) = f(\lambda)$ whenever $\lambda \in D$ and $f \in \mathbf{A}^2(D)$, then, for each fixed λ, the functional v_λ is linear. If $r = r(\lambda)$ is the radius of the largest open disc with center λ that is entirely included in D, then

$$|v_\lambda(f)| \leq \frac{1}{\sqrt{\pi r}} \|f\|.$$

187

Let D_0 be the largest open disc with center λ that is entirely included in D. Since

$$\|f\|^2 = \int_D |f(z)|^2 \, d\mu(z) \geq \int_{D_0} |f(z)|^2 \, d\mu(z)$$

$$\geq \frac{1}{\pi r^2} \left| \int_{D_0} f(z) d\mu(z) \right|^2 \qquad \text{(by the Schwarz inequality)}$$

$$= \pi r^2 \left| \frac{1}{\pi r^2} \int_{D_0} f(z) d\mu(z) \right|^2$$

$$= \pi r^2 |f(\lambda)|^2 \quad \text{(by (1))},$$

the proof of (2) is complete.

(3) The proof of the main assertion is now within reach. Suppose that $\{f_n\}$ is a Cauchy sequence in $\mathbf{A}^2(D)$. It follows from (2) that

$$|f_n(\lambda) - f_m(\lambda)| \leq \frac{1}{\sqrt{\pi \cdot r(\lambda)}} \|f_n - f_m\|$$

for every λ in D; here, as before, $r(\lambda)$ is the radius of the largest open disc with center at λ that is entirely included in D. It follows that if K is a compact subset of D, so that $r(\lambda)$ is bounded away from 0 when $\lambda \in K$, then the sequence $\{f_n\}$ of functions is uniformly convergent on K. This implies that there exists an analytic function f on D such that $f_n(\lambda) \to f(\lambda)$ for all λ in D. At the same time the completeness of the Hilbert space $\mathbf{L}^2(\mu)$ implies the existence of a complex-valued, square-integrable, but not necessarily analytic function g on D such that $f_n \to g$ in the mean of order 2. It follows that a subsequence of $\{f_n\}$ converges to g almost everywhere, and hence that $f = g$ almost everywhere. This implies that f is square-integrable, i.e., that $f \in \mathbf{A}^2(D)$, and hence that $\mathbf{A}^2(D)$ is complete.

These facts were first discussed by Bergman [11, p. 24]; the proof above is in [64]. The latter makes explicit use of the Riesz-Fischer theorem (the completeness of $\mathbf{L}^2$), instead of proving it in the particular case at hand, and consequently, from the point of view of the standard theory of Hilbert spaces, it is simpler than the analytic argument given by Bergman.

32   **Solution 32.** The evaluation of the inner products (e_n, e_m) is routine calculus. If, in fact, $D_r = \{z : |z| < r\}$, then

$$\int_{D_r} z^n z^{*m} \, d\mu(z) = \int_0^{2\pi} \int_0^r e^{i(n-m)\theta} \rho^{n+m} \rho \, d\rho \, d\theta$$

$$= 2\pi \delta_{nm} \frac{r^{n+m+2}}{n+m+2}.$$

It follows (put $r = 1$) that if $n \neq m$, then $(e_n, e_m) = 0$, and it follows also (put $n = m$) that $\|e_n\|^2 = 1$. This proves orthonormality.

To prove that the e_n's form a *complete* orthonormal set, it is tempting to argue as follows. If $f \in \mathbf{A}^2$, with Taylor series $\sum_{n=0}^{\infty} \alpha_n z^n$, then $f(z) = \sum_{n=0}^{\infty} \alpha_n \sqrt{\pi/(n+1)} \cdot e_n(z)$; this shows that each f in $\mathbf{A}^2$ is a linear combination of the e_n's, q.e.d. The argument is almost right. The trouble is that the kind of convergence it talks about is wrong. Although $\sum_{n=0}^{\infty} \alpha_n z^n$ converges to $f(z)$ at each z, and even uniformly in each compact subset of the disc, these facts by themselves do not imply that the series converges in the metric (norm) of $\mathbf{A}^2$.

There is a simple way around the difficulty: prove something else. Specifically, it is sufficient to prove that if $f \in \mathbf{A}^2$ and $f \perp e_n$ for $n = 0, 1, 2, \cdots$, then $f = 0$; and this is an immediate consequence of the second statement in Problem 32 (the statement about the relation between the Taylor and Fourier coefficients). That statement is a straightforward generalization of (1) in Solution 31 (which is concerned with e_0 only). The proof of the special case can be adapted to the general case, as follows. In each D_r, $0 < r < 1$, the series

$$f(z)z^{*m} = \sum_{n=0}^{\infty} \alpha_n z^n z^{*m}$$

converges uniformly, and, consequently, it is term-by-term integrable. This implies that

$$\int_{D_r} f(z)z^{*m} \, d\mu(z) = \sum_{n=0}^{\infty} \alpha_n \cdot 2\pi\delta_{nm} \frac{r^{n+m+2}}{n+m+2}$$
$$= \alpha_m \frac{\pi \cdot r^{2m+2}}{m+1}.$$

Since $|f \cdot e_m^*|$ is integrable over D_1, it follows that $\int_{D_r} f \cdot e_m^* \, d\mu \to \int_{D_1} f \cdot e_m^* \, d\mu = (f, e_m)$ as $r \to 1$, and the proof is complete.

Note that the argument above makes tacit use of the completeness of $\mathbf{A}^2$. The argument proves that the orthonormal set $\{e_0, e_1, e_2, \cdots\}$ is maximal; a maximal orthonormal set deserves to be called a basis only if the space is complete. The point is that in the absence of completeness the convergence of Fourier expansions cannot be guaranteed. (Cf. Problem 55.)

An alternative proof that the e_n's form a basis, which uses completeness in a less underhanded manner, is this. If $f \in \mathbf{A}^2$, with Taylor series $\sum_{n=0}^{\infty} \alpha_n z^n$, then $(f, e_n) = \sqrt{\pi/(n+1)} \cdot \alpha_n$. This implies, via the Bessel inequality, that

$$\sum_{n=0}^{\infty} \frac{\pi |\alpha_n|^2}{n+1}$$

converges. It follows that the series whose n-th term is

$$\sqrt{\pi/(n+1)} \cdot \alpha_n e_n(z)$$

converges in the mean of order 2; this conclusion squarely meets and overcomes the obstacle that stops the naive argument via power series expansions.

189

The result establishes a natural isomorphism between $\mathbf{A}^2$ and the Hilbert space of all those sequences $\langle \alpha_0, \alpha_1, \alpha_2, \cdots \rangle$ for which

$$\sum_{n=0}^{\infty} \frac{|\alpha_n|^2}{n+1} < \infty,$$

with the inner product of $\langle \alpha_0, \alpha_1, \alpha_2, \cdots \rangle$ and $\langle \beta_0, \beta_1, \beta_2, \cdots \rangle$ given by

$$\sum_{n=0}^{\infty} \frac{\pi \alpha_n \beta_n^*}{n+1}.$$

33 **Solution 33.** Formally the assertion is almost obvious. For any f in $\mathbf{L}^2$ (not only $\mathbf{H}^2$), with Fourier expansion $f = \sum_n \alpha_n e_n$, complex conjugation yields

$$f^* = \sum_n \alpha_n^* e_n^* = \sum_n \alpha_n^* e_{-n} = \sum_n \alpha_{-n}^* e_n;$$

it follows that if $f = f^*$, then $\alpha_n = \alpha_{-n}^*$ for all n. If, moreover, $f \in \mathbf{H}^2$, so that $\alpha_n = 0$ whenever $n < 0$, then it follows that $\alpha_n = 0$ whenever $n \neq 0$, and hence that $f = \alpha_0$.

The trouble with this argument is its assumption that complex conjugation distributes over Fourier expansion; that assumption must be justified or avoided. It can be justified this way: the finite subsums of $\sum_n \alpha_n e_n$ converge to f in the sense of the norm of $\mathbf{L}^2$, i.e., in the mean of order 2; it follows that a subsequence of them converges to f almost everywhere, and the desired result follows from the continuity of conjugation. The assumption can be avoided this way: since $\alpha_n = \int f e_n^* \, d\mu$, it follows that $\alpha_{-n}^* = (\int f e_{-n}^* \, d\mu)^* = \int f^* e_n^* \, d\mu$, so that if $f = f^*$, then, indeed, $\alpha_n = \alpha_{-n}^*$. It is sometimes useful to know that this last argument applies to $\mathbf{L}^1$ as well as to $\mathbf{L}^2$; it follows that the constants are the only real functions in $\mathbf{H}^1$.

34 **Solution 34.** Like the assertion (Problem 33) about real functions in $\mathbf{H}^2$, the assertion is formally obvious. If f and g are in $\mathbf{L}^2$, with Fourier expansions

$$f = \sum_n \alpha_n e_n, \qquad g = \sum_m \beta_m e_m,$$

then

$$fg = \sum_n \sum_m \alpha_n \beta_m e_n e_m = \sum_k \left(\sum_n \alpha_n \beta_{k-n} \right) e_k.$$

If, moreover, f and g are in $\mathbf{H}^2$, so that $\alpha_n = \beta_n = 0$ whenever $n < 0$, then $\sum_n \alpha_n \beta_{k-n} = 0$ whenever $k < 0$. Reason: for each term $\alpha_n \beta_{k-n}$, either $n < 0$, in which case $\alpha_n = 0$, or $n \geq 0$, in which case $k - n < 0$ and therefore $\beta_{k-n} = 0$.

The trouble with this argument is the assumption that the Fourier series of a product is equal to the formal product of the Fourier series of the factors. This assumption can be justified by appeal to the subsequence technique used in Solution 33. Alternatively the assumption can be avoided, as follows. The inner product (f, g^*) is equal to $\sum \alpha_n \beta_{-n}$ (by Parseval, and by the results of Solution 33 on the Fourier coefficients of complex conjugates); in other words, the 0-th Fourier coefficient of fg is given by

$$\int fg \, d\mu = \sum_n \alpha_n \beta_{-n}.$$

Apply this result with g replaced by the product ge_k^*. Since the Fourier coefficients γ_n of ge_k^* are given by

$$\gamma_n = \int ge_k^* e_n^* \, d\mu = \int ge_{k+n}^* \, d\mu = \beta_{k+n},$$

it follows that

$$\int fge_k^* \, d\mu = \sum_n \alpha_n \gamma_{-n} = \sum_n \alpha_n \beta_{k-n}.$$

It is an immediate corollary of this result that if $f \in \mathbf{H}^\infty$ and $g \in \mathbf{H}^2$, then $fg \in \mathbf{H}^2$. It is true also that if $f \in \mathbf{H}^\infty$ and $g \in \mathbf{H}^1$, then $fg \in \mathbf{H}^1$, but the proof requires one additional bit of analytic complication.

Solution 35. If $\varphi(z) = \sum_{n=0}^\infty \alpha_n z^n$ for $|z| < 1$, then $\varphi_r(z) = \sum_{n=0}^\infty \alpha_n r^n z^n$ for $0 < r < 1$ and $|z| = 1$. Since, for each fixed r, the latter series converges uniformly on the unit circle, it converges in every other useful sense; it follows, in particular, that the Fourier series expansion of φ_r in $\mathbf{L}^2$ is $\sum_{n=0}^\infty \alpha_n r^n e_n$, and hence that $\varphi_r \in \mathbf{H}^2$.

35

Since $\|\varphi_r\|^2 = \sum_{n=0}^\infty |\alpha_n|^2 r^{2n}$, the second (and principal) assertion reduces to this: if $\beta = \sum_{n=0}^\infty |\alpha_n|^2$ and $\beta_r = \sum_{n=0}^\infty |\alpha_n|^2 r^{2n}$, then a necessary and sufficient condition that $\beta < \infty$ is that the β_r's $(0 < r < 1)$ be bounded. In one direction the result is trivial; since $\beta_r \leq \beta$ for all r, it follows that if $\beta < \infty$, then the β_r's are bounded. Suppose now, conversely, that $\beta_r \leq \gamma$ for all r. It follows that for each positive integer k,

$$\sum_{n=0}^k |\alpha_n|^2 = \left(\sum_{n=0}^k |\alpha_n|^2 - \sum_{n=0}^k |\alpha_n|^2 r^{2n} \right) + \sum_{n=0}^k |\alpha_n|^2 r^{2n}$$

$$\leq \sum_{n=0}^k |\alpha_n|^2 (1 - r^{2n}) + \beta_r$$

$$\leq \sum_{n=0}^k |\alpha_n|^2 (1 - r^{2n}) + \gamma.$$

For k fixed, choose r so that $\sum_{n=0}^k |\alpha_n|^2 (1 - r^{2n}) < 1$; this can be done because the finite sum is a polynomial in r (and hence continuous) that vanishes when $r = 1$. Conclusion: $\sum_{n=0}^k |\alpha_n|^2 \leq 1 + \gamma$ for all k, and this implies that $\sum_{n=0}^\infty |\alpha_n|^2 \leq 1 + \gamma$.

36   **Solution 36.** Start with an arbitrary infinite-dimensional functional Hilbert space **H**, over a set X say, and adjoin a point that acts like an unbounded linear functional. To be specific: let φ be an unbounded linear functional on **H** (such a thing exists because **H** is infinite-dimensional), and write $X^+ = X \cup \{\varphi\}$. Let $\mathbf{H}^+$ be the set (pointwise vector space) of all those functions f^+ on X^+ whose restriction to X, say f, is in **H**, and whose value at φ is equal to $\varphi(f)$. (Equivalently: extend each f in **H** to a function f^+ on X^+ by writing $f^+(\varphi) = \varphi(f)$.) If f^+ and g^+ are in $\mathbf{H}^+$, with restrictions f and g, write $(f^+, g^+) = (f, g)$. (Equivalently: define the inner product of the extensions of f and g to be equal to the inner product of f and g.) The vector space $\mathbf{H}^+$ with this inner product is isomorphic to **H** with its original inner product (e.g., via the restriction mapping), and, consequently, $\mathbf{H}^+$ is a Hilbert space of functions. Since $\varphi \in X^+$ and $f^+(\varphi) = \varphi(f)$ for all f in **H**, and since φ is not bounded, it follows that $|\varphi(f)|$ can be large for unit vectors f, and therefore that $|f^+(\varphi)|$ can be large for unit vectors f^+.

37   **Solution 37.** If **H** is a functional Hilbert space, over a set X say, with orthonormal basis $\{e_j\}$ and kernel function K, write $K_y(x) = K(x, y)$, and, for each y in X, consider the Fourier expansion of K_y:

$$K_y = \sum_j (K_y, e_j)e_j = \sum_j e_j(y)^* e_j.$$

Parseval's identity implies that

$$K(x, y) = (K_y, K_x) = \sum_j e_j(x)e_j(y)^*.$$

In $\mathbf{A}^2$ the functions e_n defined by

$$e_n(z) = \sqrt{(n + 1)/\pi} \cdot z^n \quad \text{for } |z| < 1 \qquad (n = 0, 1, 2, \cdots)$$

form an orthonormal basis (see Problem 32); it follows (by the result just obtained) that the kernel function K of $\mathbf{A}^2$ is given by

$$K(x, y) = \frac{1}{\pi} \sum_{n=0}^{\infty} (n + 1)x^n y^{*n}.$$

(Note that x and y here are complex numbers in the open unit disc.) Since $\sum_{n=0}^{\infty} (n + 1)z^n = 1/(1 - z)^2$ when $|z| < 1$ (discover this by integrating the left side, or verify it by expanding the right), it follows that

$$K(x, y) = \frac{1}{\pi} \frac{1}{(1 - xy^*)^2}.$$

As for $\tilde{\mathbf{H}}^2$: by definition it consists of the functions $\tilde{f}$ on the unit disc that correspond to the elements f of $\mathbf{H}^2$. If $f = \sum_{n=0}^{\infty} \alpha_n e_n$ and if $|y| < 1$, then $\tilde{f}(y) = \sum_{n=0}^{\infty} \alpha_n y^n$ and consequently $\tilde{f}(y) = (f, K_y)$, where $K_y = \sum_{n=0}^{\infty} y^{*n} e_n$. This proves two things at once: it proves that $\tilde{f} \mapsto \tilde{f}(y)$ is

192

a bounded linear functional (so that $\tilde{\mathbf{H}}^2$ is a functional Hilbert space), and it proves that the kernel function of $\tilde{\mathbf{H}}^2$ is given by

$$K(x, y) = \sum_{n=0}^{\infty} x^n y^{*n}, \qquad |x| < 1, \; |y| < 1.$$

In closed form: $K(x, y) = 1/(1 - xy^*)$.

Solution 38. *A necessary and sufficient condition that conjugation in a* **38**
self-conjugate functional Hilbert space be isometric is that the kernel function be real.

PROOF. It is always true that

$$(f, K_y) = f(y) = (f^*(y))^* = (f^*, K_y)^*$$

for all f and all y; if, in addition, conjugation is isometric, then

$$(f^*, K_y)^* = (f, K_y^*),$$

and therefore

$$(f, K_y) = (f, K_y^*).$$

Consequence: $K_y = K_y^*$, and hence K is real. That settles necessity; sufficiency lies a little deeper.

Observe, to begin with, that the vectors K_y ($y \in X$) span **H**. Reason: if $f \perp K_y$ for all y, then $f(y) = (f, K_y) = 0$, so that $f = 0$. It follows that the norm of every vector in **H** can be calculated from its inner products with the values of K. Precisely:

$$\|f\| = \sup \frac{|(f, \sum_j \alpha_j K_{y_j})|}{\|\sum_j \alpha_j K_{y_j}\|},$$

where the supremum is extended over all non-zero finite linear combinations of the form $\sum_j \alpha_j K_{y_j}$. The numerator is equal to

$$\left| \sum_j \alpha_j^* f(y_j) \right|,$$

and the denominator is equal to

$$\left(\sum_i \sum_j \alpha_i \alpha_j^* K(y_i, y_j) \right)^{1/2}.$$

The norm of f^* is, of course, a similar supremum, with numerator

$$\left| \sum_j \alpha_j^* f^*(y_j) \right|,$$

which is equal to

$$\left| \sum_j \alpha_j f(y_j) \right|,$$

193

and denominator same as for f. The new numerator (for f^*) is obtained from the old one by replacing each α_j by its conjugate. If that replacement is made in the denominator, the result is

$$\left(\sum_i \sum_j \alpha_i{}^*\alpha_j K(y_i, y_j)\right)^{1/2}.$$

Since $K(x, y)$ is always equal to $(K(y, x))^*$, it follows that if K is assumed to be real, then the last displayed expression is equal to

$$\left(\sum_i \sum_j \alpha_i{}^*\alpha_j K(y_j, y_i)\right)^{1/2},$$

which is exactly the same as the original denominator (for f).

The set of all finite linear combinations of the form $\sum_j \alpha_j K_{y_j}$ and the set of all those of the form $\sum_j \alpha_j{}^* K_{y_j}$ are the same set; the preceding paragraph implies therefore that the suprema defining $\|f\|$ and $\|f^*\|$ are the same. The proof is complete.

In view of this solution, it is easy to construct finite-dimensional examples in which conjugation is not an isometry. If, for instance, an inner product is defined in $\mathbf{C}^2$ by the positive matrix $A = \begin{pmatrix} 2 & i \\ -i & 1 \end{pmatrix}$, and if f is the vector $\langle i, 1 \rangle$, then $\|f\| = \sqrt{5}$ and $\|f^*\| = 1$.

39   **Solution 39.** Let K be the kernel function of $\tilde{\mathbf{H}}^2$ (see Solution 37). If $f_n \to f$ in $\mathbf{H}^2$, and if $|y| < 1$, then

$$|\tilde{f}_n(y) - \tilde{f}(y)| = |(\tilde{f}_n - \tilde{f}, K_y)| \le \|f_n - f\| \cdot \|K_y\|.$$

Since

$$\|K_y\|^2 = \sum_{n=0}^{\infty} |y|^{2n} = \frac{1}{1 - |y|^2},$$

it follows that if $|y| \le r$, then

$$|\tilde{f}_n(y) - \tilde{f}(y)| \le \|f_n - f\| \cdot \frac{1}{1 - r^2}.$$

40   **Solution 40.** The function $\tilde{f}$ determines f—but how? Taylor and Fourier expansions do not reveal much about such structural properties as boundedness. The most useful way to approach the problem is to prove that the values of f (on the unit circle) are, in some sense, limits of the values of $\tilde{f}$ (on the unit disc). For this purpose, write

$$f_r(z) = \tilde{f}(rz), \qquad 0 < r < 1, \qquad |z| = 1.$$

The functions f_r are in $\mathbf{H}^2$ (see Problem 35); the assertion is that $f_r \to f$ (as $r \to 1$) in the sense of convergence in the norm of $\mathbf{H}^2$. (The boundedness of $\tilde{f}$ is not relevant yet.)

To prove the assertion, recall that if $f = \sum_{n=0}^{\infty} \alpha_n e_n$, then $f_r = \sum_{n=0}^{\infty} \alpha_n r^n e_n$, and that, consequently,

$$\| f - f_r \|^2 = \sum_{n=0}^{\infty} |\alpha_n|^2 (1 - r^n)^2.$$

It follows that for each positive integer k

$$\| f - f_r \|^2 \leq \sum_{n=0}^{k} |\alpha_n|^2 (1 - r^n)^2 + \sum_{n=k+1}^{\infty} |\alpha_n|^2.$$

The desired result ($\| f - f_r \|$ is small when r is near to 1) is now easy: choose k large enough to make the second summand small (this is independent of r), and then choose r near enough to 1 to make the first summand small.

Since convergence in $\mathbf{L}^2$ implies the existence of subsequences converging almost everywhere, it follows that $f_{r_n} \to f$ almost everywhere for a suitable subsequence $\{r_n\}$, $r_n \to 1$; the assertion about the boundedness of f is an immediate consequence.

The assertion $f_r \to f$ is true in a sense different from (better than?) convergence in the mean of order 2; in fact, it is true that $f_r \to f$ almost everywhere. This says, in other words, that if a point z in the disc tends radially to a boundary point z_0, then the function value $\tilde{f}(z)$ tends to $f(z_0)$, for almost every z_0. The result can be strengthened; radial convergence, for instance, can be replaced by non-tangential convergence. These analytic delicacies are at the center of the stage for some parts of mathematics; in the context of Hilbert space, norm convergence is enough.

Solution 41. *If $f \in \mathbf{H}^{\infty}$, then $\tilde{f}$ is bounded, and, in fact, $\| \tilde{f} \|_{\infty} = \| f \|_{\infty}$.* **41**

(The norms are the supremum of $\tilde{f}$ on the disc and the essential supremum of f on the boundary.)

PROOF. Consider the following two assertions. (1) If $f \in \mathbf{L}^{\infty}$, and, say, $|f| \leq 1$, then there exists a sequence $\{f_n\}$ of trigonometric polynomials converging to f in the norm of $\mathbf{L}^2$, such that $|f_n| \leq 1$ for all n; if, moreover, f is in $\mathbf{H}^{\infty}$, then so are the f_n's. (2) If p is a polynomial and if $|p(z)| \leq 1$ whenever $|z| = 1$, then $|p(z)| \leq 1$ whenever $|z| < 1$. Both these assertions are known parts of analysis: (1) is a consequence of Fejér's theorem about the Cesàro convergence of Fourier series, and (2) is the maximum modulus principle for polynomials. Of the two assertions, (2) seems to be far better known. In any case, (2) will be used below without any further apology; (1) will be used also, but after that it will be buttressed by the outline of a proof.

It is easy to derive the boundedness conclusion about $\tilde{f}$ from the two assertions of the preceding paragraph. Given f in $\mathbf{H}^{\infty}$, assume (this is just a matter of normalization) that $|f| \leq 1$, and, using (1), find trigonometric polynomials f_n such that $|f_n| \leq 1$ and such that $f_n \to f$ in the norm of $\mathbf{L}^2$.

Since, according to (1), the f_n's themselves are (can be chosen to be) in $\mathbf{H}^\infty$, it follows that their extensions $\tilde{f}_n$ into the interior are polynomials. Since $f_n \to f$ in the norm of $\mathbf{H}^2$, it follows from Problem 39 that $\tilde{f}_n(z) \to \tilde{f}(z)$ whenever $|z| < 1$. By (2), $|\tilde{f}_n(z)| \leqq 1$ for all n and all z. Conclusion: $|\tilde{f}(z)| \leqq 1$ for all z.

The inequality $\|\tilde{f}\|_\infty \leqq \|f\|_\infty$ is implicit in the proof above. To get the reverse inequality, use Solution 40 ($f_r \to f$ as $r \to 1$).

It remains to look at a proof of (1). If $f = \sum_n \alpha_n e_n$, write

$$s_k = \sum_{j=-k}^{+k} \alpha_j e_j \qquad (k = 0, 1, 2, \cdots).$$

Clearly $s_k \to f$ in $\mathbf{L}^2$, but this is not good enough; it does not yield the necessary boundedness results. (If $|f| \leqq 1$, it does not follow that $|s_k| \leqq 1$.) The remedy is to consider the averages

$$t_n = \frac{1}{n} \sum_{k=0}^{n-1} s_k \qquad (n = 1, 2, 3, \cdots).$$

(Note that if $f \in \mathbf{H}^2$, then so are the t_n's.) Clearly $t_n \to f$ in $\mathbf{L}^2$. (In fact it is known that $t_n \to f$ almost everywhere, but the proof is non-trivial, and, fortunately, the fact is not needed here.) This turns out to be good enough: if $|f| \leqq 1$, then it does follow that $|t_n| \leqq 1$.

For the proof, write $D_k = \sum_{j=-k}^{+k} e_j$ $(k = 0, 1, 2, \cdots)$, and

$$K_n = \frac{1}{n} \sum_{k=0}^{n-1} D_k \ (n = 1, 2, 3, \cdots);$$

the sequences of functions D_k and K_n are known as the Dirichlet and the Fejér kernels, respectively. Since $\int D_k \, d\mu = \int e_0 \, d\mu = 1$, it follows that $\int K_n \, d\mu = 1$. The principal property of the K_n's is that their values are real, and in fact positive. This is proved by computation. For $z = 1$, it is obvious; for $z \neq 1$ (but, of course, $|z| = 1$) write

$$D_k(z) = 1 + 2 \operatorname{Re} \sum_{j=1}^{k} z^j \ (k = 1, 2, 3, \cdots),$$

and apply the formula for the sum of a geometric series to get

$$D_k(z) = 2 \operatorname{Re}\left(\frac{z^k - z^{k+1}}{|1 - z|^2}\right).$$

(Computational trick: note that if $|z| = 1$, then $|1 - z|^2 = 2 \operatorname{Re}(1 - z)$.) Substitute into the expression for K_n, observe that the sum telescopes, and get

$$K_n(z) = \frac{2}{n} \operatorname{Re} \frac{1 - z^n}{|1 - z|^2}.$$

This makes it obvious that K_n is real. Since, moreover, $\operatorname{Re} z^n \leqq 1$ (recall that $|z^n| = |z|^n = 1$), i.e., $1 - \operatorname{Re} z^n \geqq 0$, it follows that $K_n(z) \geqq 0$, as asserted.

196

To apply these results to f, note that

$$s_k(z) = \sum_{j=-k}^{+k} \int f(y)y^{*j}z^j \, d\mu(y)$$

$$= \int D_k(y^*z)f(y)d\mu(y)$$

$$= \int D_k(y)f(y^*z)d\mu(y),$$

and hence that

$$t_n(z) = \int K_n(y)f(y^*z)d\mu(y);$$

this implies that if $|f| \leq 1$, then

$$|t_n(z)| \leq \int K_n(y)|f(y^*z)|d\mu(y) \leq \int K_n(y)d\mu(y) = 1.$$

The proof is over. Here is one more technical remark that is sometimes useful: under the assumptions of (1) it makes no difference whether the convergence in the conclusion is in the norm or almost everywhere. Reason: if it is in the norm, then a subsequence converges almost everywhere; if it is almost everywhere, then, by the Lebesgue bounded convergence theorem, it is also in the norm.

Solution 42. *If $f \in \mathbf{H}^\infty$, $g \in \mathbf{H}^2$, and $h = fg$, then $\tilde{h} = \tilde{f}\tilde{g}$.* **42**

PROOF. The trouble with the question as phrased is that it is easier to answer than to interpret. If f and g are in $\mathbf{H}^2$ and $h = fg$, then h is not necessarily in $\mathbf{H}^2$, and hence the definition given in Problem 35 does not apply to h; no such thing as $\tilde{h}$ is defined. The simplest way out is to assume that one factor (say f) is bounded; in that case Solution 34 shows that $h \in \mathbf{H}^2$, and the question makes sense. (There is another way out, namely to note that $h \in \mathbf{H}^1$, by Problem 34, and to extend the process of passage into the interior to $\mathbf{H}^1$. This way leads to some not overwhelming but extraneous analytic difficulties.) Once the question makes sense, the answer is automatic from Solution 34; the result there is that the Fourier coefficients of h are expressed in terms of those of f and g in exactly the same way as the Taylor coefficients of $\tilde{f} \cdot \tilde{g}$ are expressed in terms of those of $\tilde{f}$ and $\tilde{g}$. In other words: formal multiplication applies to both Fourier and Taylor series, and, consequently, the mapping from one to the other is multiplicative.

Solution 43. In order to motivate the construction of, say, f from u, it is a **43** good idea to turn the process around and to study the way u is obtained from f. Suppose therefore that $f \in \mathbf{H}^2$ with Fourier expansion $f = \sum_{n=0}^{\infty} \alpha_n e_n$, and write $u = \operatorname{Re} f$. Since $|u| \leq |f|$, the function u is in $\mathbf{L}^2$. If the Fourier expansion of u is $u = \sum_n \xi_n e_n$, then (see Solution 33)

$$u = \tfrac{1}{2}(f + f^*) = \tfrac{1}{2}\left(\sum_{n \geq 0} \alpha_n e_n + \sum_{n \geq 0} \alpha_n{}^* e_{-n}\right)$$

$$= \operatorname{Re} \alpha_0 + \sum_{n > 0} \tfrac{1}{2}\alpha_n e_n + \sum_{n < 0} \tfrac{1}{2}\alpha_{-n}{}^* e_n,$$

and therefore

$$\xi_0 = \operatorname{Re} \alpha_0 \quad \text{and} \quad \xi_n = \begin{cases} \tfrac{1}{2}\alpha_n & (n > 0), \\ \tfrac{1}{2}\alpha_{-n}{}^* & (n < 0). \end{cases}$$

It is now clear how to go in the other direction. Given $u = \sum_n \xi_n e_n$, with $\xi_n = \xi_{-n}{}^*$, and, in particular, with ξ_0 real, write

$$\alpha_0 = \xi_0 \quad \text{and} \quad \alpha_n = 2\xi_n = 2\xi_{-n}{}^* = \xi_n + \xi_{-n}{}^* \qquad (n > 0).$$

Since the sequence of α's is square-summable, an element f of $\mathbf{H}^2$ is defined by $f = \sum_{n \geq 0} \alpha_n e_n$. Write

$$f = Du$$

(D for Dirichlet); then $\operatorname{Re} Du = u$ for every real u in $\mathbf{L}^2$. It is not quite true that $D \operatorname{Re} f = f$ for every f in $\mathbf{H}^2$, but it is almost true; the difference $f - D \operatorname{Re} f$ is a purely imaginary constant that can be prescribed arbitrarily.

As for the formulation in terms of v: given u, put $v = \operatorname{Im} Du$. Since $\operatorname{Im} Du = -\operatorname{Re}(iDu)$, it is easy to get explicit expressions for the Fourier coefficients of v. If, as above, $u = \sum_n \xi_n e_n$ and $f = Du = \sum_{n \geq 0} \alpha_n e_n$, then

$$v = \operatorname{Im} f = \frac{i}{2}(f^* - f) = \frac{i}{2}\left(\sum_{n \geq 0} \alpha_n{}^* e_{-n} - \sum_{n \geq 0} \alpha_n e_n\right).$$

If $v = \sum \eta_n e_n$, then

$$\eta_0 = \operatorname{Im} \xi_0 \quad \text{and} \quad \eta_n = \begin{cases} -\dfrac{i}{2} \cdot 2\xi_n = -i\xi_n & (n > 0), \\[2mm] \dfrac{i}{2} \cdot 2\xi_n = i\xi_n & (n < 0). \end{cases}$$

If $\operatorname{Im} \xi_0 = 0$, the result can be concisely expressed:

$$\eta_n = (-i \operatorname{sgn} n)\xi_n$$

for all n.

As far as $\mathbf{L}^2$ functions on the unit circle are concerned, these algebraic trivialities are all there is to the Dirichlet problem on the unit disc. The formal expression for v in terms of u makes sense even when u is not necessarily real, and the terminology (conjugate function, Hilbert transform) remains the same. It is important to note that the Hilbert transform of a bounded function need not be bounded, or, in other words (consider extensions to the interior) that unbounded analytic functions can have bounded real parts. Standard example: $f(z) = i \log(1 - z)$.

CHAPTER 5

Infinite Matrices

Solution 44. Since **H** is the direct sum of separable subspaces that reduce A, **44** there is no loss of generality in assuming that **H** is separable in the first place. This comment, while only feebly used in the proof, eliminates the discomfort of having to worry about the pathology of the uncountable.

There is a tempting attack on the proof that is doomed to failure but is illuminating just the same. Let e_1 be an arbitrary unit vector. Since e_1 and Ae_1 span a subspace of dimension at most 2, it follows that, unless dim **H** $= 1$, there exists a unit vector e_2 orthogonal to e_1 such that $Ae_1 \in \bigvee \{e_1, e_2\}$. Since e_1, e_2, and Ae_2 span a subspace of dimension at most 3, it follows that, unless dim **H** $= 2$, there exists a unit vector e_3 orthogonal to e_1 and e_2 such that $Ae_2 \in \bigvee \{e_1, e_2, e_3\}$. An inductive repetition of this argument yields an orthonormal sequence $\{e_1, e_2, e_3, \cdots\}$ (which is finite only in case dim **H** $< \infty$) such that $Ae_n \in \bigvee \{e_1, \cdots, e_n, e_{n+1}\}$. The finite-dimensional case is transparent and, from the present point of view, uninteresting. In the infinite-dimensional case $(Ae_j, e_i) = 0$ when $i > j + 1$, and everything seems to be settled. There is a difficulty, however; there is no reason to suppose that the e_n's form a basis. If they do not, then the process of embedding them into an orthonormal basis may ruin the column-finiteness of the matrix. That is, it could happen that for some e orthogonal to all the e_n's infinitely many of the Fourier coefficients (Ae, e_i) are different from 0. If A happens to be Hermitian, then no such troubles can arise. The span of the e_n's is, in any case, invariant under A, and hence, for Hermitian A, reduces A; it follows that when the e_n's are embedded into an orthonormal basis, the new matrix elements do not interfere with the old columns. This proof, in the Hermitian case, shows more than was promised: it shows that every Hermitian operator has a Jacobi matrix. (A Jacobi matrix is a Hermitian matrix all whose non-zero entries are on

199

either the main diagonal or its two neighboring diagonals. Some authors require also that the matrix be irreducible, i.e., that none of the elements on the diagonals next to the main one vanish.) Indeed: if $(Ae_j, e_i) = 0$ when $i > j + 1$, then $(e_j, Ae_i) = 0$ when $i > j + 1$; the argument is completed by inductively enlarging the e_n's to an orthonormal basis selected the same way as the e_n's were.

In the non-Hermitian case the argument has to be refined (and the conclusion weakened to the form originally given) as follows. Let $\{f_1, f_2, f_3, \cdots\}$ be an orthonormal basis for $\mathbf{H}$. Put $e_1 = f_1$. Find a unit vector e_2 orthogonal to e_1 such that $Ae_1 \in \bigvee \{e_1, e_2\}$. (Once again restrict attention to the infinite-dimensional case.) Next find a unit vector e_3 orthogonal to e_1 and e_2 such that $f_2 \in \bigvee \{e_1, e_2, e_3\}$, and then find a unit vector e_4 orthogonal to e_1, e_2, and e_3 such that $Ae_2 \in \bigvee \{e_1, e_2, e_3, e_4\}$. Continue in this way, catching alternately one of the f_n's and the next as yet uncaught Ae_n. The selection of the needed new e is always possible. The general lemma is this: for each finite-dimensional subspace $\mathbf{M}$ and for each vector g, there exists a unit vector e orthogonal to $\mathbf{M}$ such that $g \in \mathbf{M} \vee \{e\}$. Conclusion: the sequence $\{e_1, e_2, e_3, \cdots\}$ is orthonormal by construction; it forms a basis because its span contains each f_n; and it has the property that for each n there is an i_n (calculable in case of need) such that $Ae_n \in \bigvee \{e_1, \cdots, e_{i_n}\}$. This last condition implies that $(Ae_j, e_i) = 0$ whenever $i > i_j$, and the proof is complete.

45 **Solution 45.** If $\langle \xi_0, \xi_1, \xi_2, \ldots \rangle$ is a finitely non-zero sequence of complex numbers (i.e., $\xi_n = 0$ for n sufficiently large), then

$$\sum_i \left| \sum_j \alpha_{ij} \xi_j \right|^2 = \sum_i \left| \sum_j (\sqrt{\alpha_{ij}} \sqrt{q_j}) \left(\frac{\sqrt{\alpha_{ij}} \xi_j}{\sqrt{q_j}} \right) \right|^2$$

$$\leq \sum_i \left(\sum_j \alpha_{ij} q_j \right) \left(\sum_j \frac{\alpha_{ij} |\xi_j|^2}{q_j} \right)$$

$$\leq \sum_i \gamma p_i \sum_j \frac{\alpha_{ij} |\xi_j|^2}{q_j}$$

$$= \gamma \sum_j \frac{|\xi_j|^2}{q_j} \sum_i \alpha_{ij} p_i$$

$$\leq \gamma \sum_j \frac{|\xi_j|^2}{q_j} \cdot \beta q_j = \beta \cdot \gamma \sum_j |\xi_j|^2.$$

These inequalities imply that the operator A on l^2 defined by

$$A \langle \xi_0, \xi_1, \xi_2, \cdots \rangle = \left\langle \sum_j \alpha_{0j} \xi_j, \sum_j \alpha_{1j} \xi_j, \sum_j \alpha_{2j} \xi_j, \cdots \right\rangle$$

satisfies the conditions.

Solution 46. The result is a corollary of Problem 45. For the proof, apply **46** Problem 45 with $p_i = q_i = 1/\sqrt{i - \frac{1}{2}}$. Since the Hilbert matrix is symmetric, it is sufficient to verify one of the two inequalities (with $\beta = \gamma = \pi$). The verification depends on elementary calculus, as follows:

$$\sum_i \alpha_{ij} p_i = \sum_i \frac{1}{(i + \frac{1}{2} + j + \frac{1}{2})\sqrt{i + \frac{1}{2}}}$$

$$< \int_0^\infty \frac{dx}{(x + j + \frac{1}{2})\sqrt{x}}$$

$$= 2 \int_0^\infty \frac{du}{u^2 + j + \frac{1}{2}}$$

$$= \frac{2}{\sqrt{j + \frac{1}{2}}} \int_0^\infty \frac{du}{u^2 + 1} = \frac{\pi}{\sqrt{j + \frac{1}{2}}}.$$

Solution 47. The assertion is easy to prove: since $\sum_i \sum_j (2^{-(i+j+1)})^2 < \infty$, **47** boundedness follows from the sufficient condition mentioned in the preliminary discussion of Problem 44.

The determination of the norm is made easy by the special arithmetic structure of the entries of the matrix. If f_0 is the first column,

$$f_0 = \langle \tfrac{1}{2}, \tfrac{1}{4}, \tfrac{1}{8}, \cdots \rangle,$$

then all other columns are scalar multiples of f_0, and, consequently, every vector in the range of the operator A under study is a multiple of f_0. In other words, A has rank 1; in fact, $Af = 2(f, f_0)f_0$ for every vector f. It follows that $\|A\| = 2\|f_0\|^2 = 2 \sum_{n=1}^\infty 1/4^n = 2/3$.

(Note that A is Hermitian. The spectrum of A consists of two eigenvalues: 0 with infinite multiplicity, and $\frac{2}{3}$ with multiplicity 1.)

Solution 48. The Gramian of a finite or infinite sequence $\{f_n\}$ of vectors is the **48** matrix whose $\langle i, j \rangle$ entry is the inner product (f_i, f_j). It is not difficult to prove that every positive matrix is a Gramian; it is completely trivial to prove that every Gramian is positive. To prove that the Hilbert matrix is positive, it is therefore sufficient to exhibit, in some Hilbert space, a sequence $\{f_n\}$ of vectors such that $(f_i, f_j) = 1/(i + j + 1)$ $(i, j = 0, 1, 2, \cdots)$. To do that, let the Hilbert space be $L^2(0, 1)$, and let the vectors f_n be defined by $f_n(x) = x^n$. The rest is elementary calculus.

It follows on general grounds that the Hilbert matrix has a unique positive square root. What is it? No explicit description of it seems to be known.

Solution 49. The answer is that the Gramian matrix $\langle (f_j, f_i) \rangle$ be bounded **49** (or, in other words, that there exist an operator with that matrix).

201

In one direction the proof is straightforward. If the Gramian is bounded, and if α is a finitely non-zero sequence, then

$$\left\|\sum_n \alpha_n f_n\right\|^2 = \left(\sum_j \alpha_j f_j, \sum_i \alpha_i f_i\right) = \sum_i \sum_j (f_j, f_i)\alpha_j \alpha_i^*$$

$$= (G\alpha, \alpha),$$

where G is the Gramian (considered as an operator on l^2). Consequence:

$$\left\|\sum_n \alpha_n f_n\right\|^2 \leq \|G\| \cdot \|\alpha\|^2,$$

and that implies convergence. Note, in particular, that since the Hilbert matrix is a bounded Gramian (Problem 46), this answers the question about $\mathbf{L}^2(0, 1)$.

The quickest proof of the converse is via uniform boundedness. For each $n = 1, 2, 3, \cdots$, let T_n be the linear transformation from l^2 to $\mathbf{H}$ defined by

$$T_n \alpha = \sum_{k=1}^{n} \alpha_k f_k.$$

Clearly each T_n is bounded. For a fixed α, the sequence $\{T_n \alpha\}$ consists of the partial sums of a series that is convergent by assumption, and, therefore, the sequence $\{T_n \alpha\}$ is bounded. Conclusion (see Problem 51): the sequence $\{\|T_n\|\}$ is bounded, and that implies (in view of the identity involving $\|T_n \alpha\|^2$ and the Gramian) that the Gramian is bounded.

Note that the result is a generalization of the numerical fact (Problem 29): the "Gramian" $\langle \beta_j \beta_i^* \rangle$ of a sequence of scalars is bounded if and only if the sequence is in l^2.

CHAPTER 6

Boundedness and Invertibility

Solution 50. Let $\{e_1, e_2, e_3, \cdots\}$ be an orthonormal basis for a Hilbert space **50** **H**, and find a Hamel basis for **H** that contains each e_n. Let f_0 be an arbitrary but fixed element of that Hamel basis distinct from each e_n (see Solution 7). A unique linear transformation A is defined on **H** by the requirement that $Af_0 = f_0$ and $Af = 0$ for all other elements of the selected basis: in particular, $Ae_n = 0$ for all n. If A were bounded, then its vanishing on each e_n would imply that $A = 0$. This solves parts (a) and (b) of the problem.

It is interesting to observe that unbounded examples for part (a) are not only easy to come by; in fact, it is impossible to avoid them. That is: for every linear transformation A on **H**, there exists an orthonormal basis $\{e_1, e_2, e_3, \cdots\}$ for **H** such that $\sup_n \|Ae_n\| < \infty$. Reference: [121].

The answer to (c) is no: if a linear transformation A is bounded on each orthonormal basis, then A is bounded. One way to see that is to imitate the easy beginning part of Solution 27. (If A is not bounded, then there exists a unit vector e_1 such that $\|Ae_1\| \geq 1$, and, by induction, whenever $e_1, \cdots, e_n$ are orthonormal vectors with $\|Ae_j\| \geq j$, $j = 1, \ldots, n$, then there exists a unit vector e_{n+1} orthogonal to $\{e_1, \cdots, e_n\}$ such that $\|Ae_{n+1}\| \geq n + 1$.)

For part (d), choose an arbitrary but fixed positive integer k and define an operator A (depending on k) by

$$Af = (f, e_1 + \cdots + e_k)e_1.$$

It follows that $Ae_n = e_1$ or 0 according as $n \leq k$ or $n > k$, and hence that $\|Ae_n\| \leq 1$ for all n. Since (easy computation) $A^*f = (f, e_1)(e_1 + \cdots + e_k)$ for all f, so that, in particular, $A^*e_1 = e_1 + \cdots + e_k$, it follows that

$$\|A\| = \|A^*\| \geq \|A^*e_1\| = \|e_1 + \cdots + e_k\| = \sqrt{k}.$$

A simple alternative way to say all this is to describe the matrix of A with respect to the basis $\{e_1, e_2, e_3, \cdots\}$: all the entries are 0 except the first k entries in the first row, which are 1's.

As for (e), yes, there are normal matrices, and even Hermitian ones, that have the same unboundedness properties as the ones in (d). Example: $\alpha_{ij} = 1/\sqrt{k}$ if $1 \leq i,j \leq k$ and $\alpha_{ij} = 0$ otherwise. If the matrix is divided by $\sqrt{k}$, so that each non-zero entry becomes $1/k$, then the matrix becomes idempotent. Since it is Hermitian, it is now a projection, and hence has norm 1. Conclusion: the original matrix has norm $\sqrt{k}$.

51   **Solution 51.** The conclusion can be obtained from two successive applications of the principle of uniform boundedness for vectors (Problem 27). Suppose that **Q** is a weakly bounded set of bounded linear transformations from **H** to **K**, and that, specifically, $|(Af, g)| \leq \alpha(f, g)$ for all A in **Q**. Fix an arbitrary vector g_0 and write $\mathbf{T}_0 = \{A^*g_0 : A \in \mathbf{Q}\}$. Since

$$|(f, A^*g_0)| = |(Af, g_0)| \leq \alpha(f, g_0),$$

the set $\mathbf{T}_0$ is weakly bounded in **H**, and therefore there exists a constant $\beta(g_0)$ such that $\|A^*g_0\| \leq \beta(g_0)$ for all A in **Q**.

Next, write $\mathbf{T} = \{Af : A \in \mathbf{Q}, f \in \mathbf{H}_1\}$, where $\mathbf{H}_1$ is the unit ball of **H**. Since

$$|(g, Af)| = |(A^*g, f)| \leq \beta(g) \cdot \|f\| \leq \beta(g),$$

the set **T** is weakly bounded in **K**, and therefore there exists a constant γ such that

$$\|Af\| \leq \gamma$$

whenever $A \in \mathbf{Q}$ and $f \in \mathbf{H}_1$. This implies that

$$\|A\| \leq \gamma,$$

and the proof is complete.

52   **Solution 52.** It is sufficient to prove that A^* is invertible. The range of A^* is dense in **H** (because the kernel of A is trivial), and, consequently, it is sufficient to prove that A^* is bounded from below. This means that $\|A^*g\| \geq \delta\|g\|$ for some δ (and all g in **K**). To prove it, it is sufficient to prove that if $\|A^*g\| = 1$, then $\|g\| \leq 1/\delta$ for some δ. Caution: the last reduction uses the assumption that the kernel of A^* is trivial, which is true because the range of A is dense in **K**. (The full force of the assumption that A maps **H** onto **K** will be used in a moment.) To see the difficulty, consider the transformation 0 in the role of A^*: for it the implication from $\|A^*g\| = 1$ to $\|g\| \leq 1/\delta$ is vacuously valid. Summary: it is sufficient to prove that if

$$\mathbf{S} = \{h : \|A^*h\| = 1\},$$

then the set **S** is bounded, and that can be done by proving that it is weakly bounded. To do that, take g in **K**, find f in **H** so that $Af = g$, and observe that

$$|(g, h)| = |(Af, h)| = |(f, A^*h)| \leq \|f\|$$

for all h in **S**. The proof is complete.

Solution 53. (a) The construction of non-closed vector sums is either laborious ([138, p. 21] and [50, p. 28]) or mildly sophisticated (Problem 52). All known constructions can be generalized so as to solve Problem 53 also.

Consider, for example, an operator A on a Hilbert space **H**, and let **M** be the x-axis and **N** the graph of A in $\mathbf{H} \oplus \mathbf{H}$. Question: under what conditions on A is **N** a complement of **M**? Answer: if and only if ker $A = 0$ and $\overline{\mathrm{ran}}\, A = \mathbf{H}$. Assume that these conditions are satisfied, consider a subspace $\mathbf{H}_0$ of **H** and, in an attempt to diminish **N**, consider, in the role of $\mathbf{N}_0$, that part of the graph of A that lies over $\mathbf{H}_0$. More precisely, let $\mathbf{N}_0$ be the set of all vectors in $\mathbf{H} \oplus \mathbf{H}$ that are of the form $\langle g, Ag \rangle$ for some g in $\mathbf{H}_0$. Question: under what conditions on $\mathbf{H}_0$ is $\mathbf{N}_0$ still a complement of **M**? Answer: if and only if $A\mathbf{H}_0$ is dense in **H**.

In view of all this, the construction of a diminishable complement can be accomplished by constructing an operator A on **H** and a proper subspace $\mathbf{H}_0$ of **H** such that ker $A = 0$ and $A\mathbf{H}_0$ is dense in **H**. This cannot be done if **H** is finite-dimensional; if **H** is infinite-dimensional and the kernel requirement is omitted, it is very easy to do. The only mild challenge is in the case at hand.

Let **H** be l^2 and define A (as in Problem 52) by $A\langle \xi_1, \xi_2, \xi_3, \cdots \rangle = \langle \xi_1, \frac{1}{2}\xi_2, \frac{1}{3}\xi_3, \cdots \rangle$; to define $\mathbf{H}_0$, let h be the vector $\langle 1, \frac{1}{2}, \frac{1}{3}, \cdots \rangle$ and let $\mathbf{H}_0$ be its orthogonal complement. That is: $\mathbf{H}_0$ is the set of all those vectors g ($=\langle \xi_1, \xi_2, \xi_3, \cdots \rangle$) for which (g, h) ($=\sum_n (1/n)\xi_n$) $= 0$. The only thing that needs proof is that $A\mathbf{H}_0$ is dense in l^2, and for that purpose it is sufficient to prove that each finitely non-zero vector f in l^2 can be approximated arbitrarily closely by vectors in $A\mathbf{H}_0$.

Since f is finitely non-zero, there exists a (unique and necessarily finitely non-zero) vector g such that $Ag = f$. There is no reason why the vector g should be in $\mathbf{H}_0$, but by a suitable perturbation it can be put there. Suppose, indeed, that $(g, h) = \alpha$; the plan is to find a vector g_0 such that $(g_0, h) = -\alpha$ (so that $g + g_0 \in \mathbf{H}_0$) and such that $\|Ag_0\|$ is small (so that $A(g + g_0)$ is near to f). For this purpose let p be a large positive integer, and let n be a positive integer such that all the coordinates of g after the n-th are equal to 0; define g_0 so that its coordinates with indices $n + 1$, $n + 2, \ldots, n + p$ are equal to

$$\frac{-\alpha(n + 1)}{p}, \frac{-\alpha(n + 2)}{p}, \cdots, \frac{-\alpha(n + p)}{p},$$

and all other coordinates are 0. The coordinates of Ag_0 with indices $n + 1$, $n + 2, \cdots, n + p$ are all equal to $-\alpha/p$; it follows that $(g_0, h) = -\alpha$ and

$$\|Ag_0\|^2 = p \frac{|\alpha|^2}{p^2} = \frac{|\alpha|^2}{p},$$

which is indeed small if p is large.

(b) Analytic phenomena (having to do with convergence) are likely to misbehave when infinitely repeated, but algebraic ones not. To get an infinitely multiple example of the (lattice) algebraic phenomenon observed in the preceding example, just form the direct sum of infinitely many copies of that example. That is: replace $\mathbf{H}$, A, and $\mathbf{H}_0$ by $\mathbf{H} \oplus \mathbf{H} \oplus \mathbf{H} \oplus \cdots$, $A \oplus A \oplus A \oplus \cdots$, and $\mathbf{H}_0 \oplus \mathbf{H}_0 \oplus \mathbf{H}_0 \oplus \cdots$. The verification that nothing goes wrong is straightforward.

For a different approach, see Solution 55.

54   **Solution 54.** Let $\mathbf{K}$ be a Hilbert space of dimension $\aleph_0$, let $\mathbf{E}$ be an orthonormal basis for $\mathbf{K}$ (of cardinal number $\aleph_0$), and let $\mathbf{F}$ be a subset of $\mathbf{K}$ such that $\mathbf{E} \cup \mathbf{F}$ is a Hamel basis for $\mathbf{K}$. The cardinal number of $\mathbf{F}$ is $\mathbf{c}$ (the power of the continuum). Let $\mathbf{L}$ be a Hilbert space of dimension $\mathbf{c}$, and let T be a linear transformation from $\mathbf{K}$ into $\mathbf{L}$ defined so that T maps $\mathbf{E}$ into 0 and maps $\mathbf{F}$ one-to-one onto an orthonormal basis for $\mathbf{L}$. Note that $\operatorname{ran} T$ is dense in $\mathbf{L}$.

Let $\mathbf{G}$ be the graph of T, i.e., the set of all $\langle f, Tf \rangle$ with f in $\mathbf{K}$, and let $\mathbf{H}$ be the direct sum $\mathbf{K} \oplus \mathbf{L}$. Since $\langle e, 0 \rangle \in \mathbf{G}$ for each e in $\mathbf{E}$, it follows that $\langle f, 0 \rangle \in \bar{\mathbf{G}}$ (the closure) for each f in $\mathbf{K}$. This, in turn, implies that $\langle f, Tf \rangle - \langle f, 0 \rangle = \langle 0, Tf \rangle \in \mathbf{G}$ for each f in $\mathbf{K}$. Since $\operatorname{ran} T$ is dense in $\mathbf{L}$, therefore $\bar{\mathbf{G}}$ contains every $\langle 0, g \rangle$, for g in $\mathbf{L}$, and hence $\bar{\mathbf{G}} = \mathbf{H}$; the linear manifold $\mathbf{G}$ is dense in $\mathbf{H}$.

The dimension of $\mathbf{H}$ is $\dim \mathbf{K} + \dim \mathbf{L} = \mathbf{c}$; what is the dimension of $\mathbf{G}$? Answer: $\aleph_0$. Reason: the set of all vectors of the form $\langle e, 0 \rangle$, with e in $\mathbf{E}$, is a maximal orthonormal set in $\mathbf{G}$. Indeed, if $\langle f, Tf \rangle \perp \langle e, 0 \rangle$ for all e in $\mathbf{E}$, then $f \perp \mathbf{E}$, whence $f = 0$, and therefore $Tf = 0$.

Pertinent reference: [48].

55   **Solution 55.** No, maximal orthonormal sets need not be total. One way to get an example is to use Problem 54. Suppose, indeed, that $\mathbf{G}$ is a dense linear manifold in $\mathbf{H}$ with $\dim \mathbf{G} \neq \dim \mathbf{H}$. An immediate consequence is that no orthonormal basis for $\mathbf{H}$ is included in $\mathbf{G}$. From this, in turn, it follows that there is no orthonormal set that is total for $\mathbf{G}$. The proof goes as follows. If $\mathbf{E}_0$ is such a set, then the intersection of its span in $\mathbf{H}$ with $\mathbf{G}$ is a closed subspace of $\mathbf{G}$ that includes $\mathbf{E}_0$. Hence that intersection is $\mathbf{G}$; hence the span of $\mathbf{E}_0$ in $\mathbf{H}$ is $\mathbf{H}$; hence $\mathbf{E}_0$ is an orthonormal basis for $\mathbf{H}$; and that contradicts the first consequence mentioned above.

Problems 53, 54, and 55 are concerned with the pathology of linear manifolds; in each case a transformation with dense range can be used to construct a large subspace with surprising properties. There is a much more sophisticated construction that deserves to be widely known. It is short, it does not need much machinery, and its techniques are frequently applicable. Its disadvantage is that, at this point, it borrows from the future: it needs the analytic power of the F. and M. Riesz theorem about the Hardy space $\mathbf{H}^2$ (Problem 158).

In the notation of Problem 33, consider the Hilbert space $\mathbf{L}^2$ and the subspace $\mathbf{H}^2$. Let E be a Borel subset of C (the unit circle) such that both E and $C-E$ have positive measure, and let $\mathbf{N}$ be $\mathbf{L}^2(E)$ ($=$ the set of those elements of $\mathbf{L}^2$ that vanish outside E).

Assertion: $\mathbf{N}$ is a complement of $\mathbf{H}^2$. Indeed: if $f \in \mathbf{H}^2 \cap \mathbf{N}$, then f vanishes on a set of positive measure, and, therefore, the F. and M. Riesz theorem implies that $f = 0$. If $f \perp \mathbf{H}^2 \vee \mathbf{N}$, then $f \in \mathbf{H}^{2*}$ (because $(\mathbf{H}^2)^\perp \subset \mathbf{H}^{2*}$) and f vanishes on a set of positive measure (namely E); the F. and M. Riesz theorem implies that $f = 0$.

If E_0 is a Borel subset of E such that both E_0 and $E - E_0$ have positive measure, then the result just proved applies to E_0 as well as to E. In other words $\mathbf{L}^2(E_0)$ is a complement of $\mathbf{H}^2$; since $\mathbf{N} \cap \mathbf{N}_0{}^\perp = \mathbf{L}^2(E - E_0)$, it follows that the complement $\mathbf{H}$ is infinitely diminishable.

So much for Problem 53. For a similar alternative approach to 55, let $\mathbf{P}\, (\subset \mathbf{H}^2)$ be the set of all polynomials, and consider the vector sum $\mathbf{P} + \mathbf{L}^2(E)$ as an inner product space in its own right.

The orthonormal set $\{e_0, e_1, e_2, \cdots\}$ (notation as in Problem 33) is not total in $\mathbf{P} + \mathbf{L}^2(E)$. The reason is that its span (in $\mathbf{L}^2$) is $\mathbf{H}^2$, but, since $\mu(C - E) > 0$, the F. and M. Riesz theorem implies that the only function in both $\mathbf{L}^2(E)$ and $\mathbf{H}^2$ is 0. Since $\mu(E) > 0$, there exist non-zero functions in $\mathbf{L}^2(E)$.

The orthonormal set $\{e_0, e_1, e_2, \cdots\}$ is, however, maximal in $\mathbf{P} + \mathbf{L}^2(E)$. Indeed, if $p \in \mathbf{P}, f \in \mathbf{L}^2(E)$, and $p + f \perp e_n, n = 0, 1, 2, \cdots$, then $p + f \perp \mathbf{H}^2$. This implies that $f = h - p$, where $h \in \mathbf{H}^{2*}$, and hence that $f^* = h^* - p^*$, where $h^* \in \mathbf{H}^2$. Multiply the latter equation by e_n, where n is sufficiently large to make p^*e_n belong to $\mathbf{H}^2$. The result is that $f^*e_n \in \mathbf{H}^2$, and hence, by the F. and M. Riesz theorem, $f^*e_n = 0$; therefore $f = 0$ and therefore $p = 0$ (since $p \in \mathbf{H}^2$ and $p \perp \mathbf{H}^2$).

Solution 56. If $\dim \mathbf{K} < \dim \mathbf{H}$, then there is no loss of generality in assuming that $\mathbf{K} \subset \mathbf{H}$. Suppose, accordingly, that A is an operator on $\mathbf{H}$ with range included in $\mathbf{K}$; it is to be proved that $\ker A$ is not trivial. Assume that $\dim \mathbf{K}$ is infinite; this assumption excludes trivial cases only. Let $\{f_i\}$ and $\{g_j\}$ be orthonormal bases of $\mathbf{H}$ and $\mathbf{K}$, respectively. Each A^*g_j can be expanded in terms of countably many f's; the assumed inequality between the dimensions of $\mathbf{H}$ and $\mathbf{K}$ implies the existence of an i such that $(f_i, A^*g_j) = 0$ for all j. Since $(f_i, A^*g_j) = (Af_i, g_j)$, it follows that Af_i is orthogonal to each g_j and therefore to $\mathbf{K}$. Since, however, the range of A is included in $\mathbf{K}$, it follows that $Af_i = 0$.

56

207

Consider next the statement about equality. If dim **H** is finite, all is trivial. If dim **H** is infinite, then a set of cardinal number dim **H** is dense in **H**. (Use rational linear combinations; cf. Solution 17.) It follows that a set of cardinal number dim **H** is dense in **K**, and this implies that dim **K** $\leqq$ dim **H**.

The proof above is elementary, but, for a statement that is completely natural, it is not at all completely obvious. (It is due, incidentally, to G. L. Weiss; cf. [63].) There is a quick proof, which, however, is based on a nontrivial theory (polar decomposition). It goes as follows. If A is a one-to-one linear transformation from **H** into **K**, with polar decomposition UP (see Problem 134), then, since ker A is 0, it follows that U is an isometry. As for the case of equality: if ran A is dense in **K**, then ran U is equal to **K**.

57　**Solution 57.** Observe first that no non-zero vector in the range of P is annihilated by Q. Indeed, if $Pf = f$ and $Qf = 0$, then $\|Pf - Qf\| = \|f\|$, and therefore $f \neq 0$ would imply $\|P - Q\| \geqq 1$. From this it follows that the restriction of Q to the range of P is a one-to-one bounded linear transformation from that range into the range of Q, and therefore that the rank of P is less than or equal to the rank of Q (Problem 56). The conclusion follows by symmetry.

58　**Solution 58.** Suppose that A is a linear transformation from **H** to **K**, and suppose, first, that A is bounded. Let $\{\langle f_n, Af_n \rangle\}$ be a sequence of vectors in the graph of A converging to something, say $\langle f_n, Af_n \rangle \to \langle f, g \rangle$. Since $f_n \to f$ and A is continuous, it follows that $Af_n \to Af$; since at the same time $Af_n \to g$, it follows that $g = Af$, and hence that $\langle f, g \rangle$ is in the graph of A.

The proof of the converse is less trivial; it is a trick based on Problem 52. Let **G** be the graph of A, and consider the linear transformation B from **G** to **H** defined by $B\langle f, Af \rangle = f$. Clearly B is a one-to-one mapping from **G** onto **H**; since

$$\|B\langle f, Af \rangle\|^2 = \|f\|^2 \leqq \|f\|^2 + \|Af\|^2 = \|\langle f, Af \rangle\|^2,$$

it follows that B is bounded. Since **G** is a closed subset of the complete space **H** $\oplus$ **K**, it is complete, and all is ready for an application of Problem 52; the conclusion is that B is invertible. Equivalently the conclusion says that the mapping B^{-1} from **H** into **G**, defined by $B^{-1}f = \langle f, Af \rangle$, is a bounded linear transformation. This means, by definition, that

$$\|f\|^2 + \|Af\|^2 \leqq \alpha \|f\|^2$$

for some α (and all f in **H**); the boundedness of A is an immediate consequence.

It is worth remarking that the derivation of the result from Problem 52 is reversible; the assertion there is a special case of the closed graph theorem. This, of course, is not an especially helpful comment for someone who wants to know how to prove the closed graph theorem, and not just how to bounce back and forth between it and a reformulation.

Solution 59. The answer is yes. Note first that ran $B = B\mathbf{H} = B(\ker^\perp B)$. **59**
It follows that for each f in $\mathbf{H}$ there is a unique g in $\ker^\perp B$ with $Af = Bg$;
write $Xf = g$. (There is no ambiguity here. Reason: if $Af = 0$, then $g = 0$.)
The verification that X is linear is routine, and, obviously, $Af = BXf$ for
all f, so that $A = BX$; all that needs proof is that X is bounded. That
comes quickly from the closed graph theorem. Indeed, if $\langle f_n, g_n \rangle$ is in
the graph of X, $n = 1, 2, 3, \cdots$, and $\langle f_n, g_n \rangle \to \langle f, g \rangle$, then $Af_n \to Af$ and
$Bg_n \to Bg$; since $Af_n = Bg_n$ for each n, it follows that $Af = Bg$. Since,
moreover, $\ker^\perp B$ is closed, so that g belongs to $\ker^\perp B$, it follows that $\langle f, g \rangle$
belongs to the graph of X.

Solution 60. (a) *On incomplete inner-product spaces unbounded sym-* **60**
metric transformations do exist. (b) *On a Hilbert space, every symmetric*
linear transformation is bounded.

PROOF. (a) Let $\mathbf{H}$ be the complex vector space of all finitely non-zero infinite
sequences. That is, an element of $\mathbf{H}$ is a sequence $\langle \xi_1, \xi_2, \xi_3, \cdots \rangle$, with $\xi_n = 0$
for all sufficiently large n; the "sufficiently large" may vary with the sequence.
Define inner product in $\mathbf{H}$ the natural way: if $f = \langle \xi_1, \xi_2, \xi_3, \cdots \rangle$ and
$g = \langle \eta_1, \eta_2, \eta_3, \cdots \rangle$, write $(f, g) = \sum_{n=1}^\infty \xi_n \eta_n{}^*$. Let A be the linear transformation that maps each sequence $\langle \xi_1, \xi_2, \xi_3, \cdots \rangle$ onto $\langle \xi_1, 2\xi_2, 3\xi_3, \cdots \rangle$;
in an obvious manner A is determined by the diagonal matrix whose sequence
of diagonal terms is $\langle 1, 2, 3, \cdots \rangle$. The linear transformation A is symmetric;
indeed both (Af, g) and (f, Ag) are equal to $\sum_{n=1}^\infty n\xi_n \eta_n{}^*$. The linear transformation A is not bounded; indeed if $\{f_n\}$ is the sequence whose n-th term
is 1 and all other terms are 0, then $\|f_n\| = 1$ and $\|Af_n\| = n$.

(b) This is an easy consequence of the closed graph theorem. Indeed,
if A is symmetric, and if $f_n \to f$ and $Af_n \to f'$, then, for all g,

$$(f', g) = \lim_n (Af_n, g) = \lim_n (f_n, Ag) = (f, Ag) = (Af, g),$$

and therefore $f' = Af$; this proves that A is closed, and hence that A is
bounded.

Alternatively, use the principle of uniform boundedness directly. If
$\|g\| \leq 1$, then

$$|(f, Ag)| = |(Af, g)| \leq \|Af\|$$

for all f; in other words, the image under A of the unit ball is weakly
bounded, and therefore bounded.

CHAPTER 7

Multiplication Operators

61 **Solution 61.** If A is a diagonal operator, with $Ae_j = \alpha_j e_j$, then

$$|\alpha_j| = \|\alpha_j e_j\| = \|Ae_j\| \leqq \|A\| \cdot \|e_j\| = \|A\|,$$

so that $\{\alpha_j\}$ is bounded and $\sup_j |\alpha_j| \leqq \|A\|$. The reverse inequality follows from the relations

$$\left\| A \sum_j \xi_j e_j \right\|^2 = \left\| \sum_j \alpha_j \xi_j e_j \right\|^2 = \sum_j |\alpha_j \xi_j|^2$$

$$\leqq \left(\sup_j |\alpha_j| \right)^2 \cdot \sum_j |\xi_j|^2 = \left(\sup_j |\alpha_j| \right)^2 \cdot \left\| \sum_j \xi_j e_j \right\|^2.$$

Given a bounded family $\{\alpha_j\}$, define A by $A \sum_j \xi_j e_j = \sum_j \alpha_j \xi_j e_j$; the preceding computations imply that A is an operator. Clearly A is a diagonal operator, and the diagonal of A is exactly the sequence $\{\alpha_j\}$. The proof of uniqueness is implicit in the construction: via Fourier expansions the behavior of an operator on a basis determines its behavior everywhere.

62 **Solution 62.** *If $\{\alpha_n\}$ is a sequence of complex scalars, such that $\sum_n |\alpha_n \xi_n|^2 < \infty$ whenever $\sum_n |\xi_n|^2 < \infty$, then $\{\alpha_n\}$ is bounded.*

PROOF. Expressed contrapositively, the assertion is this: if $\{\alpha_n\}$ is not bounded, then there exists a sequence $\{\xi_n\}$ such that $\sum_n |\xi_n|^2 < \infty$ but $\sum_n |\alpha_n \xi_n|^2 = \infty$. The construction is reasonably straightforward. If $\{\alpha_n\}$ is not bounded, then $|\alpha_n|$ takes arbitrarily large values. There is no loss of generality in assuming that $|\alpha_n| \geqq n$; all it takes is a slight change of notation, and, possibly, the omission of some α's. If, in that case, $\xi_n = 1/\alpha_n, n = 1, 2, 3, \cdots$, then

$$\sum_n |\xi_n|^2 \leq \sum_n \frac{1}{n^2} < \infty,$$

but $\sum_n |\alpha_n \xi_n|^2$ diverges.

Solution 63. The assertion is that if A is a diagonal operator with diagonal $\{\alpha_n\}$, then A and $\{\alpha_n\}$ are invertible together. Indeed, if $\{\beta_n\}$ is a bounded sequence such that $\alpha_n \beta_n = 1$ for all n, then the diagonal operator B with diagonal $\{\beta_n\}$ acts as the inverse of A. Conversely: if A is invertible, then $A^{-1}(\alpha_n e_n) = e_n$, so that

$$A^{-1}e_n = \frac{1}{\alpha_n} e_n;$$

since $\|A^{-1}e_n\| \leq \|A^{-1}\|$, this implies that the sequence $\{1/\alpha_n\}$ is bounded, and hence that the sequence $\{\alpha_n\}$ is invertible.

As for the spectrum: the assertion here is that $A - \lambda$ is invertible if and only if λ does not belong to the closure of the diagonal $\{\alpha_n\}$. (The purist has a small right to object. The diagonal is a *sequence* of complex numbers, and, therefore, not just a *set* of complex numbers; "the closure of the diagonal" does not make rigorous sense. The usage is an instance of a deservedly popular kind of abuse of language, unambiguous and concise; it would be a pity to let the purist have his way.) The assertion is equivalent to this: $\{\alpha_n - \lambda\}$ is bounded away from 0 if and only if λ is not in the closure of $\{\alpha_n\}$. Contrapositively: the sequence $\{\alpha_n - \lambda\}$ has 0 as a limit point if and only if the set $\{\alpha_n\}$ has λ as a cluster point. Since this is obvious, the proof is complete.

63

Solution 64. *If A is the multiplication induced by a bounded measurable function φ on a σ-finite measure space, then $\|A\| = \|\varphi\|_\infty$ ($=$the essential supremum of $|\varphi|$).*

64

PROOF. Let μ be the underlying measure. It is instructive to see how far the proof can get without the assumption that μ is σ-finite; until further notice that assumption will not be used. Since

$$\|Af\|^2 = \int |\varphi \cdot f|^2 \, d\mu \leq \|\varphi\|_\infty^2 \cdot \int |f|^2 \, d\mu = \|\varphi\|_\infty^2 \cdot \|f\|^2,$$

it follows that $\|A\| \leq \|\varphi\|_\infty$. In the proof of the reverse inequality a pathological snag is possible.

A sensible way to begin the proof is to note that if $\varepsilon > 0$, then $|\varphi(x)| > \|\varphi\|_\infty - \varepsilon$ on a set, say M, of positive measure. If f is the characteristic function of M, then

$$\|f\|^2 = \int_M 1 \cdot d\mu = \mu(M),$$

and

$$\|Af\|^2 = \int_M |\varphi|^2 \, d\mu \geq (\|\varphi\|_\infty - \varepsilon)^2 \mu(M).$$

It follows that $\|Af\| \geqq (\|\varphi\|_\infty - \varepsilon)\|f\|$, and hence that $\|A\| \geqq \|\varphi\|_\infty - \varepsilon$; since this is true for all ε, it follows that $\|A\| \geqq \|\varphi\|_\infty$. The proof is over, but it is wrong.

What is wrong is that M may have infinite measure. The objection may not seem very strong. After all, even if the measurable set $\{x : |\varphi(x)| \geqq \|\varphi\|_\infty - \varepsilon\}$ has infinite measure, the reasoning above works perfectly well if M is taken to be a measurable subset of finite positive measure. This is true. The difficulty, however, is that the measure space may be pathological enough to admit measurable sets of positive measure (in fact infinite measure) with the property that the measure of each of their measurable subsets is either 0 or ∞. There is no way out of this difficulty. If, for instance, X consists of two points x_1 and x_2 and if $\mu(\{x_1\}) = 1$ and $\mu(\{x_2\}) = \infty$, then $\mathbf{L}^2(\mu)$ is the one-dimensional space consisting of all those functions on X that vanish at x_2. If φ is the characteristic function of the singleton $\{x_2\}$, then $\|\varphi\|_\infty = 1$, but the norm of the induced multiplication operator is 0.

Conclusion: if the measure is locally finite (meaning that every measurable set of positive measure has a measurable subset of finite positive measure), then the norm of each multiplication is the essential supremum of the multiplier; otherwise the best that can be asserted is the inequality $\|A\| \leqq \|\varphi\|_\infty$. Every finite or σ-finite measure is locally finite. The most practical way to avoid excessive pathology with (usually) hardly any loss in generality is to assume σ-finiteness. If that is done, the solution (as stated above) is complete.

65   **Solution 65.** Measurability is easy. Since the measure is σ-finite, there exists an element f of $\mathbf{L}^2$ that does not vanish anywhere; since $\varphi \cdot f$ is in $\mathbf{L}^2$, it is measurable, and, consequently, so is its quotient by f.

To prove boundedness, observe that

$$\|\varphi^n \cdot f\| = \|A^n f\| \leqq \|A\|^n \cdot \|f\|$$

for every positive integer n. If $A = 0$, then $\varphi = 0$, and there is nothing to prove; otherwise write $\psi = \varphi/\|A\|$, and rewrite the preceding inequality in the form

$$\int |\psi|^{2n} \cdot |f|^2 \, d\mu \leqq \int |f|^2 \, d\mu.$$

(Here μ is, of course, the given σ-finite measure.) From this it follows that if $f \neq 0$ on some set of positive measure, then $|\psi| \leqq 1$ (i.e., $|\varphi| \leqq \|A\|$) almost everywhere on that set. If f is chosen (as above) so that $f \neq 0$ almost everywhere, then the conclusion is that $|\varphi| \leqq \|A\|$ almost everywhere.

This proof is quick, but a little too slick; it is not the one that would suggest itself immediately. A more natural (and equally quick) approach is this: to prove that $|\varphi| \leqq \|A\|$ almost everywhere, let M be a measurable set of finite measure on which $|\varphi| > \|A\|$, and prove that M must have measure 0. Indeed, if f is the characteristic function of M, then either $f = 0$ almost everywhere, or

$$\|Af\|^2 = \int |\varphi \cdot f|^2 \, d\mu = \int_M |\varphi|^2 \, d\mu > \|A\|^2 \mu(M) = \|A\|^2 \cdot \|f\|^2;$$

the latter possibility is contradictory. The proof in the preceding paragraph has, however, an advantage other than artificial polish: unlike the more natural proof, it works in a certain curious but useful situation. The situation is this: suppose that $\mathbf{H}$ is a subspace of $\mathbf{L}^2$, suppose that an operator A on $\mathbf{H}$ is such that $Af = \varphi \cdot f$ for all f in $\mathbf{H}$, and suppose that $\mathbf{H}$ contains a nowhere vanishing function. Conclusion, as before: φ is measurable and bounded (by $\|A\|$). Proof: as above.

Solution 66. *If φ is a complex-valued function such that $\varphi \cdot f \in \mathbf{L}^2$* **66**
(for a σ-finite measure) whenever $f \in \mathbf{L}^2$, then φ is essentially bounded.

PROOF. One way to proceed is to generalize the discrete (diagonal) construction (Solution 62). If φ is not bounded, then there exists a disjoint sequence $\{M_n\}$ of measurable sets of positive finite measure such that $\varphi(x) \geqq n$ whenever $x \in M_n$. (There is no trouble in proving that φ is measurable; cf. Solution 65.) Define a function f as follows: if $x \in M_n$ for some n, then

$$f(x) = \frac{1}{\sqrt{\mu(M_n) \cdot \varphi(x)}};$$

otherwise $f(x) = 0$. Since

$$\int |f|^2 \, d\mu = \sum_n \int_{M_n} |f|^2 \, d\mu$$

$$= \sum_n \int_{M_n} \frac{d\mu}{\mu(M_n)|\varphi|^2}$$

$$\leqq \sum_n \int_{M_n} \frac{d\mu}{\mu(M_n) \cdot n^2} = \sum_n \frac{1}{n^2} < \infty,$$

the function f is in $\mathbf{L}^2$; since

$$\int |\varphi \cdot f|^2 \, d\mu = \sum_n \int_{M_n} \frac{d\mu}{\mu(M_n)},$$

the function $\varphi \cdot f$ is not.

For another proof, let A be the linear transformation that multiplies each element of $\mathbf{L}^2$ by φ, and prove that A is closed, as follows. Suppose that $\langle f_n, g_n \rangle$ belongs to the graph of A (i.e., $g_n = \varphi \cdot f_n$), and suppose that $\langle f_n, g_n \rangle \to \langle f, g \rangle$ (i.e., $f_n \to f$ and $g_n \to g$). There is no loss of generality in assuming that $f_n \to f$ almost everywhere and $g_n \to g$ almost everywhere;

if this is not true for the sequence $\{f_n\}$, it is true for a suitable subsequence. Since $f_n \to f$ almost everywhere, it follows that $\varphi \cdot f_n \to \varphi \cdot f$ almost everywhere; since, at the same time, $\varphi \cdot f_n \to g$ almost everywhere, it follows that $g = \varphi \cdot f$ almost everywhere, i.e., that $\langle f, g \rangle$ is in the graph of A. Conclusion (via the closed graph theorem): A is bounded, and therefore (by Problem 65) φ is bounded.

The second proof is worth a second glance. The concept of multiplication operator can be profitably generalized to unbounded multipliers. If φ is an arbitrary (not necessarily bounded) measurable function, let $\mathbf{M}$ be the set (linear manifold) of all those f in $\mathbf{L}^2$ for which $\varphi \cdot f \in \mathbf{L}^2$. The second proof above proves that the linear transformation (from $\mathbf{M}$ into $\mathbf{L}^2$) that maps each f in $\mathbf{M}$ onto $\varphi \cdot f$ is a closed transformation. (This sort of thing is the operator analogue of a vague, but well-known and correct, measure-theoretic principle. In measure theory, every function that can be written down is measurable; in operator theory, every transformation that can be written down is closed.) Briefly: multiplications (bounded or not) are closed. The closed graph theorem can then be invoked to prove that if, in addition, a multiplication has all $\mathbf{L}^2$ for its domain, then it must be bounded.

67   **Solution 67.** For invertibility: if $\varphi \cdot \psi = 1$, then the multiplication operator induced by ψ acts as the inverse of A. Suppose, conversely, that A is invertible. This implies that φ can vanish on a set of measure 0 at most. (Otherwise take for f the characteristic function of a set of positive finite measure on which φ vanishes.) Since $\varphi \cdot A^{-1}f = f$, it follows that $A^{-1}f = (1/\varphi) \cdot f$ whenever $f \in \mathbf{L}^2$. Conclusion (from Solution 65): $|1/\varphi| \leq \|A^{-1}\|$, and therefore $|\varphi| \geq 1/\|A^{-1}\|$ almost everywhere.

The assertion about the spectrum reduces to the one about invertibility. The beginner is advised to examine the reduction in complete detail. The concept of essential range is no more slippery than other measure-theoretic concepts in which alterations on null sets are gratis, but on first acquaintance it frequently appears to be.

68   **Solution 68.** (a) *A multiplication transformation on a functional Hilbert space is necessarily bounded.*

PROOF. A proof can be based on the closed graph theorem. Suppose, indeed, that $\langle f_n, g_n \rangle$ is in the graph of A, $n = 1, 2, 3, \cdots$, and suppose that $\langle f_n, g_n \rangle \to \langle f, g \rangle$ (i.e., $f_n \to f$ and $g_n \to g$). Since convergence in $\mathbf{H}$ implies pointwise convergence (if $f_n \to f$ strongly, then $f_n \to f$ weakly), it follows that $f_n(x) \to f(x)$ and $g_n(x) \to g(x)$ for all x. Since $g_n = Af_n = \varphi \cdot f_n$, and since $\varphi(x)f_n(x) \to \varphi(x)f(x)$ for all x, it follows that $g = Af$. Conclusion: A is closed and therefore bounded.

The answer to (b) is not quite yes. The trouble is that there is nothing in the definition of a functional Hilbert space to prevent the existence of points x in

X such that $f(x) = 0$ for all f in $\mathbf{H}$. The situation can be produced at will; given $\mathbf{H}$ and X, enlarge X arbitrarily, and extend each function in $\mathbf{H}$ so as to be 0 at the new points. At the same time, "null-points" are as easy to eliminate as they are to produce; omit them all from X and restrict each function in $\mathbf{H}$ to the remaining set. As long as infinitely many null-points are present, however, the answer to (b) must be no. Reason: any function on X can be redefined at the null-points so as to become unbounded, without changing the effect that multiplication by it has on the elements of $\mathbf{H}$. Null-points play the same role for functional Hilbert spaces as atoms of infinite measure play for $\mathbf{L}^2$ spaces (cf. Solution 64).

(b) *If $\mathbf{H}$ is a functional Hilbert space with no null-points, then every (necessarily bounded) multiplication on $\mathbf{H}$ is induced by a bounded multiplier.*

PROOF. Note that

$$\| \varphi^n \cdot f \| = \| A^n f \| \leq \| A \|^n \cdot \| f \|$$

whenever n is a positive integer and f is in $\mathbf{H}$ (cf. Solution 65). If $A = 0$, then $\varphi = 0$, and there is nothing to prove; otherwise write $\psi = \varphi/\|A\|$, and rewrite the preceding inequality in the form

$$\| \psi^n \cdot f \| \leq \| f \|.$$

From this it follows that if $f(x) \neq 0$, then $|\psi(x)| \leq 1$ (i.e., $|\varphi(x)| \leq \|A\|$). Reason: $(\psi^n \cdot f)(x)$ is bounded by some multiple of $\|\psi^n \cdot f\|$. Since for each x there is an f such that $f(x) \neq 0$, it follows that $|\varphi| \leq \|A\|$ everywhere.

Here is an alternative proof that is more in the usual spirit of functional Hilbert spaces; it is due to A. L. Shields. Let K be the kernel function of the space (cf. Problem 37). Since $AK_x = \varphi \cdot K_x$ for each x, and since, at the same time, $(AK_x)(y) = (AK_x, K_y)$, it follows that

$$|\varphi(x)K(x, x)| = |(AK_x, K_x)| \leq \|A\| \cdot \|K_x\|^2.$$

Since $\|K_x\|^2 = (K_x, K_x)$, and since always $(K_x, K_y) = K_x(y)$, so that $\|K_x\|^2 = K(x, x)$, it follows that

$$|\varphi(x)K(x, x)| \leq \|A\| \cdot |K(x, x)|.$$

The relation $K(x, y) = K_y(x) = (K_y, K_x)$ implies that

$$|K(x, y)| \leq \|K_x\| \cdot \|K_y\| = \sqrt{K(x, x)}\sqrt{K(y, y)}.$$

It follows that if $K(x, x) = 0$ for some x, then $K(x, y) = 0$ for all y, i.e., $K_x = 0$, and hence $f(x) = (f, K_x) = 0$ for all f. The assumption that there are no null-points guarantees that this does not happen. Conclusion: $|\varphi(x)| \leq \|A\|$.

Solution 69. Let $\mathbf{H}$ be the set of all those absolutely continuous (complex-valued) functions on $[0, 1]$ whose derivatives belong to $\mathbf{L}^2$; define inner

69

product in H by $(f, g) = f(0)g(0)^* + \int_0^1 f'(x)g'(x)^* \, dx$. If $\|f\| = 0$, then $\int_0^1 |f'(x)|^2 \, dx = 0$, so that $f'(x) = 0$ almost everywhere, and therefore f is a constant; since, however, $f(0) = 0$, it follows that $f = 0$. This proves that the inner product is strictly positive. If $\{f_n\}$ is a Cauchy sequence in H, then $\{f_n(0)\}$ is a numerical Cauchy sequence and $\{f_n'\}$ is a Cauchy sequence in L^2. It follows that $f_n(0) \to \alpha$ and $f_n' \to g$, for some complex number α and for some g in L^2; put $f(x) = \alpha + \int_0^x g(t)dt$, and thus obtain an f such that $f_n \to f$ in H. This proves that H is complete. If $0 \leq x \leq 1$, then

$$|f(x)|^2 = \left| f(0) + \int_0^x f'(t)dt \right|^2 \leq 2\left(|f(0)|^2 + \int_0^1 |f'(t)|^2 \, dt \right) = 2\|f\|^2;$$

this proves that evaluations are bounded and hence that H is a functional Hilbert space.

If f and g are in H, then f and g are bounded; it follows that $(fg)'$ $(= fg' + f'g)$ belongs to L^2 and hence that $fg \in H$. Since 1 obviously belongs to H, all the requirements are satisfied.

This example is due to A. L. Shields.

CHAPTER 8

Operator Matrices

Solution 70. The assertion of Problem 70 is often useful in operator theory, **70** but, it turns out, the context it properly belongs to is a much more general part of algebra. If an operator is represented as a matrix whose entries are commutative operators, then it is profitable to consider the (commutative) ring with unit generated by those entries. Commutative rings have a determinant theory that is not much more frightening than in the numerical case. Indeed, if S is a finite square matrix over a commutative ring $\mathbf{M}$, then det S makes sense, as an element of $\mathbf{M}$: just apply the usual definition, according to which the determinant of a matrix of size n is a sum of $n!$ terms with appropriate signs.

If the matrix S is such that det S is an invertible element of $\mathbf{M}$, then all is well; the classical reasoning of Cramer's rule shows that S has an inverse. This remark settles the sufficiency part of Problem 70.

The difficulty of the necessity part is the presence in the background of a large non-commutative ring, namely the ring of all operators. It is conceivable that a matrix S over the small ring has an inverse whose entries are in the large ring but not in the small one, and, in that case, it is not at all obvious that anything at all follows about the element det S of the small ring. It may not be obvious, but all is well.

If $\mathbf{M}$ is a commutative subring of a ring $\mathbf{N}$ with unit, then a necessary and sufficient condition that a finite square matrix S over $\mathbf{M}$ be invertible over $\mathbf{N}$ is that det S *be invertible in* $\mathbf{N}$.

Sufficiency was discussed above.

To prove necessity, suppose that $ST = TS = 1$, where T is a matrix over $\mathbf{N}$ and 1 is the identity matrix of appropriate size.

An easy lemma is needed: if R is a matrix over $\mathbf{M}$, of the same size as S, and if $RS = SR$, then $RT = TR$. Proof: multiply the assumption fore and aft by T.

Suppose now that α is an element of $\mathbf{N}$ that commutes with each entry of S, and let R be the "scalar" matrix $\alpha \cdot 1$. Since the assumed commutativity implies that R and S commute, the lemma implies that R and T commute, and that in turn implies that α commutes with each entry of T. This is the heart of the argument; everything else is almost automatic.

Indeed, since each entry of S is fit to play the role of α, it follows that each of those entries commutes with each entry of T. Since, therefore, each entry of T is fit to play the role of α, it follows that all the entries of T commute with each other, as well as with the entries of S. That is: the ring generated by the entries of S and T together is commutative. The equation $ST = TS = 1$ (over that ring) implies that $\det S \cdot \det T = \det T \cdot \det S = 1$; q.e.d.

The statement of the theorem so proved is due to J. E. McLaughlin; the proof is due to M. A. Zorn.

71   Solution 71. Since

$$\begin{pmatrix} 1 & 0 \\ T & 1 \end{pmatrix}$$

is always invertible, with inverse

$$\begin{pmatrix} 1 & 0 \\ -T & 1 \end{pmatrix},$$

it follows that

$$\begin{pmatrix} A & B \\ C & D \end{pmatrix} \quad \text{and} \quad \begin{pmatrix} A & B \\ C & D \end{pmatrix}\begin{pmatrix} 1 & 0 \\ T & 1 \end{pmatrix}$$

are invertible together. The product works out to

$$\begin{pmatrix} A + BT & B \\ C + DT & D \end{pmatrix};$$

set $T = -D^{-1}C$ and conclude that

$$\begin{pmatrix} A & B \\ C & D \end{pmatrix}$$

is invertible if and only if

$$\begin{pmatrix} A - BD^{-1}C & B \\ 0 & D \end{pmatrix}$$

is invertible.

Introduce the temporary abbreviation $E = A - BD^{-1}C$ and proceed to consider the invertibility of

$$\begin{pmatrix} E & B \\ 0 & D \end{pmatrix}.$$

The assumption that D is invertible is still in force. If E also is invertible, then so is

$$\begin{pmatrix} E & B \\ 0 & D \end{pmatrix},$$

with inverse

$$\begin{pmatrix} E^{-1} & -E^{-1}BD^{-1} \\ 0 & D^{-1} \end{pmatrix}.$$

The converse is also true: if

$$\begin{pmatrix} E & B \\ 0 & D \end{pmatrix}$$

is invertible, then so is E. The proof is an easy computation. Suppose that

$$\begin{pmatrix} P & Q \\ R & S \end{pmatrix}$$

is the inverse of

$$\begin{pmatrix} E & B \\ 0 & D \end{pmatrix};$$

then

$$\begin{pmatrix} PE & PB + QD \\ RE & RB + SD \end{pmatrix} = \begin{pmatrix} EP + BR & EQ + BS \\ DR & DS \end{pmatrix} = \begin{pmatrix} 1 & 0 \\ 0 & 1 \end{pmatrix}.$$

Since $DR = 0$ and D is invertible, it follows that $R = 0$; since $PE = 1$ and $EP + BR = 1$, it follows that E is invertible (and, in fact, $E^{-1} = P$).

Now unabbreviate and conclude that

$$\begin{pmatrix} A & B \\ C & D \end{pmatrix}$$

is invertible if and only if $A - BD^{-1}C$ is invertible. Since D is invertible, multiplication by D does not affect any statement of invertibility; it follows that

$$\begin{pmatrix} A & B \\ C & D \end{pmatrix}$$

is invertible if and only if $AD - BD^{-1}CD$ is invertible. Up to this point the assumed commutativity of C and D was not needed; it comes in now and

219

serves to make the statement more palatable. Since C and D commute, it follows that $BD^{-1}CD = BC$. Conclusion:

$$\begin{pmatrix} A & B \\ C & D \end{pmatrix}$$

is invertible if and only if $AD - BC$ is invertible.

The unsymmetry of the hypothesis (why C and D? and why D^{-1}?) is not so ugly as first it may seem. The point is that the conclusion is equally unsymmetric. What rights does (1) $AD - BC$ have that (2) $DA - BC$, or (3) $DA - CB$, or (4) $AD - CB$ do not have? Symmetry is restored not by changing the statement but by enlarging the context. The theorem is one of four. To get a conclusion about all possible versions of the formal determinant, assume that D is invertible and make the commutativity hypothesis about (1) C and D, or (2) B and D, or, alternatively, assume that A is invertible, and make the commutativity hypothesis about (3) A and B, or (4) A and C.

It is well known and obvious that if the underlying Hilbert space is finite-dimensional, then the invertible operators are dense in the metric space of all operators. This remark (together with the result proved above) implies that in the finite-dimensional case the invertibility assumption is superfluous: if C and D commute, then a necessary and sufficient condition that

$$\begin{pmatrix} A & B \\ C & D \end{pmatrix}$$

be invertible is that $AD - BC$ be invertible. Actually the proof proves more than this: since multiplication by

$$\begin{pmatrix} 1 & 0 \\ T & 1 \end{pmatrix}$$

leaves unchanged not only the property of invertibility, but even the numerical value of the determinant, what the proof proves is that

$$\det\begin{pmatrix} A & B \\ C & D \end{pmatrix} = \det(AD - BC).$$

As for the counterexamples, an efficient place to find them is l^2. Define A and D by

$$A\langle \xi_0, \xi_1, \xi_2, \cdots \rangle = \langle \xi_1, \xi_2, \xi_3, \cdots \rangle$$

and

$$D\langle \xi_0, \xi_1, \xi_2, \cdots \rangle = \langle 0, \xi_0, \xi_1, \xi_2, \cdots \rangle,$$

and put $B = C = 0$. It follows that $AD - BC = 1$, but

$$\begin{pmatrix} A & B \\ C & D \end{pmatrix}$$

has a non-trivial kernel. (Look at $\langle f, g \rangle$, where $f = \langle 1, 0, 0, \cdots \rangle$, and $g = 0$.) If, on the other hand, B is defined by

$$B\langle \xi_0, \xi_1, \xi_2, \cdots \rangle = \langle \xi_0, 0, 0, 0, \cdots \rangle,$$

then

$$\begin{pmatrix} D & B \\ 0 & A \end{pmatrix}$$

is invertible, with inverse

$$\begin{pmatrix} A & 0 \\ B & D \end{pmatrix},$$

but the formal determinant DA has a non-trivial kernel.

Solution 72. It is convenient to begin with a lemma of some independent interest: if a finite-dimensional subspace is invariant under an invertible operator, then it is invariant under the inverse too. (Easy examples show that the assumption of finite-dimensionality is indispensable here.) To avoid the introduction of extra notation, let $\mathbf{H} \oplus \mathbf{K}$ be the space, $\mathbf{H}$ the subspace, and M the operator. (To be sure, $\mathbf{H}$ is not really a subspace of $\mathbf{H} \oplus \mathbf{K}$, but it becomes one by an obvious identification.) Since $M\mathbf{H} \subset \mathbf{H}$, and since (by invertibility) M preserves linear independence, it follows that dim $M\mathbf{H} = $ dim $\mathbf{H}$, and hence (by finite-dimensionality) that $M\mathbf{H} = \mathbf{H}$. This implies that $M^{-1}\mathbf{H} = \mathbf{H}$, and the proof of the lemma is complete.

The lemma applies to the case at hand. If

$$M = \begin{pmatrix} A & B \\ 0 & D \end{pmatrix},$$

then $\mathbf{H}$ is invariant under M; it follows from the lemma that if M is invertible, then M^{-1} has the form

$$\begin{pmatrix} A' & B' \\ 0 & D' \end{pmatrix}.$$

Finite-dimensionality has served its purpose by now; the rest of the argument is universally valid. Once it is known that a triangular matrix has a triangular inverse, then, regardless of the sizes of the entries, each diagonal entry in the matrix is invertible, and its inverse is the corresponding entry of the inverse matrix. Proof: multiply the two matrices in both possible orders and look.

CHAPTER 9

Properties of Spectra

73

Solution 73. *If A is an operator, then*

$$\Pi_0(A^*) = \Gamma(A)^* \quad and \quad \Pi(A^*) \cup \Pi(A)^* = \operatorname{spec} A^*.$$

PROOF. If $\lambda \in \Pi_0(A^*)$, then $A^* - \lambda$ has a non-zero kernel, and therefore the range of $A - \lambda^*$ has a non-zero orthogonal complement; both these implications are reversible.

The second equation is the best that can be said about the relation between Π and conjugation. The assertion is that if $A^* - \lambda$ is not invertible, then one of $A^* - \lambda$ and $A - \lambda^*$ is not bounded from below. Equivalently (with an obvious change of notation) it is to be proved that if both A^* and A are bounded from below, then A^* is invertible. The proof is trivial: if A is bounded from below, then its kernel is trivial, and therefore the range of A^* is dense; this, together with the assumption that A^* is bounded from below, implies that A^* is invertible.

Corollary. $\Pi_0(A) = \Gamma(A^*)^*$ *and* $\Pi(A) \cup \Pi(A^*)^* = \operatorname{spec} A.$

PROOF. Replace A by A^*.

74

Solution 74. *If A is an operator and p is a polynomial, then* $\Pi_0(p(A)) = p(\Pi_0(A))$, $\Pi(p(A)) = p(\Pi(A))$, *and* $\Gamma(p(A)) = p(\Gamma(A))$; *the same equations are true if A is an invertible operator and* $p(z) = 1/z$ *for* $z \neq 0$.

PROOF. It is convenient to make three elementary observations before the proof really begins. If the product of a finite number of operators (1) has a non-zero kernel, or (2) is not bounded from below, or (3) has a range that is not dense, then at least one factor must have the same property; if the factors

222

commute, then the converse of each of these statements is true. The idea of the proofs is perhaps best suggested by the following sentences. If AB sends (1) a non-zero vector onto 0, or (2) a sequence of unit vectors onto a null sequence, then argue from the right: if B does not already do so, then A must. (3) If the range of AB is not dense, argue from the left: if the range of A is dense, then the range of B cannot be.

Now for the proofs of the spectral mapping theorems. Assume, with no loss of generality, that the polynomial p has positive degree and leading coefficient 1. Since $p(\lambda) - p(\lambda_0)$ is divisible by $\lambda - \lambda_0$, it follows, by (1), that if $\lambda_0 \in \Pi_0(A)$, then $p(\lambda_0) \in \Pi_0(p(A))$, and hence that

$$p(\Pi_0(A)) \subset \Pi_0(p(A)).$$

(This part of the statement can be proved much more simply: if $Af = \lambda_0 f$, then $p(A) f = p(\lambda_0) f$. The longer sentence is adaptable to the other cases, and therefore saves time later.) If, on the other hand, $\alpha \in \Pi_0(p(A))$, then express $p(\lambda) - \alpha$ as a product of factors such as $\lambda - \lambda_0$, and apply (1); the conclusion is that $\alpha = p(\lambda_0)$ for some number λ_0 in $\Pi_0(A)$. This means that $\alpha \in p(\Pi_0(A))$, and hence that $\Pi_0(p(A)) \subset p(\Pi_0(A))$. The arguments for Π, or Γ, are exactly the same, except that (2), or (3), are used instead of (1). An alternative method is available for Γ: apply the result for Π_0 to A^*, conjugate, and apply Solution 73.

Turn now to inversion. If A is invertible and $Af = \lambda f$ with $f \neq 0$, then $\lambda \neq 0$. Apply A^{-1} to both sides of the equation, divide by λ, and obtain

$$A^{-1}f = \frac{1}{\lambda} f.$$

Conclusion:

$$\frac{1}{\Pi_0(A)} \subset \Pi_0(A^{-1}).$$

Replace A by A^{-1} and form reciprocals to get the reverse inclusion. Use the same method, but starting with $Af_n - \lambda f_n \to 0$, $\|f_n\| = 1$, to get the inversion spectral mapping theorem for Π. Derive the result for Γ by applying the result for Π_0 to the adjoint.

Solution 75. (1) If $A - \lambda$ is invertible, then so is $P^{-1}(A - \lambda)P = P^{-1}AP - \lambda$.

(2) If $Af = \lambda f$, then $P^{-1}AP(P^{-1}f) = \lambda(P^{-1}f)$.

(3) If $Af_n - \lambda f_n \to 0$, where $\|f_n\| = 1$, then $P^{-1}AP(P^{-1}f_n) - \lambda(P^{-1}f_n) = P^{-1}(Af_n - \lambda f_n) \to 0$. The norms $\|P^{-1}f_n\|$ are bounded from below by $1/\|P\|$, and, consequently, division by $\|P^{-1}f_n\|$ does not affect convergence to 0. This implies that

$$P^{-1}AP\left(\frac{P^{-1}f_n}{\|P^{-1}f_n\|}\right) - \lambda\left(\frac{P^{-1}f_n}{\|P^{-1}f_n\|}\right) \to 0.$$

(4) If g belongs to the range of $P^{-1}AP - \lambda$ $(= P^{-1}(A - \lambda)P)$, then g belongs to the image under P^{-1} of the range of $A - \lambda$; this implies that

75

223

if the closure of the range of $A - \lambda$ is not $\mathbf{H}$, then the closure of the range of $P^{-1}(A - \lambda)P$ is not $\mathbf{H}$ either.

The four proofs just given show that each named part of the spectrum of A is included in the corresponding part for $P^{-1}AP$. This assertion applied to $P^{-1}AP$ and P^{-1} (in place of A and P) implies its own converse.

76   **Solution 76.** It is to be proved that if $\lambda \neq 0$, then $AB - \lambda$ and $BA - \lambda$ are invertible or non-invertible together. Division by $-\lambda$ reduces the theorem to the general ring-theoretic assertion: if $1 - AB$ is invertible, then so is $1 - BA$. The motivation for the proof of this assertion (but not the proof itself) comes from pretending that the inverse, say C, of $1 - AB$ can be written in the form $1 + AB + ABAB + \cdots$, and that, similarly, the inverse of $1 - BA$ is $1 + BA + BABA + \cdots = 1 + B(1 + AB + ABAB + \cdots)A = 1 + BCA$. The proof itself consists of verifying that if

$$C(1 - AB) = (1 - AB)C = 1,$$

then

$$(1 + BCA)(1 - BA) = (1 - BA)(1 + BCA) = 1.$$

The verification is straightforward. It is a little easier to see if the assumption on C is rewritten in the form

$$CAB = ABC = C - 1.$$

77   **Solution 77.** *For each operator A, the approximate point spectrum $\Pi(A)$ is closed.*

PROOF. A convenient attack is to prove that the complement of $\Pi(A)$ is open. If λ_0 is not in $\Pi(A)$, then $A - \lambda_0$ is bounded from below; say

$$\|Af - \lambda_0 f\| \geq \delta \|f\|$$

for all f. Since

$$\|Af - \lambda_0 f\| \leq \|Af - \lambda f\| + \|\lambda f - \lambda_0 f\|$$

for all λ, it follows that

$$(\delta - |\lambda - \lambda_0|)\|f\| \leq \|Af - \lambda f\|.$$

This implies that if $|\lambda - \lambda_0|$ is sufficiently small, then $A - \lambda$ is bounded from below.

78   **Solution 78.** It is convenient (but not compulsory) to prove the following slightly more general assertion: if $\{A_n\}$ is a sequence of invertible operators and if A is a non-invertible operator such that $\|A_n - A\| \to 0$ as $n \to \infty$, then $0 \in \Pi(A)$. Since A is not invertible, either $0 \in \Pi(A)$ or $0 \in \Gamma(A)$. If $0 \in \Pi(A)$, there is nothing to prove. It is therefore sufficient to prove that A is not bounded from below (i.e., that $0 \in \Pi(A)$) under the assumption that ran A is not dense. Suppose then that f is a non-zero vector orthogonal to ran A, and write

$$f_n = \frac{A_n^{-1} f}{\|A_n^{-1} f\|}.$$

Since $\|f_n\| = 1$, it follows that $\|(A_n - A)f_n\| \leqq \|A_n - A\| \to 0$. Since, however, $A f_n \in$ ran A and $A_n f_n \perp$ ran A, it follows that

$$\|A_n f_n - A f_n\|^2 = \|A_n f_n\|^2 + \|A f_n\|^2 \geqq \|A f_n\|^2,$$

and hence that $\|A f_n\| \to 0$.

To derive the original spectral assertion, suppose that λ is on the boundary of spec A. It follows that there exist numbers λ_n not in spec A such that $\lambda_n \to \lambda$. The operators $A - \lambda_n$ are invertible and $A - \lambda$ is not; since

$$\|(A - \lambda_n) - (A - \lambda)\| = |\lambda_n - \lambda| \to 0,$$

it follows from the preceding paragraph that $\lambda \in \Pi(A)$.

CHAPTER 10

Examples of Spectra

79 **Solution 79.** Normality says that $\|Af\| = \|A^*f\|$ for every vector f. It follows that $\|(A - \lambda)f\| = \|(A^* - \lambda^*)f\|$ for every λ, and hence that $\Pi_0(A) = (\Pi_0(A^*))^*$. The conclusion follows from Solution 73.

80 **Solution 80.** *If A is a diagonal operator, then both $\Pi_0(A)$ and $\Gamma(A)$ are equal to the diagonal, and $\Pi(A)$ ($=$ spec A) is the closure of the diagonal.*

PROOF. Suppose that $\{e_j\}$ is an orthonormal basis such that $Ae_j = \alpha_j e_j$. The first assertion is that a number is an eigenvelue of A if and only if it is equal to one of the α_j's. "If" is trivial: each α_j is an eigenvalue of A. By an obvious subtraction, the "only if" is equivalent to this: if A has a non-zero kernel, then at least one of the α_j's vanishes. Contrapositively: if $\alpha_j \neq 0$ for all j, then $Af = 0$ implies $f = 0$. Indeed: if $f = \sum_j \xi_j e_j$, then $Af = \sum_j \alpha_j \xi_j e_j$, so that $Af = 0$ is equivalent to $\alpha_j \xi_j = 0$ for all j; since no α_j vanishes, every ξ_j must.

Now that $\Pi_0(A)$ is known, the result of Problem 79 applies. Since a diagonal operator is normal, it follows that $\Gamma(A)$ also is equal to the diagonal, and that the approximate point spectrum is the same as the entire spectrum.

81 **Solution 81.** *If A is the multiplication induced by a multiplier φ (over a σ-finite measure space), then both $\Pi_0(A)$ and $\Gamma(A)$ are equal to the set of those complex numbers λ for which $\varphi^{-1}(\{\lambda\})$ has positive measure, and $\Pi(A)$ ($=$ spec A) is the essential range of φ.*

PROOF. If $f \in \mathbf{L}^2$ and $\varphi(x)f(x) = \lambda f(x)$ almost everywhere, then $\varphi(x) = \lambda$ whenever $f(x) \neq 0$. This implies that in order for λ to be an eigenvalue of A, the function φ must take the value λ on a set of positive measure. If, conversely, $\varphi(x) = \lambda$ on a set M of positive measure, and if f is the charac-

226

teristic function of a measurable subset of M of positive finite measure, then $f \in L^2$, $f \neq 0$, and $Af = \lambda f$, so that λ is an eigenvalue of A. The remaining assertions are proved just as in Solution 80.

Solution 82. *If U is the unilateral shift, then* spec $U = D$ ($=$ *the closed unit disc*), $\Pi_0(U) = \varnothing$, $\Pi(U) = C$ ($=$ *the unit circle*), *and* $\Gamma(U) = D - C$ ($=$ *the interior of the unit disc*). *For the adjoint*: spec $U^* = D$, $\Pi_0(U^*) = D - C$, $\Pi(U^*) = D$, *and* $\Gamma(U^*) = \varnothing$. **82**

PROOF. It is wise to treat U and U^* together; each gives information about the other. To treat U^*, whether together with U or separately, it is advisable to know what it is. Since (for $i, j = 0, 1, 2, \cdots$)

$$(U^*e_i, e_j) = (e_i, Ue_j) = (e_i, e_{j+1}) = \delta_{i, j+1},$$

it follows that

$$U^*e_0 = 0;$$

if $i > 0$, then

$$\delta_{i, j+1} = \delta_{i-1, j} = (e_{i-1}, e_j),$$

and therefore

$$U^*e_i = e_{i-1} \qquad (i = 1, 2, 3, \cdots).$$

In terms of coordinates the result is that

$$U^*\langle \xi_0, \xi_1, \xi_2, \cdots \rangle = \langle \xi_1, \xi_2, \xi_3, \cdots \rangle.$$

The functional representation of U (i.e., its representation as a multiplication on $\mathbf{H}^2$) is deceptive; since the adjoint of a multiplication operator is multiplication by the complex conjugate function, it is tempting to think that if $f \in \mathbf{H}^2$, then $(U^*f)(z) = z^*f(z)$. This is not only false, it is nonsense; $\mathbf{H}^2$ is not invariant under multiplication by e_{-1}. The correspondence between adjunction and conjugation works for L^2, but there is no reason to assume that it will work for a subspace of L^2. The correct expression for U^* on $\mathbf{H}^2$ is given by

$$(U^*f)(z) = z^*(f(z) - (f, e_0)).$$

Now for the spectrum and its parts. Since U is an isometry, so that $\|U\| = 1$, it follows that the spectrum of U is included in the closed unit disc, and hence the same is true of U^*.

If $Uf = \lambda f$, where $f = \langle \xi_0, \xi_1, \xi_2, \cdots \rangle$, then

$$\langle 0, \xi_0, \xi_1, \xi_2, \cdots \rangle = \langle \lambda\xi_0, \lambda\xi_1, \lambda\xi_2, \cdots \rangle,$$

so that $0 = \lambda\xi_0$, and $\xi_n = \lambda\xi_{n+1}$ for all n. This implies that $\xi_n = 0$ for all n (look separately at the cases $\lambda = 0$ and $\lambda \neq 0$), and hence that $\Pi_0(U) = \varnothing$. Consequence: $\Gamma(U^*) = \varnothing$.

Here is an alternative proof that U has no eigenvalues, which has some geometric merit. It is a trivial fact, true for every operator A, that if f is an eigenvector belonging to a non-zero eigenvalue, then f belongs to ran A^n for every positive integer n. (Proof by induction. Trivial for $n = 0$; if $f = A^n g$, then $f = (1/\lambda)Af = (1/\lambda)A^{n+1}g$.) The range of U^n consists of all vectors orthogonal to all the e_j's with $0 \leqq j < n$, and, consequently, $\bigcap_n$ ran U^n consists of 0 alone. This proves that U has no eigenvalues different from 0. The eigenvalue 0 is excluded by the isometric property of U: if $Uf = 0$, then $0 = \|Uf\| = \|f\|$.

If $U^*f = \lambda f$, then

$$\langle \xi_1, \xi_2, \xi_3, \cdots \rangle = \langle \lambda\xi_0, \lambda\xi_1, \lambda\xi_2, \cdots \rangle,$$

so that $\xi_{n+1} = \lambda\xi_n$, or $\xi_{n+1} = \lambda^n\xi_0$, for all n. If $\xi_0 = 0$, then $f = 0$; otherwise a necessary and sufficient condition that the resulting ζ's be the coordinates of a vector (i.e., that they be square-summable) is that $|\lambda| < 1$. Conclusion: $\Pi_0(U^*)$ is the open unit disc (and consequently $\Gamma(U)$ is the open unit disc). Each λ in that disc is a simple eigenvalue of U^* (i.e., it has multiplicity 1); the corresponding eigenvector f_λ (normalized so that $(f_\lambda, e_0) = 1$) is given by

$$f_\lambda = \langle 1, \lambda, \lambda^2, \lambda^3, \cdots \rangle.$$

Since spectra are closed, it follows that both spec U and spec U^* include the closed unit disc, and hence that they are equal to it. All that remains is to find $\Pi(U)$ and $\Pi(U^*)$. Since the boundary of the spectrum of every operator is included in the approximate point spectrum, it follows that both $\Pi(U)$ and $\Pi(U^*)$ include the unit circle. If $|\lambda| < 1$, then

$$\|Uf - \lambda f\| \geqq |\|Uf\| - \|\lambda f\|| = |1 - |\lambda|| \cdot \|f\|$$

for all f, so that $U - \lambda$ is bounded from below; this proves that $\Pi(U)$ is exactly the unit circle. For U^* the situation is different: since Π_0 is always included in Π, and since $\Pi_0(U^*)$ is the open unit disc, it follows that $\Pi(U^*)$ is the closed unit disc.

83 **Solution 83.** *If the set of eigenvectors of an operator A has a non-empty interior, then A is a scalar.*

Suppose indeed that $\mathbf{E}$ is a non-empty open ball consisting of eigenvectors of an operator A, and let g be an element of $\mathbf{E}$ ($g \neq 0$). Assertion: if $f \in \mathbf{E}$, then f and g belong to the same eigenvalue. This is obvious if f and g are linearly dependent. If they are linearly independent, and if $Af = \alpha f$, $Ag = \beta g$, and $A(\frac{1}{2}(f + g)) = \gamma(\frac{1}{2}(f + g))$, then $\alpha f + \beta g = Af + Ag = A(f + g) = \gamma f + \gamma g$, and, because of linear independence, $\alpha = \gamma$ and $\beta = \gamma$.

Consequence: throughout the open set $\mathbf{E}$ the operator A agrees with the (scalar) operator α. Since the set of points of equality of two operators is a

228

subspace, and since a subspace with a non-empty interior is the whole space, it follows that $A = \alpha$.

Observe that if $Af = \lambda f$ with $f \neq 0$, then $(Af, f) = \lambda(f, f)$, so that $\lambda = (Af, f)/\|f\|^2$. The eigenvalue equation can therefore be rewritten in the form $\|f\|^2 Af = (Af, f)f$; the equation holds, of course, when $f = 0$ also. Since both sides of the equation depend continuously on f, it follows that the set of eigenvectors (0 included) is always a closed set.

Solution 84. *If W is the bilateral shift, then* spec $W = C$ ($=$ *the unit circle*), $\Pi_0(W) = \varnothing$, $\Pi(W) = C$, *and* $\Gamma(W) = \varnothing$. *The same equations are true for the adjoint W^*.*

84

PROOF. The determination of the spectrum of W, and of the fine structure of that spectrum, can follow the pattern indicated in the study of the unilateral shift U (Solution 82), but there is also another way to do it, a better way. Corresponding to the functional representation of U on $\mathbf{H}^2$, the bilateral shift W has a natural functional representation on $\mathbf{L}^2(\mu)$ (where μ is normalized Lebesgue measure on the unit circle; see Problem 33). Since the functions e_n defined by $e_n(z) = z^n$ ($n = 0, \pm 1, \pm 2, \cdots$) form an orthonormal basis for $\mathbf{L}^2$, and since the effect on them of shifting forward by one index is the same as the effect of multiplying by e_1, it follows that the bilateral shift is the same as the multiplication operator on $\mathbf{L}^2$ defined by

$$(Wf)(z) = zf(z).$$

This settles everything for W; everything follows from Solution 81.

As for W^*, its study can be reduced to that of W. Indeed, since W is unitary, its adjoint is the same as its inverse. The calculation of W^{-1} takes no effort at all; clearly W^{-1} shifts backward the same way as W shifts forward. There is a thoroughgoing symmetry between W and W^*; to obtain one from the other, just replace n by $-n$. In more pedantic language: W and W^* are unitarily equivalent, and, in particular, the unitary operator R determined by the conditions $Re_n = e_{-n}$ ($n = 0, \pm 1, \pm 2, \cdots$) transforms W onto W^* (i.e., $R^{-1}WR = W^*$). Conclusion: the spectrum of W^* is equal to the spectrum of W, and the same is true, part for part, for each of the usual parts of the spectrum.

Solution 85. Suppose first that the eigenvectors of A^* span $\mathbf{H}$. Let X be an index set such that corresponding to each x in X there is an eigenvector K_x of A^*, and such that the K_x's span $\mathbf{H}$; denote the eigenvalue corresponding to K_x by $\varphi(x)^*$. (The conjugation has no profound significance here; it is just a notational convenience.) It follows that $A^*K_x = \varphi(x)^*K_x$. For each f in $\mathbf{H}$, let $\hat{f}$ be the function on X defined by $\hat{f}(x) = (f, K_x)$. The correspondence $f \mapsto \hat{f}$ is linear. If $\hat{f} = 0$, i.e., if $(f, K_x) = 0$ for all x, then $f = 0$ (since the K_x's span $\mathbf{H}$). This justifies the definition $(\hat{f}, \hat{g}) = (f, g)$. With this definition of inner product, the set $\tilde{\mathbf{H}}$ of all functions of the form

85

$\tilde{f}$ (with f in $\mathbf{H}$) becomes a functional Hilbert space. [Note: $|\tilde{f}(x)| = |(f, K_x)| \leqq \|f\| \cdot \|K_x\| = \|\tilde{f}\| \cdot \|K_x\|$.] Let $\tilde{A}$ be the image of A under the isomorphism $f \mapsto \tilde{f}$ (i.e., $\tilde{Af} = (Af)^\sim$); then

$$(\tilde{A}\tilde{f})(x) = (Af)^\sim(x) = (Af, K_x) = (f, A^*K_x)$$
$$= (f, \varphi(x)^*K_x) = \varphi(x)(f, K_x)$$
$$= \varphi(x)\tilde{f}(x),$$

so that $\tilde{A}$ is a multiplication.

The converse is proved by retracing the steps of the last computation. In detail, if A is a multiplication (with multiplier φ, say) on a functional Hilbert space $\mathbf{H}$ with domain X and kernel function K, so that $(Af)(x) = \varphi(x)f(x)$, then $(Af, K_x) = \varphi(x)(f, K_x)$ (where $K_x(y) = K(y,x)$), and therefore $(f, A^*K_x - \varphi(x)^*K_x) = 0$ for all f. It follows that $A^*K_x = \varphi(x)^*K_x$; since in a functional Hilbert space the set of K_x's always spans the space, the proof is complete.

Compare the construction with what is known about the unilateral shift (Solution 82).

CHAPTER 11

Spectral Radius

Solution 86. A standard trick for proving operator functions analytic is the $\qquad$ **86**
identity

$$(1 - A)^{-1} = 1 + A + A^2 + \cdots.$$

If $\|A\| < 1$, then the series converges (with respect to the operator norm),
and obvious algebraic manipulations prove that its sum indeed acts as the
inverse of $1 - A$. (Replace A by $1 - A$ and recapture the assertion that if
$\|1 - A\| < 1$, then A is invertible. Cf. [50, p. 52]; see also Problem 99.)

Suppose now that λ_0 is not in the spectrum of A, so that $A - \lambda_0$ is invertible. To prove that $(A - \lambda)^{-1}$ is analytic in λ, for λ near λ_0, express
$A - \lambda$ in terms of $A - \lambda_0$:

$$A - \lambda = (A - \lambda_0) - (\lambda - \lambda_0)$$
$$= (A - \lambda_0)(1 - (A - \lambda_0)^{-1}(\lambda - \lambda_0)).$$

If $|\lambda - \lambda_0|$ is sufficiently small, then $\|(A - \lambda_0)^{-1}(\lambda - \lambda_0)\| < 1$, and the
series trick can be applied. The conclusion is that if $|\lambda - \lambda_0|$ is sufficiently
small, then $A - \lambda$ is invertible, and

$$(A - \lambda)^{-1} = (A - \lambda_0)^{-1} \sum_{n=0}^{\infty} ((A - \lambda_0)^{-1}(\lambda - \lambda_0))^n.$$

It follows that if f and g are in **H**, then

$$(\rho(\lambda)f, g) = \sum_{n=0}^{\infty} ((A - \lambda_0)^{-n-1}f, g)(\lambda - \lambda_0)^n$$

in a neighborhood of λ_0, and hence that ρ is analytic at λ_0.

As for $\lambda = \infty$, note that

$$A - \frac{1}{\lambda} = -\frac{1}{\lambda}(1 - \lambda A)$$

whenever $\lambda \neq 0$, and hence that $A - 1/\lambda$ is invertible whenever $|\lambda|$ is sufficiently small (but different from 0). Since

$$\rho\left(\frac{1}{\lambda}\right) = \left(A - \frac{1}{\lambda}\right)^{-1} = -\lambda(1 - \lambda A)^{-1},$$

the series trick applies again:

$$\tau(\lambda) = \rho\left(\frac{1}{\lambda}\right) = -\lambda(1 + \lambda A + \lambda^2 A^2 + \cdots).$$

The parenthetical series converges for small λ, and the factor $-\lambda$ in front guarantees that $\tau(0) = 0$.

Analyticity doesn't have to be treated via power series. The definition by differentiability can be used too, but it requires a slight loan from the future. Thus, for instance, to prove that ρ is analytic at ∞, note that $1 - \lambda A \to 1$ as $\lambda \to 0$, and hence (from the continuity of inversion, Problem 100) that $\rho(1/\lambda) \to 0$ as $\lambda \to 0$.

87 **Solution 87.** Proceed by contradiction. If the spectrum of A is empty, then $(\rho_A f, g)$ (i.e., the function $\lambda \mapsto ((A - \lambda)^{-1} f, g))$ is an entire function for each f and g; since $\rho_A(\infty) = 0$, the function $(\rho_A f, g)$ is bounded in a neighborhood of ∞ and therefore in the whole plane. Liouville's theorem implies that $(\rho_A f, g)$ is a constant; since $\rho_A(\infty) = 0$, it follows that

$$((A - \lambda)^{-1} f, g) = 0$$

identically in f, g, and λ. Since this is absurd (replace f and g by $(A - \lambda)f$ and f), the assumption of empty spectrum is untenable.

88 **Solution 88.** Since $(r(A))^n = r(A^n) \leq \|A^n\|$, so that $r(A) \leq \|A^n\|^{1/n}$ for all n, it follows that

$$r(A) \leq \liminf_n \|A^n\|^{1/n}.$$

The reverse inequality leans on the analytic character of the resolvent (Problem 86). If

$$\tau(\lambda) = \rho\left(\frac{1}{\lambda}\right) = \left(A - \frac{1}{\lambda}\right)^{-1},$$

then $\tau(\lambda) = -\lambda(1 - \lambda A)^{-1}$ whenever $\lambda \neq 0$ and $1/\lambda$ is not in the spectrum of A. Since, for each f and g, the numerical function $(\tau f, g)$ is analytic as long as $|1/\lambda| > r(A)$ (i.e., $|\lambda| < 1/r(A)$), it follows that its Taylor series

$$-\lambda \sum_{n=0}^{\infty} \lambda^n (A^n f, g)$$

converges whenever $|\lambda| < 1/r(A)$. This implies that the sequence $\{((\lambda A)^n f, g)\}$ is bounded for each such λ. The principle of uniform boundedness yields the conclusion that the sequence $\{|\lambda|^n \cdot \|A^n\|\}$ is bounded. If $|\lambda|^n \cdot \|A^n\| \leqq \alpha$ for all n, then

$$|\lambda| \cdot \|A^n\|^{1/n} \leqq \alpha^{1/n},$$

and therefore

$$|\lambda| \cdot \limsup_n \|A^n\|^{1/n} \leqq 1.$$

Since this is true whenever $|\lambda| < 1/r(A)$, it follows that

$$\limsup_n \|A^n\|^{1/n} \leqq r(A).$$

The proof is complete.

Solution 89. The asserted unitary equivalence can be implemented by a **89** diagonal operator. To see which diagonal operator to use, work backwards. Assume that D is a diagonal operator with diagonal $\{\delta_n\}$, and assume that $AD = DB$. It follows (apply both sides to e_n) that

$$\alpha_n \delta_n = \beta_n \delta_{n+1}$$

for each n. Put $\delta_0 = 1$, and determine the other δ's by recursion. Consider first the positive n's. If $\beta_n \neq 0$, put $\delta_{n+1} = (\alpha_n/\beta_n)\delta_n$. If $\beta_n = 0$, then $\alpha_n = 0$ (since, by assumption, $|\alpha_n| = |\beta_n|$); in that case put $\delta_{n+1} = 1$. For negative n's (if there are any) apply the same process in the other direction. That is, if $\alpha_n \neq 0$, put $\delta_n = (\beta_n/\alpha_n)\delta_{n+1}$; if $\alpha_n = 0$, then put $\delta_n = 1$. The result is a sequence $\{\delta_n\}$ of complex numbers of modulus 1. The steps leading to this sequence are reversible. Given the sequence, let it induce a diagonal operator D; note that since $|\delta_n| = 1$ for all n, the operator D is unitary; and, finally, note that since $ADe_n = DBe_n$ for all n, the operator D transforms A onto B.

Solution 90. Suppose first that S is an invertible operator such that $A = $ **90** $S^{-1}BS$. It follows that $A^* = S^*B^*S^{*-1}$, and hence that $A^{*n} = S^*B^{*n}S^{*-1}$. Use the argument that worked for unitary equivalence to infer that S^* sends ker B^{*n} onto ker A^{*n}. This implies that the matrix $\langle \sigma_{ij} \rangle$ of S is lower triangular. Consider the equation $SA = BS$, and evaluate the matrix entries in row $n + 1$, column n ($n = 0, 1, 2, \cdots$) for both sides. The result is that $\sigma_{n+1,n+1}\alpha_n = \beta_n\sigma_{n,n}$, and hence that

$$\left| \frac{\beta_0 \cdots \beta_n}{\alpha_0 \cdots \alpha_n} \right| = \left| \frac{\sigma_{n+1,n+1}}{\sigma_{0,0}} \right| = \left| \frac{(Se_{n+1}, e_{n+1})}{\sigma_{0,0}} \right| \leqq \frac{\|S\|}{|\sigma_{0,0}|}.$$

Consequence: $\{|\alpha_0 \cdots \alpha_n/\beta_0 \cdots \beta_n|\}$ is bounded away from 0. To get boundedness (away from ∞), work with S^{-1} (instead of S) and with the equation $AS^{-1} = S^{-1}B$ (instead of $SA = BS$).

If, conversely, the boundedness conditions are satisfied, then write $\sigma_0 = 1$, $\sigma_{n+1} = \beta_0 \cdots \beta_n/\alpha_0 \cdots \alpha_n$, let S be the (invertible) diagonal operator with diagonal sequence $\{\sigma_0, \sigma_1, \sigma_2, \cdots\}$, and verify that $SA = BS$.

233

91   **Solution 91.** *If A is a weighted shift with weights α_n, then $\|A\| = \sup_n |\alpha_n|$, and $r(A) = \lim_k \sup_n |\prod_{i=0}^{k-1} \alpha_{n+i}|^{1/k}$.*

The expression for r looks mildly complicated, but there are cases when it can be used to compute something.

PROOF. Since A is the product of a shift and the diagonal operator with diagonal $\{\alpha_n\}$, and since a shift is an isometry, it follows that the norm of A is equal to the norm of the associated diagonal operator.

To prove the assertion about the spectral radius, evaluate the powers of A. If $Ae_n = \alpha_n e_{n+1}$, then $A^2 e_n = \alpha_n \alpha_{n+1} e_{n+2}$, $A^3 e_n = \alpha_n \alpha_{n+1} \alpha_{n+2} e_{n+3}$, etc. What this shows is that A^k is the product of an isometry (namely the k-th power of the associated shift) and a diagonal operator (namely the one whose n-th diagonal term is the product of k consecutive α's starting with α_n). Conclusion: the norm of A^k is the supremum of the moduli of the "sliding products" of length k, or, explicitly,

$$\|A^k\| = \sup_n \left| \prod_{i=0}^{k-1} \alpha_{n+i} \right| \qquad (k = 1, 2, 3, \cdots).$$

The expression for the spectral radius follows immediately.

92   **Solution 92.** *If $p_k = \|A^k\|$, $k = 0, 1, 2, \cdots$, then $p_{j+k} \leqq p_j p_k$, and, conversely, if a sequence $\{p_k\}$ satisfies these inequalities, then it is the sequence of power norms of some operator A. The operator A can, in fact, be chosen as a weighted shift.* Indeed, if A is a weighted shift, with (positive) weight sequence $\{\alpha_0, \alpha_1, \alpha_2, \cdots\}$, then $\|A^k\| = \sup_n(\alpha_n \alpha_{n+1} \cdots \alpha_{n+k-1})$. Given the p's define the α's: put $\alpha_0 = p_0$ and $\alpha_n = p_n/p_{n-1}$, $n = 1, 2, 3, \cdots$. Since $a_n \leqq p_{n-1} p_1/p_{n-1}$, the α's are bounded, as they must be to define a weighted shift. The rest is a natural computation:

$$\|A^k\| = \sup_n (\alpha_n \alpha_{n+1} \cdots \alpha_{n+k-1})$$

$$= \sup_n \frac{p_{n+k-1}}{p_{n-1}} \leqq p_k = \alpha_0 \cdots \alpha_{k-1} \leqq \|A^k\|.$$

(The reasoning assumed that none of the p's is 0. If some of them are 0, the proof can be modified in an obvious way.) The result and the proof are due to L. J. Wallen.

93   **Solution 93.** *If A is the unilateral weighted shift with weights $\{\alpha_0, \alpha_1, \alpha_2, \cdots\}$, and if $\alpha_n \neq 0$ for $n = 0, 1, 2, \cdots$, then $\Pi_0(A) = \varnothing$, and $\Pi_0(A^*)$ is a disc with center 0 and radius $\liminf_n |\prod_{i=0}^{n-1} \alpha_i|^{1/n}$. The disc may be open or closed, and it may degenerate to the origin only.*

PROOF. The proof for A is the same as for the unweighted unilateral shift (Solution 82). In sequential (coordinate) notation, if $Af = \lambda f$, where $f = \langle \xi_0, \xi_1, \xi_2, \cdots \rangle$, then $Af = \langle 0, \alpha_0 \xi_0, \alpha_1 \xi_1, \alpha_2 \xi_2, \cdots \rangle$, so that $0 = \lambda \xi_0$ and

234

$\alpha_n \xi_n = \lambda \xi_{n+1}$ for all n. This implies that $\xi_n = 0$ for all n; look separately at the cases $\lambda = 0$ and $\lambda \neq 0$.

To treat A^*, it is advisable to know what it is. That can be learned by looking at matrices (the diagonal just below the main one flips over to the one just above), by imitating the procedure used to find U^* (Solution 82), or by writing A as the product of U and a diagonal operator and applying the known result for U^*. The answer is that $A^* e_n = 0$ if $n = 0$ and $A^* e_n = \alpha_{n-1}^* e_{n-1}$ if $n > 0$. Sequentially: if $f = \langle \xi_0, \xi_1, \xi_2, \cdots \rangle$, then $A^* f = \langle \alpha_0^* \xi_1, \alpha_1^* \xi_2, \alpha_2^* \xi_3, \cdots \rangle$. It follows that $A^* f = \lambda f$ if and only if

$$\alpha_n^* \xi_{n+1} = \lambda \xi_n$$

for all n. This implies that if $n > 1$, then ξ is the product of ξ_0 by

$$\frac{\lambda^n}{\prod_{i=0}^{n-1} \alpha_i^*}.$$

Since a sequence of numbers defines a vector if and only if it is square-summable, it follows that $\lambda \in \Pi_0(A^*)$ if and only if

$$\sum_{n=1}^{\infty} \left| \frac{\lambda^n}{\prod_{i=0}^{n-1} \alpha_i} \right|^2 < \infty.$$

The condition is that a certain power series in λ^2 be convergent; that proves that the λ's satisfying it form a disc. The radius of the disc can be obtained from the formula for the radius of convergence of a power series.

If $\alpha_n = 1$ $(n = 0, 1, 2, \cdots)$, then $\prod_{i=0}^{n-1} \alpha_i = 1$ $(n = 1, 2, 3, \cdots)$, and therefore the power series converges in the open unit disc; cf. Solution 82. If

$$\alpha_n = \left(1 + \frac{1}{n+1} \right)^2 \quad \left(= \left(\frac{n+2}{n+1} \right)^2 \right),$$

then $\prod_{i=0}^{n-1} \alpha_i = (n + 1)^2$, and therefore the power series converges in the closed unit disc, which in this case happens to be the same as the spectrum; cf. Solution 91. If $\alpha_n = 1/(n + 1)$, then $\prod_{i=0}^{n-1} \alpha_i = 1/n!$, and therefore the power series converges at the origin only.

Solution 94. The answer is yes: the approximate point spectrum of a weighted   **94**
shift can fill an annulus.

The main idea in the proof is that if one of the weights is repeated very often, then it is very nearly an eigenvalue. Suppose, to be precise, that A is a weighted shift with weight sequence $\{\alpha_n : n = 0, 1, 2, \cdots\}$; if

$$\alpha_{m+1} = \alpha_{m+2} = \cdots = \alpha_{m+k} = \lambda,$$

then there exists a unit vector f such that $\|Af - \lambda f\| \leq 2/\sqrt{k}$. Reason: consider the orthonormal basis $\{e_n : n = 0, 1, 2, \cdots\}$ that is being shifted, and write $f = (1/\sqrt{k}) \sum_{j=1}^{k} e_{m+j}$.

It follows that if a weight λ occurs infinitely often, in arbitrarily long blocks, then λ must be in the approximate point spectrum. To prove the

SOLUTIONS

desired result, consider a sequence $\{\lambda_n\}$ dense in some interval, and let $\{\alpha_n\}$ be a sequence in which each λ occurs infinitely often in arbitrarily long blocks. (Take one λ_1, then two λ_1's and two λ_2's, then three λ_1's, three λ_2's, and three λ_3's, etc.) The resulting weighted shift will have each λ in its approximate point spectrum; since the approximate point spectrum is closed and circularly symmetric, the proof is complete.

There is a valuable discussion of many properties of weighted shifts in [133].

95   **Solution 95.** *If $p = \{p_n\}$ is a sequence of positive numbers such that $\{p_{n+1}/p_n\}$ is bounded, then the shift S on $l^2(p)$ is unitarily equivalent to the weighted shift A, with weights $\{\sqrt{p_{n+1}/p_n}\}$, on l^2.*

PROOF. If $f = \langle \xi_0, \xi_1, \xi_2, \cdots \rangle \in l^2(p)$ write

$$Uf = \langle \sqrt{p_0}\xi_0, \sqrt{p_1}\xi_1, \sqrt{p_2}\xi_2, \cdots \rangle.$$

The transformation U maps $l^2(p)$ into l^2; it is clearly linear and isometric. If $\langle \eta_0, \eta_1, \eta_2, \cdots \rangle \in l^2$, and if $\xi_n = \eta_n/\sqrt{p_n}$, then $\sum_{n=0}^{\infty} p_n|\xi_n|^2 = \sum_{n=0}^{\infty} |\eta_n|^2$; this proves that U maps $l^2(p)$ onto l^2.

Assertion: U transforms S onto A. Computation:

$$\begin{aligned} USU^{-1}\langle \eta_0, \eta_1, \eta_2, \cdots \rangle &= US\langle \eta_0/\sqrt{p_0}, \eta_1/\sqrt{p_1}, \eta_2/\sqrt{p_2}, \cdots \rangle \\ &= U\langle 0, \eta_0/\sqrt{p_0}, \eta_1/\sqrt{p_1}, \eta_2/\sqrt{p_2}, \cdots \rangle \\ &= \langle 0, \sqrt{p_1/p_0}\eta_0, \sqrt{p_2/p_1}\eta_1, \sqrt{p_3/p_2}\eta_2, \cdots \rangle \\ &= A\langle \eta_0, \eta_1, \eta_2, \cdots \rangle. \end{aligned}$$

Conclusion: the transform of the ordinary shift on a weighted sequence space is a weighted shift on the ordinary sequence space.

In view of this result, all questions about weighted sequence spaces can be answered in terms of weighted shifts. The spectral radius of S, for instance, is $\lim_k \sup_n (\prod_{i=0}^{k-1} \sqrt{p_{n+i+1}/p_{n+i}})^{1/k}$ (see Solution 91).

96   **Solution 96.** *If A is a unilateral weighted shift with positive weights α_n such that $\alpha_n \to 0$, then* spec $A = \{0\}$ *and* $\Pi_0(A) = \varnothing$.

PROOF. Use Solution 91 to evaluate $r(A)$. In many special cases that is quite easy to do. If, for instance, $\alpha_n = 1/2^n$, then the supremum (over all n) of $(\prod_{i=0}^{k-1} 1/2^{n+i})^{1/k}$ is attained when $n = 0$. It follows that that supremum is

$$\left(\prod_{i=0}^{k-1} \frac{1}{2^i} \right)^{1/k} = \frac{1}{2^m},$$

where

$$m = \frac{1}{k} \sum_{i=0}^{k-1} i = \frac{1}{2}(k-1).$$

This implies that the supremum tends to 0 as k tends to ∞.

236

In the general case, observe first that $(\prod_{i=0}^{k-1} \alpha_i)^{1/k} \to 0$ as $k \to \infty$. (This assertion is the multiplicative version of the one according to which convergence implies Cesàro convergence. The additive version is that if $\alpha_n \to 0$, then $(1/k) \sum_{i=0}^{k-1} \alpha_i \to 0$. The proofs are easy and similar. It is also easy to derive the multiplicative one from the additive one.) Since $\alpha_{n+1} \to 0$, it follows equally that $(\prod_{i=0}^{k-1} \alpha_{1+i})^{1/k} \to 0$; more generally, $(\prod_{i=0}^{k-1} \alpha_{n+i})^{1/k} \to 0$ as $k \to \infty$ for each n.

The problem is to prove that $\sup_n (\prod_{i=0}^{k-1} \alpha_{n+i})^{1/k}$ is small when k is large. Given ε (>0) and given n ($=0, 1, 2, \cdots$), find $k_0(n, \varepsilon)$ so that $(\prod_{i=0}^{k-1} \alpha_{n+i})^{1/k} < \varepsilon$ whenever $k \geqq k_0(n, \varepsilon)$. If n_0 is such that $\alpha_n < \varepsilon$ for $n \geqq n_0$, then $\max(k_0(0, \varepsilon), k_0(1, \varepsilon), \cdots, k_0(n_0 - 1, \varepsilon))$ is "large" enough; if k is greater than or equal to this maximum, then $\sup_n (\prod_{i=0}^{k-1} \alpha_{n+i})^{1/k} < \varepsilon$. Indeed, if $n < n_0$, then $(\prod_{i=0}^{k-1} \alpha_{n+i})^{1/k} < \varepsilon$ just because $k \geqq k_0(n, \varepsilon)$; if $n \geqq n_0$, then $(\prod_{i=0}^{k-1} \alpha_{n+1})^{1/k} < \varepsilon$ because each factor in the product is less than ε.

To see that $\Pi_0(A)$ is empty, apply Solution 93.

Solution 97. *A quasinilpotent operator is analytic if and only if it is nilpotent.*   **97**

"If" is trivial. To prove "only if", suppose that A is quasinilpotent and $f(A) = 0$, where f is analytic in a neighborhood of 0. Write $f(z) = z^n g(z)$, where $n \geqq 0$, g is analytic in a neighborhood of 0, and $g(0) \neq 0$. The last condition implies that g is invertible, i.e., that there exists a function h analytic in a neighborhood of 0 and such that $g(z)h(z) = 1$. Consequence:

$$0 = f(A) = f(A)h(A) = A^n g(A)h(A) = A^n.$$

Conclusion: the answer to the question of Problem 97 is no. If A is quasinilpotent but not nilpotent, then A is not analytic.

Solution 98. *There exists a countable set of operators, each with spectrum*   **98**
$\{0\}$, *whose direct sum has spectral radius 1.*

PROOF. Here is an example described in terms of weighted shifts. Consider the (unilateral) sequence

$$\{1, 0, 1, 1, 0, 1, 1, 1, 0, 1, 1, 1, 1, 0, \cdots\},$$

and let A be the unilateral weighted shift with these weights. The 0's in the sequence guarantee that A is the direct sum of the operators given by

$$\begin{pmatrix} 0 & 0 \\ 1 & 0 \end{pmatrix}, \quad \begin{pmatrix} 0 & 0 & 0 \\ 1 & 0 & 0 \\ 0 & 1 & 0 \end{pmatrix}, \quad \begin{pmatrix} 0 & 0 & 0 & 0 \\ 1 & 0 & 0 & 0 \\ 0 & 1 & 0 & 0 \\ 0 & 0 & 1 & 0 \end{pmatrix}, \cdots,$$

and hence it is the direct sum of operators each of which has spectrum $\{0\}$. Since, however, the sequence of weights has arbitrarily long blocks of 1's, the formula for the spectral radius of a weighted shift (Solution 91) implies that $r(A) = 1$.

What makes such examples possible is the misbehavior of the approximate point spectrum. For the point spectrum the best possible assertion is true (and easy to prove): the point spectrum of a direct sum is the union of the point spectra of the summands. Passage to adjoints implies that the same best possible assertion is true for the compression spectrum.

CHAPTER 12

Norm Topology

Solution 99. *The metric space of operators on an infinite-dimensional Hilbert space is not separable.* **99**

PROOF. Since every infinite-dimensional Hilbert space has a separable infinite-dimensional subspace, and since every separable infinite-dimensional space is isomorphic to $L^2(0, 1)$, there is no loss of generality in assuming that the underlying Hilbert space is $L^2(0, 1)$ to begin with. That granted, let φ_t be the characteristic function of $[0, t]$, and let P_t be the multiplication operator induced by φ_t, $0 \leqq t \leqq 1$. If $s < t$, then $P_t - P_s$ is the multiplication operator induced by the characteristic function of $(s, t]$, and therefore $\|P_t - P_s\| = 1$. Conclusion: there exists an uncountable set of operators such that the distance between any two of them is 1; the existence of such a set is incompatible with separability. For an alternative example of the same thing, consider diagonal operators whose diagonals consist of 0's and 1's only.

Solution 100. *The set of invertible operators is open and inversion is* **100** *continuous.*

PROOF. Recall first that if $\|1 - A\| < 1$, then A is invertible and $A^{-1} = \sum_{n=0}^{\infty} (1 - A)^n$ (cf. Solution 86); it follows that

$$\|A^{-1}\| \leqq \sum_{n=0}^{\infty} \|1 - A\|^n = \frac{1}{1 - \|1 - A\|}.$$

Suppose now that A_0 is an invertible operator. Since

$$1 - AA_0^{-1} = (A_0 - A)A_0^{-1}$$

for each A, it follows that if $\|A_0 - A\| < 1/\|A_0^{-1}\|$, then $\|1 - AA_0^{-1}\| < 1$.

239

This implies that if $\|A_0 - A\| < 1/\|A_0^{-1}\|$, then A is invertible (because AA_0^{-1} is) and

$$\|A^{-1}\| = \|((AA_0^{-1})A_0)^{-1}\| \leqq \|A_0^{-1}\| \cdot \|A_0 A^{-1}\|$$

$$\leqq \frac{\|A_0^{-1}\|}{1 - \|A_0 - A\| \cdot \|A_0^{-1}\|}.$$

Conclusion: not only is the set of invertible operators open, but so long as an operator stays in a sufficiently small neighborhood of one of them, it is not only invertible, but its inverse remains bounded.

The result of the preceding paragraph makes the continuity proof accessible. Observe that

$$\|A_0^{-1} - A^{-1}\| = \|A_0^{-1}(A - A_0)A^{-1}\| \leqq \|A_0^{-1}\| \cdot \|A - A_0\| \cdot \|A^{-1}\|.$$

If A_0 is fixed and if A is sufficiently near to A_0, then the middle factor on the right makes the outcome small, and the other two factors remain bounded.

As for the puzzle: if $A^n \to B$, then multiply by A, infer that $A^{n+1} \to AB$, and conclude that $AB = B$; since B is invertible, it follows that $A = B = 1$.

101   **Solution 101.** On a Hilbert space of positive dimension a conjugate class cannot even have a non-empty interior. Suppose, indeed, that the operator A is such that $\|X\| < \varepsilon$ implies that $A + X$ is similar to A, for some positive number ε. It follows that $A + \varepsilon/2$ is similar to A, and hence that $S^{-1}AS + \varepsilon/2 = A$ for some invertible S. Consideration of the spectra of the two sides of this equation shows its impossibility.

102   **Solution 102.** The sequence of weights for A_k is

$$\left\{ \cdots, 1, 1, 1, \left(\frac{1}{k}\right), 1, 1, 1, \cdots \right\}.$$

Since $1/k \leqq 1$, it follows that the supremum of the sliding products that enter the formula for the spectral radius of a weighted shift (see Solution 91) is equal to 1, and hence that $r(A_k) = 1$. Conclusion: the spectrum of A_k is included in the closed unit disc, and this is true for $k = 1, 2, 3, \cdots, \infty$.

If $k < \infty$, then A_k is invertible, and, in fact, A_k^{-1} itself is a weighted shift. Since $A_k^{-1}e_n$ is e_{n-1} or ke_{n-1} according as $n \neq 1$ or $n = 1$, it follows that A_k^{-1} shifts the e_n's backwards (and weights them as just indicated). Backwards and forwards are indistinguishable to within unitary equivalence (cf. Solution 84), and, consequently, the theory of weighted shifts is applicable to A_k^{-1}. The sequence of weights for A_k^{-1} is

$$\{\cdots, 1, 1, 1, (1), k, 1, 1, 1, \cdots\}.$$

The supremum of the sliding products of length m is now equal to k; it follows that $r(A_k^{-1}) = \lim_m k^{1/m} = 1$. Conclusion: the spectrum of A_k^{-1} is included in the closed unit disc, and this is true for $k = 1, 2, 3, \cdots$ (but not for ∞).

The conclusions of the preceding two paragraphs, together with the spectral mapping theorem for inverses, imply that the spectrum of A_k (and also the spectrum of A_k^{-1}) is included in the unit circle (perimeter). This, together with the circular symmetry of the spectra of weighted shifts (see Problem 89), implies that the spectrum of A_k is equal to the unit circle ($k = 1, 2, 3, \cdots$).

The spectrum of A_∞ is clearly not the unit circle; since $A_\infty e_0 = 0$, the spectrum of A_∞ contains the origin. This shows that the spectrum of A_∞ is discontinuously different from the spectra of the other A_k's. (Note that $A_k \to A_\infty$, i.e., $\|A_k - A_\infty\| \to 0$, as $k \to \infty$.) The spectrum of A_∞ is, in fact, equal to the unit disc. The quickest way to prove this is to note that the span of the e_n's with $n > 0$ reduces A_∞ (both it and its orthogonal complement are invariant under A_∞), and that the restriction of A_∞ to that span is the unilateral shift. Since the spectrum of every operator includes the spectrum of each direct summand, the proof is complete.

This example is due to G. Lumer.

Solution 103. Let **T** be the set of all singular operators (on a fixed Hilbert space), and given an operator A, fixed from now on, let $\varphi(\lambda)$ be the distance (in the metric space of operators) from $A - \lambda$ to **T**. The function φ is continuous. (This is an elementary fact about metric spaces; it does not even depend on **T** being closed.) If Λ_0 is an open set that includes spec A, if Δ is the closed disc with center 0 and radius $1 + \|A\|$, and if $\lambda \in \Delta - \Lambda_0$, then $\varphi(\lambda) > 0$. (This does depend on **T** being closed; if $\varphi(\lambda) = 0$, i.e., $d(A - \lambda, \mathbf{T}) = 0$, then $A - \lambda \in \mathbf{T}$, i.e., $\lambda \in$ spec A.) Since $\Delta - \Lambda_0$ is compact, there exists a positive number ε such that $\varphi(\lambda) \geqq \varepsilon$ for all λ in $\Delta - \Lambda_0$; there is clearly no loss of generality in assuming that $\varepsilon < 1$. Suppose now that $\|A - B\| < \varepsilon$. It follows that if $\lambda \in \Delta - \Lambda_0$, then

$$\|(A - \lambda) - (B - \lambda)\| < \varepsilon \leqq d(A - \lambda, \mathbf{T}).$$

This implies that $B - \lambda$ is not in **T**, and hence that λ is not in spec B. Conclusion: spec B is disjoint from $\Delta - \Lambda_0$. At the same time, if $\lambda \in$ spec B, then

$$|\lambda| \leqq \|B\| \leqq \|A\| + \|A - B\| < 1 + \|A\|,$$

so that spec $B \subset \Delta$. These two properties of spec B say exactly that spec $B \subset \Lambda_0$; the proof is complete.

A different proof can be based on the known properties of resolvents. If $\varphi(\lambda) = \|(A - \lambda)^{-1}\|$, then φ is defined and continuous outside Λ_0; since it vanishes at ∞, it is bounded (cf. Problem 86). If, say, $\varphi(\lambda) < \alpha$ whenever $\lambda \notin \Lambda_0$, put $\varepsilon = 1/\alpha$. If $\|A - B\| < \varepsilon$ and $\lambda \notin \Lambda_0$, then

$$\|(A - \lambda) - (B - \lambda)\| = \|A - B\| < \varepsilon < \frac{1}{\|(A - \lambda)^{-1}\|};$$

it follows as in Solution 100 that $B - \lambda$ is invertible.

241

Perhaps the simplest proof is a sequential one. Suppose that $A_n \to A$ and $\lambda \in \text{limsup}_n \text{ spec } A_n$; choose λ_n in spec A_n so that for a suitable subsequence $\lambda_{n_k} \to \lambda$. Since $A_{n_k} - \lambda_{n_k}$ is singular and $A_{n_k} - \lambda_{n_k} \to A - \lambda$, it follows from Problem 100 that $A - \lambda$ is singular, so that $\lambda \in \text{spec } A$.

The metric space proof is due to C. Wasiutynski; the resolvent proof is due to E. A. Nordgren.

104   **Solutions 104.** *There exists a convergent sequence of nilpotent operators such that the spectral radius of the limit is positive.*

PROOF. The construction is based on a sequence $\{\varepsilon_n\}$ of positive numbers converging to 0. The question of what the ε's can be will be answered after the question of what they are expected to do. Begin by defining a sequence $\{\alpha_n\}$ as follows: every second α is equal to ε_0 (i.e., $\alpha_0 = \varepsilon_0$, $\alpha_2 = \varepsilon_0$, $\alpha_4 = \varepsilon_0$, $\cdots$); every second one of the remaining α's is equal to ε_1 (i.e., $\alpha_1 = \varepsilon_1$, $\alpha_5 = \varepsilon_1$, $\alpha_9 = \varepsilon_1$, $\cdots$); every second one of the still remaining α's is equal to ε_2; and so on ad infinitum. The sequence of α's looks like this:

$$\varepsilon_0, \varepsilon_1, \varepsilon_0, \varepsilon_2, \varepsilon_0, \varepsilon_1, \varepsilon_0, \varepsilon_3, \varepsilon_0, \varepsilon_1, \varepsilon_0, \varepsilon_2, \varepsilon_0, \varepsilon_1, \varepsilon_0, \cdots.$$

Let A be the weighted unilateral shift whose weights are the α's and, for each non-negative integer k, let A_k be the weighted unilateral shift whose weights are what the α's become when each ε_k is replaced by 0. Thus, for instance, the sequence of weights for A_2 is

$$\varepsilon_0, \varepsilon_1, \varepsilon_0, 0, \varepsilon_0, \varepsilon_1, \varepsilon_0, \varepsilon_3, \varepsilon_0, \varepsilon_1, \varepsilon_0, 0, \varepsilon_0, \varepsilon_1, \varepsilon_0, \cdots.$$

Two things are obvious from this construction: A_k is nilpotent of index 2^{k+1}, and the norm of $A_k - A$ (which is a weighted shift) is ε_k.

All that remains is to prove that the ε's can be chosen so as to make $r(A) > 0$. For this purpose note that

$$\alpha_0 = \varepsilon_0,$$

$$\alpha_0 \alpha_1 \alpha_2 = \varepsilon_0^2 \varepsilon_1,$$

$$\alpha_0 \alpha_1 \alpha_2 \alpha_3 \alpha_4 \alpha_5 \alpha_6 = \varepsilon_0^4 \varepsilon_1^2 \varepsilon_2,$$

and, in general, if $n = 2^p - 2$ $(p = 1, 2, 3, \cdots)$, then

$$\alpha_0 \cdots \alpha_n = \varepsilon_0^{2^{p-1}} \cdots \varepsilon_{p-1}.$$

Hence

$$\log(\alpha_0 \cdots \alpha_n) = \sum_{k=0}^{p-1} 2^{p-1-k} \log \varepsilon_k = 2^p \sum_{k=0}^{p-1} \frac{\log \varepsilon_k}{2^{k+1}},$$

or

$$\log(\alpha_0 \cdots \alpha_n)^{1/n+1} = \frac{2^p}{2^p - 1} \sum_{k=0}^{p-1} \frac{\log \varepsilon_k}{2^{k+1}}.$$

This implies that if the series

$$\sum_{k=0}^{\infty} \frac{\log \varepsilon_k}{2^{k+1}}$$

is convergent (which happens if, for instance, $\varepsilon_k = 1/2^k$), then

$$\liminf_n \log(\alpha_0 \cdots \alpha_n)^{1/n+1} > -\infty,$$

and therefore

$$\liminf_n (\alpha_0 \cdots \alpha_n)^{1/n+1} > 0.$$

The desired conclusion follows from Solution 91.

In some concrete cases the spectral radius can be computed; if, for instance, $\varepsilon_k = 1/2^k$, then $r = (r(A)) = 1$. Here is how the computation goes. Since

$$\lim_{p \to \infty} \frac{2^p}{2^p - 1} \sum_{k=0}^{p-1} \frac{\log \varepsilon_k}{2^{k+1}} = \sum_{k=0}^{\infty} \frac{\log \varepsilon_k}{2^{k+1}},$$

it follows that

$$\log r = -\frac{\log 2}{2} \sum_{k=0}^{\infty} \frac{k}{2^k}.$$

If $f(z) = \sum_{k=0}^{\infty} z^k/2^k$, then $\log r = -((\log 2)/2)f'(1)$. Since

$$f'(z) = -2 \log(2 - z),$$

it follows that $f'(1) = 0$, and hence that $r = 1$.

This example is due to S. Kakutani; see [112, p. 282].

Solutions 105. *The restriction of* spec *to the set of normal operators is* **105** *continuous.*

To prove the statement, it is to be proved that if $\{A_n\}$ is a sequence of normal operators and $A_n \to A$, then

$$\text{spec } A \subset \liminf_n \text{spec } A_n.$$

(Note that A is necessarily normal, but that fact does not explicitly enter into the proof.)

Question: when is a number λ *not* in $\liminf_n$ spec A_n? Answer: exactly when the distance from λ to the set spec A_n does not tend to 0 as $n \to \infty$; in other words, exactly when there exists a positive number ε such that

$$d(\lambda, \text{spec } A_n) \geqq \varepsilon$$

for infinitely many values of n. The inequality says that

$$\frac{1}{|\lambda - \lambda'|} \leqq \frac{1}{\varepsilon}$$

whenever $\lambda' \in \text{spec } A_n$. This, in turn, says that not only is λ absent from spec A_n, so that $A_n - \lambda$ is invertible, but, in fact, $r((A_n - \lambda)^{-1}) \leqq 1/\varepsilon$.

No use was made of normality so far; here it comes. The way normality is used is via the observation that for normal operators the spectral radius is the same as the norm. Accordingly, the contrapositive of what is to be proved is that if a number λ and a subsequence $\{A_{n_k}\}$ are such that

$$\|(A_{n_k} - \lambda)^{-1}\| \leqq \frac{1}{\varepsilon}$$

(for some ε), then $A - \lambda$ is invertible. There is no loss in simplifying the notation and assuming that $\|(A_n - \lambda)^{-1}\| \leqq 1/\varepsilon$ for all n.

Everything is now prepared for the final argument. Since

$$\|(A_n - \lambda)^{-1} - (A_m - \lambda)^{-1}\|$$
$$= \|(A_n - \lambda)^{-1}((A_m - \lambda) - (A_n - \lambda))(A_m - \lambda)^{-1}\|$$
$$\leqq \|(A_n - \lambda)^{-1}\| \cdot \|A_m - A_n\| \cdot \|(A_m - \lambda)^{-1}\|$$
$$\leqq \frac{1}{\varepsilon^2} \|A_m - A_n\|,$$

and since $A_n \to A$, it follows that the sequence $\{(A_n - \lambda)^{-1}\}$ converges, to some operator B, say. Since

$$(A - \lambda)B = \lim_n (A_n - \lambda) \cdot \lim_n (A_n - \lambda)^{-1}$$
$$= \lim_n (A_n - \lambda)(A_n - \lambda)^{-1} = 1,$$

and similarly, of course, $B(A - \lambda) = 1$, the proof is complete.

106 **Solution 106.** The answer is yes and no: yes if the underlying Hilbert space $\mathbf{H}$ is finite-dimensional and no otherwise.

Since, by assumption, $\text{spec}(B + A/n) = \text{spec } A/n$, and spec A/n is included in the disc with center 0 and radius $\|A\|/n$, it follows that if $B_n = B + A/n$, then $r(B_n) \to 0$ (and, of course, $B_n \to B$). If dim $\mathbf{H} < \infty$, then spec B_n is a finite set of eigenvalues (each counted as often as its algebraic multiplicity requires); the elementary symmetric functions of those eigenvalues determine the coefficients of the (monic) characteristic polynomial $\det(\lambda - B_n)$. Since, on the other hand, even the largest (in absolute value) of those eigenvalues tends to 0 as n tends to ∞, it follows that the elementary symmetric functions tend to 0, and hence that $\det(\lambda - B_n)$ tends to $\lambda^{\dim \mathbf{H}}$. Since, finally, $\det(\lambda - B_n) \to \det(\lambda - B)$, it follows indeed that B is nilpotent.

So much for the finite-dimensional case. (What just happened was, in effect, a proof that in the finite-dimensional case spectrum is continuous. Cf. Problem 105.)

A counterexample in the infinite-dimensional case resembles the example in Solution 104 of a convergent sequence of nilpotent operators with a limit that has large spectrum. Let A be a weighted shift with weight sequence

$$0, 0, -1, 0, 0, -1, 0, 0, -1, \cdots,$$

so that A is nilpotent of index 3. Let B be the weighted shift in whose weight sequence every second term is 1, every second one of the remaining terms is $\frac{1}{2}$, every second one of the still remaining terms is $\frac{1}{3}$, and so on (with $\frac{1}{4}, \frac{1}{5}, \cdots, 1/n, \cdots$). Since each weight of B is greater than or equal to the corresponding weight of the Kakutani shift with $\varepsilon_k = 1/2^k$ (Solution 104), it follows that the spectral radius of B is greater than or equal to that of the Kakutani shift (and is therefore positive). Consequence: B is not quasinilpotent.

For each positive integer n, the weight $1/n$ occurs in the weight sequence of B at position 2^{n-1} the first time, and, from then on, periodically with period 2^n. In the weight sequence of nB, therefore, the weight 1 occurs at positions $2^{n-1} + 2^n \cdot k$, $k = 0, 1, 2, \cdots$. Since 3 and 2^n are relatively prime, the weight $0\ (= -1 + 1)$ must occur in the weight sequence of $A + nB$ at least once, and, from then on, periodically with period $3 \cdot 2^n$. Conclusion: $A + nB$ is nilpotent, and, therefore, $\mathrm{spec}(A + nB) = \{0\} = \mathrm{spec}\ A$ for all n.

These results are due to K. J. Harrison.

CHAPTER 13

Operator Topologies

107 **Solution 107.** The first assertion involving uniformity has nothing to do with operators; it is just a statement about uniform weak convergence of vectors on the unit sphere. The proofs of the two assertions are very similar; what follows is a proof of the second one. For that purpose, assume $A = 0$; this loses no generality. The assumption in this case is that, for each positive number ε, if n is sufficiently large, then

$$\|A_n f\| < \varepsilon \quad \text{whenever} \quad \|f\| = 1;$$

the uniformity manifests itself in that the size of n does not depend on f. It follows that if n is sufficiently large, then

$$\|A_n\| \leqq \varepsilon.$$

The argument is general; it applies to all nets, not only to sequences.

108 **Solution 108.** *The norm is continuous with respect to the uniform topology and discontinuous with respect to the strong and weak topologies.*

PROOF. The proof for the uniform topology is contained in the inequality

$$\big| \|A\| - \|B\| \big| \leqq \|A - B\|.$$

This is just a version of the subadditivity of the norm, and it implies that the norm is a uniformly continuous function in the norm topology. The proof says nothing about the continuity of the norm in any other topology. A norm is always continuous with respect to the topology it defines; other topologies take their chances.

To show that the norm is not continuous with respect to the strong topology (not even sequentially), and, a fortiori, it is not continuous with

246

respect to the weak topology, consider the following example. Let $\{\mathbf{M}_n\}$ be a decreasing sequence of non-zero subspaces with intersection 0, and let $\{P_n\}$ be the corresponding sequence of projections. The sequence $\{P_n\}$ converges to 0 strongly. (To see this, form an orthonormal basis for $\mathbf{M}_1^\perp$, one for $\mathbf{M}_1 \cap \mathbf{M}_2^\perp$, another for $\mathbf{M}_2 \cap \mathbf{M}_3^\perp$, etc., and string them together to make a basis for the whole space. Cf. also Solution 120.) The sequence $\{\|P_n\|\}$ of norms does not converge to the number 0; indeed $\|P_n\| = 1$ for all n.

Solution 109. If $A_n \to A$ (weak), then, for each f and g,

$$|(Af, g) - (A_n f, g)|$$

is small when n is large, and therefore, for each $\varepsilon > 0$,

$$|(Af, g)| \leqq |(A_n f, g)| + \varepsilon$$

for n sufficiently large. It follows that

$$|(Af, g)| \leqq \|A_n\| \cdot \|f\| \cdot \|g\| + \varepsilon$$

for n sufficiently large, and hence that

$$|(Af, g)| \leqq \left(\liminf_n \|A_n\|\right) \cdot \|f\| \cdot \|g\| + \varepsilon.$$

Let ε tend to 0, and then recall that the supremum of $|(Af, g)|$ over all unit vectors f and g is $\|A\|$.

109

Solution 110. *The adjoint is continuous with respect to the uniform and the weak topologies and discontinuous with respect to the strong topology.*

110

PROOF. The proof for the uniform topology is contained in the identity

$$\|A^* - B^*\| = \|A - B\|.$$

If a function from one space to another is continuous, then it remains so if the topology of the domain is made larger, and it remains so if the topology of the range is made smaller. (This is the reason why the strong discontinuity of the norm implies its weak discontinuity.) If, however, a mapping from a space to itself is continuous, then there is no telling how it will behave when the topology is changed; every change works both ways. In fact, everything can happen, and the adjoint proves it. As the topologies march down (from norm to strong to weak), the adjoint changes from being continuous to being discontinuous, and back again.

To prove the strong discontinuity of the adjoint, let U be the unilateral shift, and write $A_k = U^{*k}$, $k = 1, 2, 3, \cdots$. Assertion: $A_k \to 0$ strongly, but the sequence $\{A_k^*\}$ is not strongly convergent to anything. Indeed:

$$\|A_k\langle \xi_0, \xi_1, \xi_2, \cdots \rangle\|^2 = \|\langle \xi_k, \xi_{k+1}, \xi_{k+2}, \cdots \rangle\|^2$$
$$= \sum_{n=k}^{\infty} |\xi_n|^2,$$

247

so that $\|A_k f\|^2$ is, for each f, the tail of a convergent series, and therefore $A_k f \to 0$. The negative assertion about $\{A_k{}^*\}$ can be established by proving that if $f \neq 0$, then $\{A_k{}^*f\}$ is not a Cauchy sequence. Indeed:

$$\|A_{m+n}{}^*f - A_n{}^*f\|^2 = \|U^{m+n}f - U^n f\|^2 = \|U^m f - f\|^2$$
$$= \|U^m f\|^2 - 2\,\mathrm{Re}(U^m f, f) + \|f\|^2$$
$$= 2(\|f\|^2 - \mathrm{Re}(f, U^{*m}f)).$$

Since $\|U^{*m}f\| \to 0$ as $m \to \infty$, it follows that $\|A_{m+n}{}^*f - A_n{}^*f\|$ refuses to become small as m and n become large; in fact if m is large, then

$$\|A_{m+n}{}^*f - A_n{}^*f\|$$

is nearly equal to $\sqrt{2}\,\|f\|$.

As for the weak continuity of the adjoint, that is implied by the identity

$$|(A^*f, g) - (B^*f, g)| = |(f, Ag) - (f, Bg)| = |(Ag, f) - (Bg, f)|.$$

111 **Solution 111.** The proof for the uniform topology is contained in the inequalities

$$\|AB - A_0 B_0\| \leqq \|AB - AB_0\| + \|AB_0 - A_0 B_0\|$$
$$\leqq \|A\| \cdot \|B - B_0\| + \|A - A_0\| \cdot \|B_0\|$$
$$\leqq (\|A - A_0\| + \|A_0\|)\|B - B_0\| + \|A - A_0\| \cdot \|B_0\|.$$

An elegant counterexample for the strong topology depends on the assertion that the set of all nilpotent operators of index 2 (i.e., the set of all operators A such that $A^2 = 0$) is strongly dense. (The idea is due to Arnold Lebow.) To prove this, suppose that

$$\{A: \|A_0 f_i - A f_i\| < \varepsilon, i = 1, \cdots, k\}$$

is an arbitrary basic strong neighborhood. There is no loss of generality in assuming that the f's are linearly independent (or even orthonormal); otherwise replace them by a linearly independent (or even orthonormal) set with the same span, and, at the same time, make ε as much smaller as is necessary. For each $i\,(=1, \cdots, k)$ find a vector g_i such that $\|A_0 f_i - g_i\| < \varepsilon$ and such that the span of the g's has only 0 in common with the span of the f's; so long as the underlying Hilbert space is infinite-dimensional, this is possible. Let A be the operator such that

$$A f_i = g_i \quad \text{and} \quad A g_i = 0 \qquad (i = 1, \cdots, k)$$

and

$$A h = 0 \quad \text{whenever } h \perp f_i \text{ and } h \perp g_i \qquad (i = 1, \cdots, k).$$

Clearly A is nilpotent of index 2, and, just as clearly, A belongs to the prescribed neighborhood. (For a different proof, see Problem 225.)

If squaring were strongly continuous, then the set of nilpotent operators of index 2 would be strongly closed, and therefore every operator would be nilpotent of index 2, which is absurd.

This result implies, of course, that multiplication is not jointly strongly continuous. Since the strong topology is larger than the weak, so that a strongly dense set is necessarily weakly dense, the auxiliary assertion about nilpotent operators holds for the weak topology as well as for the strong. Conclusion: squaring is not weakly continuous, and, consequently, multiplication is not jointly weakly continuous.

There is another way to see that squaring is not strongly continuous; it is less geometric but, in recompense, it is also less computational.

The idea is to use a badly unbounded set that contains 0 in its weak closure. If $\{e_1, e_2, e_3, \cdots\}$ is an orthonormal set, then the set $\mathbf{E}$ of all vectors of the form $\sqrt{k}e_k$ will serve; see Solution 28. For each k, define an operator A_k by $A_k f = (f, \sqrt{k}e_k)e_k$; it follows that $\|A_k f\| = |(f, \sqrt{k}e_k)|$. (Observe that A_k is a positive operator of rank 1.) Let $\{k_n\}$ be a net of positive integers such that $\sqrt{k_n}e_{k_n} \to 0$ weakly. (It cannot be a sequence.) It follows that

$$(f, \sqrt{k_n}e_{k_n}) \to 0$$

for each f, and therefore that $\|A_{k_n}f\| \to 0$. As for the squares:

$$A_k^2 f = A_k(f, \sqrt{k}e_k)e_k = (f, \sqrt{k}e_k)(e_k, \sqrt{k}e_k)e_k = k(f, e_k)e_k.$$

If, in particular, $f = \sum_{k=1}^{\infty} (1/k)e_k$, then $A_k^2 f = e_k$, so that $\|A_k^2 f\| = 1$. Conclusion: $\|A_{k_n}^2 f\|$ cannot converge to 0, so that squaring is not strongly continuous.

Solution 112. The easiest proof uses convergence. The convergence of **112** sequences is sometimes misleading in general topology, but the convergence of nets (generalized sequences) is good enough. Suppose therefore that $A_j \to A$ strongly, i.e., that $A_j f \to Af$ for each f. It follows, in particular, that $A_j B f \to AB f$ for each f, and this settles strong continuity in A. If, on the other hand, $B_j \to B$ strongly, i.e, if $B_j f \to Bf$ for each f, then apply A to conclude that $AB_j f \to AB f$ for each f; this settles strong continuity in B. Weak continuity can be treated the same way. If $(A_j f, g) \to (Af, g)$ for each f and g, then, in particular, $(A_j B f, g) \to (AB f, g)$ for each f and g; if $(B_j f, g) \to (Bf, g)$ for each f and g, then, in particular, $(AB_j f, g) = (B_j f, A^*g) \to (Bf, A^*g) = (AB f, g)$ for each f and g.

Solution 113. (a) The crux of the matter is boundedness. Assume first that **113** the sequence $\{\|A_n\|\}$ of norms is bounded. Since, for each f,

$$\|A_n B_n f - AB f\| \leqq \|A_n B_n f - A_n B f\| + \|A_n B f - AB f\|$$
$$\leqq \|A_n\| \cdot \|(B_n - B)f\| + \|(A_n - A)B f\|,$$

the assumed boundedness implies, as desired, that $A_n B_n f \to AB f$.

Now what about the boundedness assumption? The answer is that it need not be assumed at all; it can be proved. It is, in fact, an immediate consequence of the principle of uniform boundedness for operators: if a sequence of operators is weakly convergent (and all the more if it is strongly convergent), then it is weakly bounded, and therefore bounded.

(b) *Multiplication is not weakly sequentially continuous.*

PROOF. Let U be the unilateral shift, and write $A_n = U^{*n}$, $B_n = U^n$, $n = 1, 2, 3, \cdots$. Since $A_n \to 0$ strongly, it follows that $A_n \to 0$ weakly, and hence that $B_n \to 0$ weakly (cf. Solution 110). Since, however, $A_n B_n = 1$ for all n, it is not true that $A_n B_n \to 0$ weakly.

114 **Solution 114.** Let A_n be the infinite matrix (or, by a slight abuse of language, the operator) whose entries at the positions $\langle 1, n \rangle$ and $\langle n, 1 \rangle$ are 1, and whose remaining entries are all 0 ($n = 2, 3, 4, \cdots$). The matrix A_n^2 has its only two non-zero entries at $\langle 1, 1 \rangle$ and $\langle n, n \rangle$. The weak asymptotic properties of both sequences $\{A_n\}$ and $\{A_n^2\}$ are easy to obtain: $A_n \to 0$ weakly and $A_n^2 \to A$ weakly, where A is the matrix whose only non-zero entry is 1 at the position $\langle 1, 1 \rangle$.

This proves the weak sequential discontinuity of squaring at one point, namely 0. The same thing happens everywhere. Indeed, for any operator B, put $B_n = B + A_n$. Clearly $B_n \to B$ weakly, and

$$B_n^2 = B^2 + BA_n + A_nB + A_n^2 \to B^2 + 0 + 0 + A$$

weakly. (Cf. Problem 112 on separate continuity.) Since $A \neq 0$, it follows that squaring is weakly sequentially discontinuous at B. (The latter part of the argument is due to C. A. McCarthy.)

115 **Solution 115.** There exist weakly convergent sequences of projections that are not strongly convergent. A simple way of describing an example is to use matrices (as in Solution 114). Let P_n be the infinite matrix whose entries at the positions $\langle 1, 1 \rangle$, $\langle 1, n \rangle$, $\langle n, 1 \rangle$, and $\langle n, n \rangle$ are $\frac{1}{2}$, and whose remaining entries are all 0 ($n = 2, 3, 4, \cdots$). Each P_n is a projection (Hermitian and idempotent). If A is the matrix whose only non-zero entry is $\frac{1}{2}$ at the position $\langle 1, 1 \rangle$, then $P_n \to A$ weakly. The sequence $\{P_n\}$ cannot, however, converge to anything strongly. Reason: if it did, then the limit would have to coincide with A (because strong convergence implies weak), and therefore A would have to be a projection (because a strong limit of projections is a projection).

CHAPTER 14

Strong Operator Topology

Solution 116. The assertion is that if $\{A_n\}$ is a net of normal operators and A **116**
is a normal operator such that $A_n \to A$ (strong), then $A_n{}^* \to A^*$ (strong).
What is easy and known (Solution 110) is that $A_n{}^* \to A^*$ (weak). Since (by
the assumed normality and the assumed strong convergence)

$$\|A_n{}^*f\| = \|A_n f\| \to \|Af\| = \|A^*f\|$$

for each f, Problem 20 is applicable and yields the result.
   Reference: [83].
   Caution: the assertion does not mean that if $\{A_n\}$ is a net of normal
operators and $A_n \to A$ (strong), then $A_n{}^* \to A^*$; the normality of A must
be explicitly assumed. Suppose, indeed, that $\{e_0, e_1, e_2, \cdots\}$ is an orthonormal basis, and suppose that U_n ($n = 1, 2, 3, \cdots$) is an operator such
that $U_n e_k = e_{k+1}$ when $0 \leqq k < n - 1$, $U_n e_{n-1} = e_0$, and $U_n e_k = 0$
when $k \geqq n$.
   It follows that $U_n \to U$ (strong), where U is the unilateral shift. Since,
however, $U_n{}^*e_0 = e_{n-1}$, the sequence $\{U_n{}^*e_0\}$ is not strongly convergent.

Solution 117. The answer is no. One good way to construct a counter- **117**
example is to use the unilateral shift U and exploit the fact that $U^*U \neq UU^*$.
The details can be arranged as follows.
   Let $\mathbf{N}$ be the set of all pairs $\langle p, \mathbf{E} \rangle$ where p is a (strictly) positive integer
and $\mathbf{E}$ is a strong neighborhood of 0; write $\langle p, \mathbf{E} \rangle \leqq \langle q, \mathbf{F} \rangle$ in case $p \leqq q$
and $\mathbf{E} \supset \mathbf{F}$. The relation so defined makes $\mathbf{N}$ a directed set.
   The next step is to define a net of operators on $\mathbf{N}$. For each $n = \langle p, \mathbf{E} \rangle$ in
$\mathbf{N}$, find a positive integer $k = k(p, \mathbf{E})$ so that $pU^{*k} \in \mathbf{E}$ (which can be done
because, for p fixed, pU^{*k} tends strongly to 0 as k becomes large); write
$A_n = pU^{*k}$ and $B_n = (1/p)U^k$.

The net $\{B_n\}$ converges to 0, not only strongly, but in the norm; that is, $\|B_n\|$ can be made arbitrarily small by making n sufficiently large. Indeed: given ε (>0), find p_0 so that $1/p_0 < \varepsilon$, let $\mathbf{E}_0$ be an arbitrary strong neighborhood of 0, and put $n_0 = \langle p_0, \mathbf{E}_0 \rangle$. If $n \geqq n_0$, so that $n = \langle p, \mathbf{E} \rangle$ with $p \geqq p_0$ and $\mathbf{E} \subset \mathbf{E}_0$, then $\|B_n\| < \varepsilon$. Proof:

$$\left\| \frac{1}{p} U^{k(p, \mathbf{E})} \right\| = \frac{1}{p}.$$

Note that $\{B_n\}$ is bounded.

The net $\{A_n\}$ also converges to 0 strongly. Proof: given a strong neighborhood $\mathbf{E}_0$ of 0, write $n_0 = \langle 1, \mathbf{E}_0 \rangle$. If $n \geqq n_0$, so that $n = \langle p, \mathbf{E} \rangle$ with $p \geqq 1$ and $\mathbf{E} \subset \mathbf{E}_0$, then $A_n \in \mathbf{E}_0$ (because $A_n \in \mathbf{E}$, by definition).

The product net $\{A_n B_n\}$ does not converge to 0 strongly; in fact, $A_n B_n$ is identically equal to 1.

118 **Solution 118.** No. For an example, consider an orthonormal basis

$$\{e_0, e_1, e_2, \cdots\},$$

and write $A_n = U^{*n}$ (where U is the unilateral shift defined by $Ue_n = e_{n+1}$, $n = 0, 1, 2, \cdots$). Since $e_n \to 0$ weakly and $U^{*n} \to 0$ strongly, the assumptions are satisfied; since, however, $U^{*n} e_n = e_0$, the conclusion is not.

119 **Solution 119.** Suppose that $\{e_0, e_1, e_2, \cdots\}$ is an orthonormal basis and let P_n be the projection onto the span of $\{e_0, \cdots, e_n\}$, $n = 1, 2, \cdots$. Clearly $P_n \to 1$ strongly, spec $P_n = \{0, 1\}$, and spec $1 = \{1\}$. Consequence: arbitrarily near to the operator 1, in the sense of the strong topology, there are operators with a (relatively) much larger spectrum. More simply:

$$\limsup_n \operatorname{spec} P_n \nsubseteq \operatorname{spec} 1.$$

Conclusion: the spectrum is not strongly upper semicontinuous.

In this example the spectral radius does not misbehave: $r(P_n) = 1$ for all n, and $r(P) = 1$. It is easy to modify the example so as to prove that r is not continuous: just consider the corresponding unilateral shift U

$$(Ue_n = e_{n+1}, n = 0, 1, 2, \cdots),$$

and write $A_n = UP_n$. The result is that $A_n \to U$ strongly; each A_n is nilpotent, so that $r(A_n) = 0$ for all n, but $r(U) = 1$. Consequence: arbitrarily near to the operator U, in the sense of the strong topology, there are operators with a much smaller spectral radius; the spectral radius (and hence the spectrum) is not strongly lower semicontinuous.

Could it be that despite all this the spectral radius is upper semicontinuous? No, but to prove that a different example is needed. The powers U^{*n} will do. Since $U^{*n} \to 0$ strongly and $r(U^{*n}) = 1$ for all n, it follows that arbitrarily near to 0, in the sense of the strong topology, there are operators with much larger spectral radius.

Solution 120. *A bounded increasing sequence of Hermitian operators is* **120** *always convergent with respect to the strong topology, but not necessarily with respect to the uniform topology.*

PROOF. One way to prove the assertion about the strong topology is to make use of the weak version. Let $\{A_n\}$ be a bounded increasing sequence of Hermitian operators, and let A be its weak limit. Since $A_n \leqq A$, the operator $A - A_n$ is positive, and therefore it has a positive square root, say B_n (see Problem 121). Since

$$\|B_n f\|^2 = (B_n f, B_n f) = (B_n^2 f, f) = ((A - A_n)f, f) \to 0,$$

the sequence $\{B_n\}$ tends strongly to 0. Since $\{\|A - A_n\|\}$ is bounded, so is $\{\|B_n\|\}$; say $\|B_n\| \leqq \beta$ for all n. The asserted strong convergence now follows from the relation

$$\|(A - A_n)f\| = \|B_n^2 f\| \leqq \beta \|B_n f\|.$$

As once before (cf. Solution 1) sequences play no essential role here; nets would do just as well.

There is sometimes a technical advantage in not using the theorem about the existence of positive square roots. The result just obtained can be proved without that theorem, if it must be, but the proof with square roots shows better what really goes on. Here is how a proof without square roots goes. Assume, with no loss of generality, that $A \leqq 1$. If $m < n$, then

$$\|(A_n - A_m)f\|^4 = ((A_n - A_m)f, (A_n - A_m)f)^2$$
$$\leqq ((A_n - A_m)f, f)((A_n - A_m)^2 f, (A_n - A_m)f),$$

by the Schwarz inequality for the inner product determined by the positive operator $A_n - A_m$. Since $A_n - A_m \leqq 1$, so that $\|A_n - A_m\| \leqq 1$, it follows that

$$\|A_n f - A_m f\|^4 \leqq ((A_n f, f) - (A_m f, f))\|f\|^2.$$

A frequently used consequence of the strong convergence theorem is about projections. If $\{M_n\}$ is an increasing sequence of subspaces, then the corresponding sequence $\{P_n\}$ of projections is an increasing (and obviously bounded) sequence of Hermitian operators. It follows that there exists a Hermitian operator P such that $P_n \to P$ strongly. Assertion: P is the projection onto the span, say M, of all the M_n's (cf. Solution 108). Reason: if f belongs to some M_n, then $Pf = f$, and if f is orthogonal to all M_n's, then $Pf = 0$; these two comments together imply that there is a dense set on which P agrees with the projection onto M.

Increasing sequences of projections serve also to show that the monotone convergence assertion is false for the uniform topology. Indeed, if the sequence $\{M_n\}$ is strictly increasing, then the sequence $\{P_n\}$ cannot converge to P (or, for that matter, to anything at all) in the norm, because it is not even a Cauchy sequence. In fact, a monotone sequence of projections can be a Cauchy sequence in trivial cases only; $\|P_n - P_m\| = 1$ unless $P_n = P_m$.

121 **Solution 121.** It is convenient (for purposes of reference) to break up the proof into small steps, as follows.

(1) All the positive integral powers of a positive operator are positive. Indeed $(A^{2n}f, f) = \|A^n f\|^2$ and $(A^{2n+1}f, f) = (A \cdot A^n f, A^n f)$; the former is positive because norms are, and the latter is positive because A is. In the sequel the result is needed not for A but for $1 - A$. (Note: the assertion is a trivial consequence of the spectral theorem.)

(2) Each B_n is a polynomial in $1 - A$ with positive coefficients (by induction), and hence (by (1)) each B_n is a positive operator.

(3) By (2), all the B_n's commute with one another, and it follows that

$$B_{n+2} - B_{n+1} = \tfrac{1}{2}(B_{n+1}{}^2 - B_n{}^2) = \tfrac{1}{2}(B_{n+1} - B_n)(B_{n+1} + B_n).$$

This implies (by (2) and induction) that $B_{n+1} - B_n$ is a polynomial in $1 - A$ with positive coefficients, and hence positive; it follows that the sequence $\{B_n\}$ is increasing.

(4) The definition of B_{n+1} in terms of B_n implies (induction) that $\|B_n\| \le 1$ for all n; the sequence $\{B_n\}$ is bounded.

(5) By (3) and (4), $\{B_n\}$ is a bounded increasing sequence of positive operators, and therefore it is strongly convergent to some (necessarily positive) operator B. Note that the argument needs Solution 120. Since the point of what is now going on is to avoid square roots, it is necessary to use the version of Solution 120 that does not use square roots.

Convergence is proved; it remains only to evaluate the limit. This is easy from Problem 113; since $B_n \to B$ (strongly), it follows that $B_n{}^2 \to B^2$ (strongly), and hence that

$$B = \tfrac{1}{2}((1 - A) + B^2).$$

This says that

$$A = 1 - 2B + B^2 = (1 - B)^2,$$

and the proof is complete.

The proof is standard; cf. [114].

122 **Solution 122.** Even a small amount of experience with non-commutative projections shows that the familiar algebraic operations are not likely to suffice to express $E \wedge F$ in terms of E and F. The following quite pretty and geometrical consideration shows how topology comes in, and motivates the actual proof. Suppose that the underlying Hilbert space $\mathbf{H}$ is two-dimensional real Euclidean space, and suppose that $\mathbf{M}$ and $\mathbf{N}$ are two distinct but not orthogonal lines through the origin. Take an arbitrary point f in $\mathbf{H}$, project it on $\mathbf{M}$ (i.e., form Ef), project the result on $\mathbf{N}$ (FEf), then project on $\mathbf{M}$ ($EFEf$), and continue so on ad infinitum; it looks plausible that the sequence so obtained converges to 0, which, in this case, is $(E \wedge F)f$. This suggests the formation of the sequence

$$E, FE, EFE, FEFE, EFEFE, \cdots.$$

The proof itself works with the subsequence

$$EFE, EFEFE, EFEFEFE, \cdots;$$

this is a matter of merely technical convenience.

Since $\|EFE\| \leqq 1$, the powers of EFE form a decreasing (and even commutative) sequence of positive operators. It follows that $(EFE)^n$ is weakly convergent to, say, G; since in this case weak and strong convergence are equivalent, G belongs to the given von Neumann algebra. Assertion: $G = E \wedge F$. Clearly G is Hermitian. Since $(EFE)^m G = G$ for all m, therefore $G^2 = G$, so that G is a projection. Since $(EFE)^m FG = G$ for all m, therefore $GFG = G$; this implies that $G \leq F$. (Proof: $0 = G - GFG = G(1 - F)G = G(1 - F)(1 - F)G$, and $(1 - F)G = (G(1 - F))^*$.) Since $E(EFE)^n = (EFE)^n$ for all n, therefore $EG = G$ or $G \leq E$. If, finally, G_0 is a projection such that $G_0 \leq E$ and $G_0 \leq F$, then $G_0(EFE)^n = G_0$, whence $G_0 G = G_0$, so that $G_0 \leq G$. The proof is complete.

The theorem has its own dual for an easy corollary. The assertion is that the projection $E \vee F$ on the subspace $\mathbf{M} \vee \mathbf{N}$ belongs to any von Neumann algebra containing E and F. Since

$$E \vee F = 1 - ((1 - E) \wedge (1 - F)),$$

the proof is immediate.

An examination of the proof shows that not all the defining properties of von Neumann algebras were used; all that was needed was a sequentially strongly closed set of operators such that if A and B are in the set, then so is ABA (for the theorem about $E \wedge F$) or

$$1 - (1 - A)(1 - B)(1 - A)$$

(for the theorem about $E \vee F$). Observe that even in the latter case it is not required that 1 belong to the set; an expression such as

$$1 - (1 - A)(1 - B)(1 - A)$$

is a convenient way of writing something that can obviously (though clumsily) be written without 1 if so desired.

255

CHAPTER 15

Partial Isometries

123 **Solution 123.** Use the spectral theorem to represent A as a multiplication by, say, φ. If $\lambda \in \operatorname{spec} A$ and if N is an arbitrary neighborhood of $F(\lambda)$, then $F^{-1}(N)$ is a neighborhood of λ, and therefore $\varphi^{-1}(F^{-1}(N))$ has positive measure. Since $\varphi^{-1}(F^{-1}(N)) = (F \circ \varphi)^{-1}(N)$, it follows that $F(\lambda)$ is in the essential range of $F \circ \varphi$, so that $F(\lambda) \in \operatorname{spec} F(A)$. This proves that $F(\operatorname{spec} A) \subset \operatorname{spec} F(A)$.

To prove the reverse inclusion is the same as to prove that if $\lambda \notin F(\operatorname{spec} A)$, then $\lambda \notin \operatorname{spec} F(A)$. The set $F(\operatorname{spec} A)$ is compact. (It is the image under the continuous function F of the compact set $\operatorname{spec} A$.) Since λ is not in it, λ has a neighborhood disjoint from it. If N is such a neighborhood, then $F^{-1}(N) = \varnothing$, and therefore not only does $(F \circ \varphi)^{-1}(N)$ $(= \varphi^{-1}(F^{-1}(N)))$ have measure zero, but, in fact, it is empty. Consequence: λ does not belong to the spectrum of $F(A)$. This completes the proof.

124 **Solution 124.** Perhaps one reason why the non-spectral proof is elusive here is that square roots seem to be needed, and square roots without the spectral theorem are known to require a little effort; see Problem 121. The necessity for square roots is not surprising in a theorem that is explicitly about squares.

If $0 \leqq A \leqq 1$, then $(Af, f) \leqq (f, f)$ for all f, and, in particular,

$$(A\sqrt{Af}, \sqrt{Af}) \leqq (\sqrt{Af}, \sqrt{Af}),$$

or

$$(A^2 f, f) \leqq (Af, f);$$

this means that $A^2 \leqq A$. That's the easier direction.

256

In the other direction, note that if A is Hermitian, then

$$\|Af\|^2 = (Af, Af) = (A^2f, f),$$

and hence, in particular, if $A \geqq 0$, then

$$\|\sqrt{A}f\|^2 = (Af, f).$$

It follows that if $A^2 \leqq A$, then

$$\|Af\| \leqq \|\sqrt{A}f\| \leqq \|\sqrt{A}\| \cdot \|f\|$$

for all f, and hence that $\|A\| \leqq \|\sqrt{A}\|$. Consequence: $\|A\|^2 \leqq \|A\|$, or $\|A\| \leqq 1$; since A is Hermitian, it follows that $A \leqq 1$.

Solution 125. The answer is yes, but the formulation of the problem is mis- **125** leading. The fact is that if $p(A)$ is diagonal for any non-zero polynomial p, then A also must be diagonal.

Use the spectral theorem to represent A as multiplication by φ on $\mathbf{L}^2(X)$ for some measure space X with σ-finite measure μ. By assumption $p(A)$ has a set Λ_0 of eigenvalues λ such that the corresponding eigenspaces $\mathbf{M}_\lambda = (\{f : p(A)f = \lambda f\})$, $\lambda \in \Lambda_0$, span the whole space. (Normality implies that if $\lambda_1 \neq \lambda_2$, then $\mathbf{M}_{\lambda_1} \perp \mathbf{M}_{\lambda_2}$.)

Question: which functions in $\mathbf{L}^2(X)$ belong to $\mathbf{M}_\lambda$? Answer: those functions f for which

$$(p(\varphi(x)) - \lambda)f(x) = 0 \quad \text{a.e.},$$

that is, those whose support is a subset of

$$E_\lambda = \varphi^{-1}\{z : p(z) - \lambda = 0\}.$$

The separability of the space implies that $\mathbf{M}_\lambda \neq 0$ for only countably many values of λ, and, correspondingly, that $\mu(E_\lambda) \neq 0$ for only countably many values of λ. The sets $\{E_\lambda : \lambda \in \Lambda_0\}$ are pairwise (almost) disjoint, and their union is (almost) equal to X. If $\mathbf{E}_\lambda$ is an orthonormal basis for $\mathbf{L}^2(E_\lambda)$ ($=$ the subspace of those functions in $\mathbf{L}^2(X)$ whose support is a subset of E_λ), then $\bigcup_{\lambda \in \Lambda_0} \mathbf{E}_\lambda$ is an orthonormal basis for $\mathbf{L}^2(X)$.

There is an intelligent way of choosing a basis $\mathbf{E}_\lambda$ for $\mathbf{L}^2(E_\lambda)$. Since the set $\{z : p(z) - \lambda = 0\}$ is finite, it follows that the essential range of the restriction of φ to E_λ is finite. That is: E_λ is the union of a finite number of sets on each of which φ is a constant. Choose $\mathbf{E}_\lambda$ as the union of a corresponding finite number of orthonormal bases, one for each of those subsets of E_λ. Each vector in each such basis is an eigenvector of A, and the union of all the bases so obtained is a basis for the whole space.

The answer is the same even if the underlying Hilbert space is not assumed to be separable, but the proof involves some fuss necessitated by the possible pathology of large (meaning, non-σ-finite) measure spaces.

Solution 126. The answer is yes. Suppose, indeed, that A and A_n are normal **126** operators ($n = 1, 2, 3, \cdots$) with $A_n \to A$. Let c be a positive number such that $\|A_n\| \leqq c$ for all n (and hence such that $\|A\| \leqq c$). By the Weierstrass polynomial approximation theorem in the plane there exists a sequence $\{p_m\}$ of polynomials in two variables such that $p_m(z, z^*) \to F(z)$ uniformly in the disc with center 0 and radius c. Since, for every normal operator X, the norm $\|p_m(X, X^*) - F(X)\|$ is dominated by the supremum norm of $p_m(z, z^*) - F(z)$ on the disc with center 0 and radius c, it follows that $p_m(X, X^*) \to F(X)$ uniformly over all normal operators X with $\|X\| \leqq c$.

Now note that

$$\|F(A_n) - F(A)\| \leqq \|F(A_n) - p_m(A_n, A_n^*)\|$$
$$+ \|p_m(A_n, A_n^*) - p_m(A, A^*)\| + \|p_m(A, A^*) - F(A)\|$$

for all m and n. Choose m so large that the first and last summands are small for all n (this is the uniformity mentioned above), and then, for m fixed, choose n so as to make the middle summand small.

The method of proof can be generalized so as to yield similar results for functions that are not defined on the entire plane. A typical and important example is the square root function on the positive half of the real line: the conclusion is that the mapping $A \mapsto \sqrt{A}$, defined for positive operators, is continuous.

127   **Solution 127.** Suppose that **H** and **K** are Hilbert spaces and suppose that U is a partial isometry from **H** into **K** with initial space **M**. (For a discussion of such transformations and their adjoints, see Problem 51.) If E is the projection from **H** onto **M**, and if $f \in \mathbf{M}$, then

$$(U^*Uf, f) = \|Uf\|^2 = \|f\|^2 = (Ef, f);$$

if $f \perp \mathbf{M}$, then

$$(U^*Uf, f) = 0 = (Ef, f).$$

It follows that $(U^*Uf, f) = (Ef, f)$ for all f in **H**, and this implies that $U^*U = E$.

Suppose, conversely, that U is a bounded linear transformation from **H** into **K** such that U^*U is a projection with domain **H** and range **M**, say. It follows that

$$\|Uf\|^2 = (U^*Uf, f) = (Ef, f) = \|Ef\|^2$$

for all f, and hence that $\|Uf\| = \|f\|$ or $Uf = 0$ according as $f \in \mathbf{M}$ or $f \perp \mathbf{M}$.

To prove Corollary 1, observe that $\ker U^*U = \ker U$ (this is true for every bounded linear transformation U). The proof of Corollary 2 is a trick. If U^*U is idempotent, then $(UU^*)^3 = U(U^*UU^*U)U^* = (UU^*)^2$; the spectral theorem implies that a Hermitian operator A with $A^3 = A^2$ is idempotent. The assertion about initial and final spaces follows from the observation that $\ker^\perp UU^* = \ker^\perp U^* = \operatorname{ran} U$ (since $\operatorname{ran} U$ is closed). As for Corollary 3: if U is a partial isometry, then the product of U and the projection U^*U agrees with U on both ker U and its orthogonal complement; if, conversely, $U = UU^*U$, then premultiply by U^* and conclude that U^*U is idempotent.

Solution 128. *If U is an isometry and if $U\mathbf{M} = \mathbf{M}$, then $\mathbf{M}$ reduces U; if U is a co-isometry and if $\mathbf{M}$ reduces U, then $U\mathbf{M} = \mathbf{M}$. The first implication is false for co-isometries; the second implication is false for isometries.*   **128**

PROOF. If $U^*U = 1$ and $U\mathbf{M} = \mathbf{M}$, then $U^*\mathbf{M} = U^*U\mathbf{M} = \mathbf{M}$. If $UU^* = 1$ and both $U\mathbf{M} \subset \mathbf{M}$ and $U^*\mathbf{M} \subset \mathbf{M}$, then apply U to the second inclusion to obtain the reverse of the first.

The first implication is false if U is the adjoint of the unilateral shift and $\mathbf{M}$ is the (one-dimensional) subspace of eigenvectors belonging to a non-zero eigenvalue (see Solution 82). In that case $U\mathbf{M} = \mathbf{M}$, but $\mathbf{M}$ does not reduce U. The second implication is false if U is the unilateral shift and $\mathbf{M}$ is the whole space. In that case $\mathbf{M}$ reduces U, but $U\mathbf{M} \neq \mathbf{M}$.

Solution 129. The assertion about closure is obvious; the reason is that (1)   **129**
the mapping $A \mapsto AA^*A$ is continuous, and (2) the equation $A = AA^*A$ characterizes partial isometries.

An even easier version of the same proof shows that the set of all isometries is closed; consider the mapping (1) $A \mapsto A^*A$ and the equation (2) $A^*A = 1$. This comment is pertinent to the question concerning the connectedness of the set of all non-zero partial isometries. One way to prove that the answer to that question is no is to prove that the set of all isometries is not only closed but also open in the set of all partial isometries (in the relative topology of the latter). The fact is that if a partial isometry is sufficiently near to an isometry, then it is an isometry. More precisely, if U is a partial isometry, if V is an isometry, and if $\|U - V\| < 1$, then U is an isometry. It is sufficient to prove that if $Uf = 0$, then $f = 0$. Indeed, since

$$\|f\| = \|Vf\| = \|Uf - Vf\| \leq \|U - V\| \cdot \|f\|,$$

it follows that if $f \neq 0$, then $\|U - V\| \geq 1$, which contradicts the assumption that $\|U - V\| < 1$.

The same argument shows that if the underlying Hilbert space is infinite-dimensional, then the set of all isometries is not connected. Reason: the set of all unitary operators is a non-empty proper (!) subset that is simultaneously open and closed.

Solution 130. The kernel of U and the initial space of V can have only 0   **130**
in common. Indeed, if f is a non-zero vector such that $Uf = 0$ and $\|Vf\| = \|f\|$, then $\|Uf - Vf\| = \|f\|$, and this contradicts the hypothesis $\|U - V\| < 1$. It follows that the restriction of U to the initial space of V is one-to-one, and hence (Problem 56) the dimension of the initial space of V is less than or equal to the dimension of the entire range of U. In other

words, the result is that $\rho(V) \leqq \rho(U)$; the assertion about ranks follows by symmetry.

The assertion about nullities can be phrased this way: if $v(U) \neq v(V)$, then $\|U - V\| \geqq 1$. Indeed, if $v(U) \neq v(V)$, say, for definiteness, $v(U) < v(V)$, then there exists at least one unit vector f in the kernel of V that is orthogonal to the kernel of U. To say that f is orthogonal to the kernel of U is the same as to say that f belongs to the initial space of U. It follows that $1 = \|f\| = \|Uf\| = \|Uf - Vf\| \leqq \|U - V\|$, and the proof of the assertion about nullities is complete.

The assertion about co-ranks is an easy corollary: if $\|U - V\| < 1$, then $\|U^* - V^*\| < 1$, and therefore $\rho'(U) = v(U^*) = v(V^*) = \rho'(V)$.

The result appears in [114, § 105] for the special case of projections (which is, in fact, Problem 57). The present statement is a generalization, and, at the same time, the proof is a considerable simplification. The proof in [114] is, however, more constructive; it not only proves that two subspaces have the same dimension, but it exhibits a partial isometry that has the first for initial space and the second for final space. The generalization appears in [65].

131　**Solution 131.** Suppose that V_1 and V_2 are partial isometries with the same rank, co-rank, and nullity; let $\mathbf{N}_1$ and $\mathbf{N}_2$ be their kernels, $\mathbf{M}_1$ and $\mathbf{M}_2$ their initial spaces, and $\mathbf{R}_1$ and $\mathbf{R}_2$ their ranges. Let U be an arbitrary unitary operator that maps $\mathbf{N}_1$ onto $\mathbf{N}_2$ and $\mathbf{M}_1$ onto $\mathbf{M}_2$. Let W be a linear transformation that maps $\mathbf{R}_1^\perp$ isometrically onto $\mathbf{R}_2^\perp$; for f in $\mathbf{R}_1$, define $Wf = V_2 U V_1^* f$. Since it is easy to verify that this definition yields a linear transformation W that maps $\mathbf{R}_1$ isometrically onto $\mathbf{R}_2$, it follows that there exists a unitary operator W that maps $\mathbf{R}_1$ onto $\mathbf{R}_2$ and $\mathbf{R}_1^\perp$ onto $\mathbf{R}_2^\perp$ as indicated. If $g \in \mathbf{N}_1$, then

$$WV_1 g = 0 = V_2 Ug;$$

if $g \in \mathbf{M}_1$, then

$$WV_1 g = V_2 U V_1^* V_1 g = V_2 Ug.$$

It follows that $WV_1 = V_2 U$, or $WV_1 U^* = V_2$. If $t \mapsto W_t$ and $t \mapsto U_t$ are continuous curves of unitary operators that join 1 to W and to U, then $t \mapsto W_t V_1 U_t^*$ is a continuous curve of partial isometries all with the same rank, co-rank, and nullity, that joins V_1 to V_2.

This proof is a simplification of the one in [65]; it is due to R. G. Douglas.

132　**Solution 132.** Suppose that A and B are unitarily equivalent. If U is a unitary operator that transforms A onto B, then U transforms A^* onto B^*, and therefore U transforms $A' = \sqrt{1 - AA^*}$ onto $B' = \sqrt{1 - BB^*}$; it follows that

$$\begin{pmatrix} U & 0 \\ 0 & U \end{pmatrix}$$

transforms $M(A)$ onto $M(B)$.

If, conversely, a unitary operator matrix of size 2 transforms $M(A)$ onto $M(B)$, then it transforms $M(A)^*$ onto $M(B)^*$, and therefore it transforms $M(A)M(A)^* \left(= \begin{pmatrix} 1 & 0 \\ 0 & 0 \end{pmatrix} \right)$ onto itself. A unitary operator matrix that commutes with $\begin{pmatrix} 1 & 0 \\ 0 & 0 \end{pmatrix}$ is necessarily of the form $\begin{pmatrix} U & 0 \\ 0 & V \end{pmatrix}$, with U and V unitary. The assumed relation between $M(A)$ and $M(B)$ implies that $U^*AU = B$.

Solution 133. *If a compact subset Λ of the closed unit disc contains the origin, then there exists a partial isometry with spectrum Λ.* **133**

PROOF. Let A be a normal contraction with spectrum Λ (see Problem 63). If, as in Problem 132,

$$M = \begin{pmatrix} A & A' \\ 0 & 0 \end{pmatrix},$$

where $A' = \sqrt{1 - AA^*}$, then M is a partial isometry; what is its spectrum? The question reduces to this: for which values of λ is the operator matrix

$$M - \lambda = \begin{pmatrix} A - \lambda & A' \\ 0 & -\lambda \end{pmatrix}$$

not invertible? By Problem 70, $M - \lambda$ is invertible if and only if $-\lambda(A - \lambda)$ is invertible, which happens when $\lambda \neq 0$ and $A - \lambda$ is invertible. Conclusion: spec $M = \{0\} \cup \Lambda = \Lambda$.

In the finite-dimensional case more can be said. If Λ is a finite subset of the closed unit disc, with 0 in Λ, and if each element of Λ is assigned a positive integral multiplicity, then there exists a partial isometry with spectrum Λ whose eigenvalues have exactly the prescribed algebraic multiplicities; see [65].

CHAPTER 16

Polar Decomposition

134 **Solution 134.** Begin with the construction of P. Since A^*A is a positive operator on $\mathbf{H}$, it has a (unique) positive square root; call it P. Since

$$\|Pf\|^2 = (Pf, Pf) = (P^2f, f) = (A^*Af, f) = \|Af\|^2$$

for all f in $\mathbf{H}$, it follows that the equation

$$UPf = Af$$

unambiguously defines a linear transformation U from the range $\mathbf{R}$ of P into the space $\mathbf{K}$, and that U is isometric on $\mathbf{R}$. Since U is bounded on $\mathbf{R}$, it has a unique bounded extension to the closure $\overline{\mathbf{R}}$, and, from there, a unique extension to a partial isometry from $\mathbf{H}$ to $\mathbf{K}$ with initial space $\overline{\mathbf{R}}$. The equation $A = UP$ holds by construction. The kernel of a partial isometry is the orthogonal complement of its initial space, and the orthogonal complement of the range of a Hermitian operator is its kernel. This implies that ker U = ker P and completes the existence proof.

To prove uniqueness, suppose that $A = UP$, where U is a partial isometry, P is positive, and ker U = ker P. It follows that $A^* = PU^*$ and hence that

$$A^*A = PU^*UP = PEP,$$

where E is the projection from $\mathbf{H}$ onto the initial space of U. Since that initial space is equal to $\text{ker}^\perp U$, and hence to $\overline{\text{ran}}\, P$, it follows that $EP = P$, and hence that $A^*A = P^2$. Since the equation $UPf = Af$ uniquely determines U for f in ran P, and since $Uf = 0$ when f is in ker P, it follows that U also is uniquely determined by the stated conditions.

To deduce Corollary 1, multiply $A = UP$ on the left by U^*, and use the equation $U^*U = E$; cf. Solution 127. For Corollary 2, observe that ker U = ker P = ker A^*A = ker A, and ker U^* = $\text{ran}^\perp U$ = $\text{ran}^\perp A$.

262

Solution 135. Suppose that A is a bounded linear transformation from a **135** Hilbert space **H** to a Hilbert space **K**, let $A = UP$ be the polar decomposition of A, let **M** ($\subset$ **H**) be the initial space of the partial isometry U, and let **R** ($\subset$ **K**) be the range of U (or, equivalently, the closure of the range of A). If dim $\mathbf{M}^{\perp} \leqq$ dim $\mathbf{R}^{\perp}$, then there exist isometries from **H** into **K** that agree with U on **M** (many of them); all that is needed is to map $\mathbf{M}^{\perp}$ isometrically into $\mathbf{R}^{\perp}$ and to combine such a mapping with what U does on **M**. If, on the other hand, dim $\mathbf{R}^{\perp} \leqq$ dim $\mathbf{M}^{\perp}$, then there exist isometries from **K** into **H** that agree with U^{*} on **R**; the adjoint of each such isometry is a co-isometry from **H** into **K** that agrees with U on **M**. In either case there exists a linear transformation V from **H** into **K** such that either V or V^{*} is an isometry and such that V agrees with U on **M**. Since the range of P is included in **M**, it follows that $VP = UP = A$.

Solution 136. *The extreme points of the unit ball in the space of operators* **136** *on a Hilbert space are the maximal partial isometries.*

PROOF. Suppose first that U is an isometry and that $U = \alpha A + \beta B$, with $\alpha > 0$, $\beta > 0$, $\alpha + \beta = 1$, $\|A\| \leqq 1$, and $\|B\| \leqq 1$. If f is a unit vector, then so is Uf, and $Uf = \alpha Af + \beta Bf$, where $\|Af\| \leqq 1$ and $\|Bf\| \leqq 1$. Since the closed unit ball of a Hilbert space is strictly convex (Problem 4), it follows that $Af = Bf = Uf$, and hence that $A = B = U$. Conclusion: isometries are extreme points. The result for co-isometries is an immediate consequence.

The converse can be proved by showing that every operator A, with $\|A\| \leqq 1$, is equal to a convex combination (in fact, to the average) of two extreme points of the kind already found. Here the theory of polar decompositions (or, rather, a consequence of it) is useful. By Problem 135, it is possible to write $A = VP$, where V is a maximal partial isometry and $0 \leqq P \leqq 1$. (The justification for the upper bound on P is that $\|A\| \leqq 1$.) Assertion: there exists a unitary operator W such that $P = \frac{1}{2}(W + W^{*})$. (The assertion is true and the proof below is valid whenever $-1 \leqq P \leqq 1$; in case the underlying Hilbert space is one-dimensional, then both the assertion and its proof make simple geometric sense.) To prove the assertion, just write

$$W = P + i\sqrt{1 - P^{2}},$$

and verify that everything works. Now, since $A = VP$ and $P = \frac{1}{2}(W + W^{*})$, it follows that $A = \frac{1}{2}(VW + VW^{*})$. Since the product of a maximal partial isometry and a unitary operator is a maximal partial isometry, the proof is complete.

Kadison [82] has proved that, for certain operator algebras, the extreme points in the unit ball of the algebra are those partial isometries U that satisfy the identity

$$(1 - U^{*}U)A(1 - UU^{*}) = 0$$

for all A in the algebra. For the algebra of all operators on a Hilbert space this is consistent with what was just proved. It is, indeed, clear that if either U or U^* is an isometry, then the Kadison condition is satisfied. Suppose, conversely, that the condition is satisfied, and assume that $1 - UU^* \neq 0$; it is to be proved that $1 - U^*U = 0$. In other words, it is to be proved that if $(1 - UU^*)f \neq 0$ for some f, then $(1 - U^*U)g = 0$ for each g. That is easy: given g, find an operator A such that $A(1 - UU^*)f = g$.

137 **Solution 137.** Write $UP = A$. If U commutes with P, then U commutes with P^2; since P also commutes with P^2, it follows that $A \ (=UP)$ commutes with $A^*A \ (=P^2)$. .

The converse is harder. If A is quasinormal, then A commutes with $P^2 \ (=A^*A)$. It follows from the most elementary aspects of the functional calculus that A commutes with P. (Compare Problem 121, which shows that the positive square root of a positive operator is the weak limit of a sequence of polynomials in that operator. Alternatively, apply the Weierstrass theorem on the approximation of continuous functions by polynomials to prove that "weak" can be replaced by "uniform".) This says that $(UP - PU)P = 0$, so that $UP - PU$ annihilates ran P. Since ker $P =$ ker U, it is trivial that $UP - PU$ annihilates ker P also, and it follows that $UP - PU = 0$.

138 **Solution 138.** If $A = \begin{pmatrix} 0 & 0 \\ 1 & 0 \end{pmatrix}, f = \begin{pmatrix} 1 \\ 0 \end{pmatrix}$, and $g = \begin{pmatrix} 0 \\ 1 \end{pmatrix}$, then $(Af, g) = 1$, but if each of X and Y is one of $\sqrt{A^*A}$ and $\sqrt{AA^*}$, then, three times out of four, $(Xf, f) \cdot (Y g, g) = 0$. The fourth possibility is a "mixed" Schwarz inequality that is always true.

To prove it, let $A = UP$ be the polar decomposition, with $P = \sqrt{A^*A}$, and put $Q = \sqrt{AA^*}$. Since $AA^* = (UP)(PU^*) = U(A^*A)U^*$, or, in other words,

$$Q^2 = UP^2U^*,$$

it follows that

$$Q = UPU^*.$$

Reason: note first that

$$U^*UP^2 = U^*AP = P^2$$

(by Corollary 1 of Problem 134), and therefore

$$Q^4 = UP^2U^*UP^2U^* = UP^4U^*,$$

and, inductively,

$$Q^{2n} = UP^{2n}U^*$$

for all $n = 0, 1, 2, \cdots$, and then infer that $F(Q^2) = UF(P^2)U^*$ for every polynomial F. (Compare Solution 137.) The rest is straightforward:

$$|(Af, g)|^2 = |(UPf, g)|^2 = |(Pf, U^*g)|^2$$
$$\leq (Pf, f) \cdot (PU^*g, U^*g) = (Pf, f) \cdot (UPU^*g, g)$$
$$= (Pf, f) \cdot (Qg, g).$$

This elegant application of polar decomposition was shown me by John Duncan.

Solution 139. Is the question about unilateral or bilateral shifts? It turns out that it makes no difference.

139

Suppose that $\{e_n\}$ is an orthonormal basis and $Ae_n = \alpha_n e_{n+1}$ for all n. Assume with no loss that $\alpha_n \geq 0$ (see Problem 89). Since $A^*Ae_n = \alpha_n^2 e_n$, it follows that $(A^*A)Ae_n = \alpha_n \alpha_{n+1}^2 e_{n+1}$ and $A(A^*A)e_n = \alpha_n^3 e_{n+1}$. Consequence: A is quasinormal if and only if $\alpha_n \alpha_{n+1}^2 = \alpha_n^3$ for all n. If $\alpha_n \neq 0$ for some n, then $\alpha_{n+1}^2 = \alpha_n^2$, and therefore $\alpha_{n+1} = \alpha_n$; it follows by induction that $\alpha_{n+k} = \alpha_n$ for all positive k. This implies that either there is a first n for which $\alpha_n \neq 0$, and in that case the α's are constantly equal to that α_n from then on, and 0 before, or else $\alpha_n = 0$ for all n. If, in particular, α_n is never 0, then A is just a scalar multiple of the unweighted shift.

Solution 140. By Problem 135, every operator has the form VP, where V is a maximal partial isometry and P is positive. Given a positive number ε, find an invertible operator Q (which can be made positive, if so desired) such that $\|P - Q\| < \varepsilon$. It follows that $\|VP - VQ\| < \varepsilon$. The proof of the density theorem for unilaterally invertible operators is completed by observing that if V is a maximal partial isometry, then V is unilaterally invertible (left-invertible if V is an isometry and right-invertible if V^* is one), and that the product of a unilaterally invertible operator and an invertible operator is unilaterally invertible.

140

To obtain the negative conclusion, consider an operator A that is left-invertible but not right-invertible. (Example: the unilateral shift.) Assertion: there is a neighborhood of A that contains no right-invertible operators. Assume (with no loss of generality) that A has a left inverse B, with $\|B\| \leq 1$. In the presence of this normalization, the assertion can be made more precise: the open ball with center A and radius 1 contains no right-invertible operators.

Now, for the proof, observe first that B is right-invertible (with right inverse A), but not left-invertible. (If both $BA = 1$ and $CB = 1$, then B is invertible and $C = A = B^{-1}$.) It is to be proved that if $\|A - T\| < 1$, then T is not right-invertible. Indeed:

$$\|1 - BT\| = \|B(A - T)\| \leq \|A - T\| < 1,$$

and hence BT is invertible; this implies that T is left-invertible. If T is right-invertible also, then it is invertible; the invertibility of both BT and T implies that of B, which is false.

141   **Solution 141.** One way to approach the proof is to show that for each invertible operator A there is a continuous curve that connects it to the identity. For this purpose, consider the polar decomposition UP of A. Since A is invertible, so also are U and P. It follows that U is unitary and P is strictly positive. Join U to 1 by a continuous curve $t \mapsto U_t$ of unitary operators (cf. Problem 131), and, similarly, join P to 1 by a continuous curve $t \mapsto P_t$ of strictly positive operators. (The latter does not even need the spectral theorem; consider $tP + (1 - t)$, $0 \leqq t \leqq 1$.) If $A_t = U_t P_t$, then $t \mapsto A_t$ is a continuous curve of invertible operators that joins A $(= A_1)$ to 1 $(= A_0)$.

CHAPTER 17

Unilateral Shift

Solution 142. If $\mathbf{H}$ is not separable, then it is the direct sum of separable **142**
infinite-dimensional subspaces that reduce A, and, consequently, there is no
loss of generality in assuming that $\mathbf{H}$ is separable in the first place. In a
separable Hilbert space all infinite-dimensional subspaces have the same
dimension; the assertion, therefore, is just that $\mathbf{H}$ is the direct sum of $\aleph_0$
infinite-dimensional subspaces that reduce A. It is sufficient to prove the
assertion for 2 in place of $\aleph_0$. In other words, it is sufficient to prove that
*for each normal operator on a separable infinite-dimensional Hilbert space
there exists a reducing subspace such that both it and its orthogonal complement are infinite-dimensional.* Indeed, if this is true, then there exists a re-
ducing subspace $\mathbf{H}_1$ of $\mathbf{H}$ such that both $\mathbf{H}_1$ and $\mathbf{H}_1^\perp$ are infinite-dimensional.
Another application of the same result (consider the restriction of A to $\mathbf{H}_1^\perp$)
implies that there exists a reducing subspace $\mathbf{H}_2$ of $\mathbf{H}_1^\perp$ such that both $\mathbf{H}_2$
and $\mathbf{H}_1^\perp \cap \mathbf{H}_2^\perp$ are infinite-dimensional. Proceed inductively to obtain an
infinite sequence $\{\mathbf{H}_n\}$ of pairwise orthogonal infinite-dimensional reducing
subspaces. If the intersection $\bigcap_{n=1}^\infty \mathbf{H}_n^\perp$ is not zero, adjoin it to, say, $\mathbf{H}_1$.

It remains to prove the assertion italicized above. The spectral theorem
shows that there is no loss of generality in restricting attention to a multiplication operator A induced by a bounded measurable function φ on some
measure space. For each Borel subset M of the complex plane, let $E(M)$
be the multiplication operator induced by the characteristic function of
$\varphi^{-1}(M)$; the operator $E(M)$ is the projection onto the subspace of functions that vanish outside $\varphi^{-1}(M)$. Clearly each $E(M)$ commutes with A,
i.e., the range of each $E(M)$ reduces A. If, for some M, both $E(M)$ and
$1 - E(M)$ have infinite-dimensional ranges, the desired assertion is true.

In the contrary case what must happen is that for each M either $E(M)$ or
$1 - E(M)$ has finite rank. Draw a sequence of finer and finer square grids on

the plane, and let each square in each grid play the role of M; it follows that if $E(M)$ has positive rank, then M contains at least one point λ such that $E(\{\lambda\})$ has positive rank. There cannot be more than finitely many λ's like that, for then they could be separated into two infinite subsets, and that would contradict the main assumption of this paragraph. Conclusion: there exists at least one point λ such that the dimension of the range of $E(\{\lambda\})$ is infinite; let $\mathbf{M}$ be that range. The restriction of A to $\mathbf{M}$ is a scalar and is therefore reduced by every subspace of $\mathbf{M}$. Split $\mathbf{M}$ into two infinite-dimensional subspaces $\mathbf{M}_0$ and $\mathbf{M}_1$; if $\mathbf{H}_0 = \mathbf{M}_0$ and $\mathbf{H}_1 = \mathbf{M}_1 \vee \mathbf{M}^\perp$, then $\mathbf{H}_0$ and $\mathbf{H}_1$ do everything that is required.

143   **Solution 143.** *Every unitary operator on an infinite-dimensional Hilbert space is the product of four symmetries; three is not always enough.*

If the underlying Hilbert space $\mathbf{H}$ is finite-dimensional, then the concept of determinant makes sense. Since the determinant of a symmetry is ± 1, it follows that no unitary operator with a non-real determinant can be the product of symmetries.

PROOF. Suppose that $\mathbf{H}$ is an infinite-dimensional Hilbert space, and begin by representing $\mathbf{H}$ as the direct sum of a sequence $\{\mathbf{H}_n\}$ of equi-dimensional subspaces each of which reduces the given unitary operator U (Problem 142). It is convenient to let the index n run through all integers, positive, negative, and zero.

Relative to the fixed direct sum decomposition $\mathbf{H} = \sum_n \mathbf{H}_n$, define a *right shift* as a unitary operator S such that $S\mathbf{H}_n = \mathbf{H}_{n+1}$, and define a *left shift* as a unitary operator T such that $T\mathbf{H}_n = \mathbf{H}_{n-1}$, $n = 0, \pm 1, \pm 2, \cdots$. The equi-dimensionality of all the $\mathbf{H}_n$'s guarantees the existence of shifts. If S is an arbitrary right shift, write $T = S^*U$. Since $T\mathbf{H}_n = S^*U\mathbf{H}_n = S^*\mathbf{H}_n = \mathbf{H}_{n-1}$ for all n, it follows that T is a left shift. Since $U = ST$, it follows that every unitary operator is the product of two shifts; the proof will be completed by showing that every shift is the product of two symmetries.

Since the inverse (equivalently, the adjoint) of a left shift is a right shift, it is sufficient to consider right shifts. Suppose then that S is a right shift; let P be the operator that is equal to S^{1-2n} on $\mathbf{H}_n$ and let Q be the operator that is equal to S^{-2n} on $\mathbf{H}_n$ ($n = 0, \pm 1, \pm 2, \cdots$). If $f \in \mathbf{H}_n$, then $Qf = S^{-2n}f \in S^{-2n}\mathbf{H}_n = \mathbf{H}_{-n}$, so that $PQf = PS^{-2n}f = S^{1-2(-n)}S^{-2n}f = Sf$. The existence proof is complete.

To prove that on every Hilbert space there exists a unitary operator that is not the product of three symmetries, let ω be a non-real cube root of unity, and let U be $\omega 1$. The operator U belongs to the center of the group of all unitary operators; the order of U in that group is exactly three. The remainder of the proof has nothing to do with operator theory; the point is that in no group can a central element of order 3 be the product of three elements of order 2. Suppose indeed that u is central and that $u = xyz$, where $x^2 = y^2 = z^2 = 1$; it follows that

$$u^4 = uxuyuz = u(xu)y(uz) = u(yz)y(xy)$$
$$= y(uz)y(xy) = yxy \cdot yxy = 1.$$

Reference: [62].

Solution 144. (a) *The unilateral shift is not the product of a finite number of normal operators.* (b) *The norm of both the real and the imaginary part of the unilateral shift is 1.* (c) *The distance from the unilateral shift to the set of normal operators is 1.*

144

PROOF. (a) The principal tool is the observation that if a normal operator has a one-sided inverse, then it has an inverse. (Proof: for every operator, left invertibility is the same as boundedness from below; boundedness from below for a normal operator is the same as boundedness from below for its adjoint.) Suppose, indeed, that $U = A_1 \cdots A_n$, where U is the unilateral shift and $A_1, \cdots, A_n$ are normal. Since $U^* = A_n^* \cdots A_1^*$, it follows that

$$A_n^* \cdots A_1^* A_1 \cdots A_n = 1,$$

and hence that A_n is left invertible. In view of the preceding comments, this implies that A_n is invertible, and therefore so is A_n^*. Invertible operators can be peeled off either end of a product without altering its invertibility character. It follows by an obvious inductive repetition of the argument that each of the A's is invertible, and so therefore is U. This is a contradiction, and the proof is complete.

(b) If U is the unilateral shift, and if $A = \frac{1}{2}(U + U^*)$, then it is clear that $\|A\| \leqq 1$. Since 1 is an approximate eigenvalue of U, there exists a sequence $\{f_n\}$ of unit vectors such that $Uf_n - f_n \to 0$. Apply U^* and change sign to get $U^*f_n - f_n \to 0$. Add and divide by 2 to get $Af_n - f_n \to 0$. Conclusion: 1 is an approximate eigenvalue of A, and therefore $\|A\| \geqq 1$. To get the result for the imaginary part, note that if $U = A + iB$, then $-iU = B - iA$, and $-iU$ is unitarily equivalent to U (cf. Problem 89).

(c) It is trivial that there is a normal operator (namely 0) within 1 of U; the less trivial part of the assertion is that if A is normal, then $\|U - A\| \geqq 1$. If A is invertible, this follows from Solution 140; the assertion there implies that the open ball with center U and radius 1 contains no invertible operators. The general case is now immediate: the set of invertible normal operators is dense in the set of all normal operators.

Solution 145. *The unilateral shift has no square root.*

145

PROOF. It turns out that U^* is easier to treat than U, and, of course, it comes to the same thing. Suppose therefore that $V^2 = U^*$, and let $\mathbf{N}_0$ be the (one-dimensional) kernel of U^*. Since $\ker V \subset \ker V^2 = \mathbf{N}_0$, it follows that

269

dim ker $V \leqq 1$. If the kernel of V were trivial (zero-dimensional), then the same would be true of U^*; it follows that dim ker $V = 1$, and hence that ker $V = N_0$. Since U^* maps the underlying Hilbert space onto itself, the same must be true of V. It follows in particular that N_0 is included in the range of V, and hence that there exists a vector f such that Vf is a non-zero element of N_0. Since N_0 is the kernel of V, this implies that $V^2f = 0$, i.e., that $U^*f = 0$, and hence that $f \in N_0$. Do it again: since N_0 is the kernel of V, this implies that $Vf = 0$, in contradiction to the way f was chosen in the first place. Conclusion: there is no such V.

Similar negative results were first obtained in [64]; the techniques used there would serve here too. The very much simpler proof given above is due to J. G. Thompson. Further interesting contributions to the square root problem were made in [34] and [126]. See also Problem 151.

146   **Solution 146.** It is obvious that every multiplication operator on L^2 commutes with W. If A is the multiplication operator induced by a bounded measurable function φ, then

$$Ae_0 = \varphi \cdot e_0 = \varphi.$$

This shows that in any attempt to prove that some operator A is a multiplication on L^2 there is no choice in the determination of the multiplier; if there is one, it must be Ae_0.

Suppose now that $AW = WA$, and put $\varphi = Ae_0$. The first (and in fact the major) difficulty is to prove that φ is bounded; one way to do it is this. If ψ is an arbitrary bounded measurable function, and if B is the multiplication operator it induces, then, in the usual sense of the functional calculus for normal operators, $B = \psi(W)$. Since W commutes with A, every function of W commutes with A, and hence, in particular, B commutes with A; it follows that

$$\varphi \cdot \psi = \psi \cdot \varphi = B\varphi = BAe_0 = ABe_0 = A\psi.$$

The statement that every function of W commutes with A is not trivial; it is the Fuglede commutativity theorem for normal operators. (See [50, p. 68] and Problem 192.) It is not necessary in this argument to use all bounded measurable functions; it would be sufficient to use trigonometric polynomials (i.e., finite linear combinations of the e_n's). That way the Fuglede theorem can be avoided; all that is needed is to observe that if W commutes with A, then $W^* (= W^{-1})$ commutes with A.

At this point Problem 65 is almost applicable. The hypothesis there was that A is an operator on L^2 such that $Af = \varphi \cdot f$ for all f in L^2; the situation here is that A is an operator on L^2 such that $A\psi = \varphi \cdot \psi$ for all bounded measurable ψ. The difference is large enough to invalidate one of the proofs that worked there, but not large enough to invalidate the second, more "natural" proof. Conclusion: φ is bounded.

The rest of the proof is trivial. Since φ is bounded, it induces a multiplication operator; since that multiplication operator agrees with A on the dense set of all bounded functions, it agrees with A everywhere.

To prove the corollary, note that if a multiplication is a projection, then the multiplier is a characteristic function.

Solution 147. As in Solution 146, it is inevitable to put $\varphi = Ae_0$ and to try to prove that φ is the desired multiplier. Since, for each n, multiplication by e_n leaves $\mathbf{H}^2$ invariant ($n = 0, 1, 2, \cdots$), it follows that $\varphi \cdot e_n \in \mathbf{H}^2$. Since, moreover,

$$\varphi \cdot e_n = e_n \cdot \varphi = U^n \varphi = U^n A e_0 = A U^n e_0 = A e_n,$$

it follows that, for each polynomial p, the product $\varphi \cdot p$ belongs to $\mathbf{H}^2$ and $\varphi \cdot p = Ap$. If φ were known to be bounded, the proof would be over (the multiplication operator induced by φ agrees with A on a dense set), and, if it were known that $\varphi \cdot f = Af$ for all f in $\mathbf{H}^2$, then φ would be bounded (cf. the last comment in Solution 65). Since at the moment neither of these ifs is known, there is nothing for it but to prove something. The least troublesome way seems to be to adapt (or, to put it bluntly, to repeat) the second proof used in Solution 66.

If $f \in \mathbf{H}^2$, then there exist polynomials p_n such that $p_n \to f$ in $\mathbf{H}^2$; it follows, of course, that $Ap_n \to Af$ in $\mathbf{H}^2$. There is no loss of generality in assuming that $p_n \to f$ almost everywhere and $Ap_n \to Af$ almost everywhere; if this is not true for the sequence $\{p_n\}$, it is true for a suitable subsequence. Since $p_n \to f$ almost everywhere, it follows that $\varphi \cdot p_n \to \varphi \cdot f$ almost everywhere; since, at the same time, $\varphi \cdot p_n \to Af$ almost everywhere, it follows that $\varphi \cdot f = Af$ almost everywhere.

There are two ideas in this twice used proof: (1) if a closed transformation agrees with a bounded one on a dense set, then it is bounded, and (2) multiplications are always closed.

The corollary is equivalent to this: if E is a projection that commutes with U, then $E = 0$ or $E = 1$. The result proved above implies that E is the restriction to $\mathbf{H}^2$ of a multiplication, where the multiplier itself is in $\mathbf{H}^\infty$. Since an idempotent multiplication on $\mathbf{H}^2$ must be induced by an idempotent multiplier (apply to e_0), the multiplier must be the characteristic function of a set, and hence, in particular, real; the desired conclusion follows from Problem 33.

The corollary, incidentally, does not have to be deduced from the main assertion; for an easy direct proof see [50, p. 41].

Solution 148. Let U be the unilateral shift, represented as the restriction to $\mathbf{H}^2$ of the multiplication induced by e_1; see, for instance, Problem 147. If A commutes with U, then (by Problem 147) there exists a function φ in $\mathbf{H}^\infty$ such that $Af = \varphi \cdot f$ for all f in $\mathbf{H}^2$. The crucial tool is that φ is the limit almost everywhere of a sequence $\{p_n\}$ of polynomials such that $\|p_n\|_\infty \leq \|\varphi\|_\infty$ for every n; cf. Solution 41. It follows that if $f \in \mathbf{H}^2$, then

$$|p_n(z)f(z)| \leq \|\varphi\|_\infty \cdot |f(z)|$$

almost everywhere. Since $p_n \cdot f \to \varphi \cdot f$ almost everywhere, the Lebesgue dominated convergence theorem applies; the conclusion is that $p_n \cdot f \to \varphi \cdot f$ in $\mathbf{H}^2$. Since multiplication by p_n is a polynomial in U (namely $p_n(U)$), the proof is complete.

149 **Solution 149.** The only way an isometry V on a Hilbert space $\mathbf{H}$ can fail to be unitary is to map $\mathbf{H}$ onto a proper subspace of $\mathbf{H}$. This suggests that the extent to which $V\mathbf{H}$ differs from $\mathbf{H}$ is a useful measure of the non-unitariness of V. One application of V compresses $\mathbf{H}$ into $V\mathbf{H}$, another application of V compresses $V\mathbf{H}$ into $V^2\mathbf{H}$, and so on. The incompressible core of $\mathbf{H}$ seems to be what is common to all the $V^n\mathbf{H}$'s. This is true, and it is the crux of the matter; the main thing to prove is that that incompressible core reduces the operator V. A slightly sharper result is sometimes useful; it is good to know exactly what the orthogonal complement of that core is. Write $\mathbf{N} = (V\mathbf{H})^\perp$; in terms of $\mathbf{N}$ the main result is that

$$\bigcap_{n=0}^\infty V^n\mathbf{H} = \bigcap_{n=0}^\infty (V^n\mathbf{N})^\perp.$$

Both the statement (and the proof below) become intuitively obvious if orthogonal complements are replaced by ordinary set-theoretic complements. (A picture helps.)

Begin with the observation that $V\mathbf{M}^\perp \subset (V\mathbf{M})^\perp$ for all subspaces $\mathbf{M}$. (Indeed, if $f \in \mathbf{M}^\perp$, so that Vf is a typical element of $V\mathbf{M}^\perp$, and if $g \in \mathbf{M}$, so that Vg is a typical element of $V\mathbf{M}$, then $Vf \perp Vg$ follows, since V is an isometry, from $f \perp g$.) This implies that

$$V^{n+1}\mathbf{H} = V^n(V\mathbf{H}) = V^n\mathbf{N}^\perp \subset (V^n\mathbf{N})^\perp,$$

and that settles half the proof. For the reverse inclusion, assume that $f \in \bigcap_{n=0}^\infty (V^n\mathbf{N})^\perp$ and prove by induction that $f \in V^n\mathbf{H}$ for all n. If $n = 0$, this is trivial. If $f \in V^n\mathbf{H}$, so that $f = V^n g$ for some g, then $V^n g \perp V^n\mathbf{N}$ (since $f \in (V^n\mathbf{N})^\perp$), and therefore $g \perp \mathbf{N}$. This implies that $g \in V\mathbf{H}$, and hence that $f \in V^{n+1}\mathbf{H}$, as desired. The proof of the asserted equation is complete.

The rest is easy. Obviously $\bigcap_{n=0}^\infty V^n\mathbf{H}$ is invariant under V; since, by the result just proved, its orthogonal complement is equal to $\bigvee_{n=0}^\infty V^n\mathbf{N}$, which is also invariant under V, it follows that $\bigcap_{n=0}^\infty V^n\mathbf{H}$ reduces V. The restriction of V to this reducing subspace is unitary (because it is an isometry whose range is equal to its domain). The restriction of V to the orthogonal complement $\bigvee_{n=0}^\infty V^n\mathbf{N}$ is a direct sum of copies of the unilateral shift; the number of copies is dim $\mathbf{N}$.

Solution 150. *If U is the unilateral shift, then $\|U - V\| = 2$ for each* **150**
unitary operator V.

PROOF. The proof begins with the observation that if -1 belongs to the
spectrum of an operator A, then -2 belongs to the spectrum of $A - 1$.
It follows that if A is a non-normal (i.e., non-unitary) isometry, then

$$r(A - 1) \geqq 2,$$

and hence $\|A - 1\| \geqq 2$. (Use Problem 149, and recall that the spectrum
of the unilateral shift is the closed unit disc.) If V is unitary, then $\|U - V\| =$
$\|V^*U - 1\|$. Since V^*U is a non-normal isometry, it follows that
$\|U - V\| \geqq 2$; the reverse inequality is trivial.

This is a geometrically very peculiar result. The unilateral shift is on the
unit sphere of the space of operators, and so also is each unitary operator.
What was just proved can be expressed in geometric language by saying that
if V is unitary, then U and V are diametrically opposite; they are as far from
each other as if they were at the opposite ends of a diameter. What is peculiar
is that this is true for every V.

Solution 151. *A unilateral shift of multiplicity m has a square root if* **151**
and only if m is even. (For the purpose of this assertion every infinite
cardinal number m is even; recall that $m = 2m$.)

The "if" is obvious: if $m = 2n$, then the square of a unilateral shift of
multiplicity n is a unilateral shift of multiplicity m, whether n is finite or
infinite.

The "only if" part of the statement is more delicate; it is the negative
assertion that if U is a unilateral shift of (finite) odd multiplicity, then U has
no square root. For the proof it is convenient ro recall and generalize a
classical and important finite-dimensional fact.

If, for every operator A, null A (the nullity of A) is dim ker A (cf. Problem
130), then the assertion of Sylvester's law of nullity is that

$$\text{null}(AB) \leqq \text{null } A + \text{null } B.$$

The proof can be arranged so as to work equally well whether the dimension
of the underlying space is finite or infinite. Indeed:

$$B^{-1}(\text{ker } A) = \text{ker } B + (B^{-1}(\text{ker } A) \cap \text{ker}^{\perp} B),$$

and the restriction of B to the second summand maps that summand one-to-
one into ker A.

Suppose now that U is a unilateral shift of multiplicity m and that V is a
square root of U^*. Since $m = $ null U^*, the preceding paragraph implies that

$$m \leqq 2 \text{ null } V.$$

273

The idea of the rest of the proof is to show that if m is finite, then equality must hold, and hence that m cannot be odd. What is wanted is therefore a special case of the following statement, which is an odd kind of infinite-dimensional reverse of Sylvester's law.

Lemma. *If A is an operator of finite nullity on a Hilbert space* $\mathbf{H}$, *with* ran $A = \mathbf{H}$, *then* null $A^2 \geqq 2$ null A.

PROOF. If $\{f_1, \cdots, f_m\}$ is a linear basis for ker A, then, for each $j (= 1, \cdots, m)$, there exists a vector g_j with $Ag_j = f_j$. Since each of $f_1, \cdots, f_m, g_1, \cdots, g_m$ is in ker A^2, the proof can be completed by proving that that set of $2m$ vectors is linearly independent. To do that, suppose that

$$\alpha_1 f_1 + \cdots + \alpha_m f_m + \beta_1 g_1 + \cdots + \beta_m g_m = 0,$$

apply A, infer that

$$\beta_1 f_1 + \cdots + \beta_m f_m = 0,$$

so that $\beta_1 = \cdots = \beta_m = 0$, and then use the linear independence of the f_j's once more to conclude that $\alpha_1 = \cdots = \alpha_m = 0$.

152

Solution 152. Given: a Hilbert space $\mathbf{H}$ and on it a contraction A such that $A^n \to 0$ strongly. To construct: a Hilbert space $\tilde{\mathbf{H}}$ and on it a shift U with the stated unitary equivalence property. The construction is partially motivated by the following observation: if a vector f in $\tilde{\mathbf{H}}$ is replaced by Af, then the sequence

$$\langle f, Af, A^2 f, \cdots \rangle$$

is shifted back by one step, i.e., it is replaced by

$$\langle Af, A^2 f, A^3 f, \cdots \rangle.$$

What this suggests is that $\tilde{\mathbf{H}}$ be something like the direct sum

$$\mathbf{H} \oplus \mathbf{H} \oplus \mathbf{H} \oplus \cdots.$$

That does not work. There is no reason why the sequence $\langle f, Af, A^2 f, \cdots \rangle$ should belong to the direct sum (the series $\sum_{n=0}^{\infty} \| A^n f \|^2$ need not converge), and, even if it does, the correspondence between f and $\langle f, Af, A^2 f, \cdots \rangle$ may fail to be norm-preserving (even if $\sum_{n=0}^{\infty} \| A^n f \|^2$ converges, its sum will be equal to $\| f \|^2$ only in case $Af = 0$).

The inspiration that removes these difficulties is to transform each term of the sequence $\langle f, Af, A^2 f, \cdots \rangle$ by an operator T so that the resulting series of square norms converges to $\| f \|^2$ the easy way, by telescoping. That is: replace $\langle f, Af, A^2 f, \cdots \rangle$ by $\langle Tf, TAf, TA^2 f, \cdots \rangle$, so that

$$\| Tf \|^2 = \| f \|^2 - \| Af \|^2,$$

$$\| TAf \|^2 = \| Af \|^2 - \| A^2 f \|^2,$$

$$\| TA^2 f \|^2 = \| A^2 f \|^2 - \| A^3 f \|^2,$$

etc.

The first of these equations alone, if required to hold for all f, implies that $T^*T = 1 - A^*A$, and, conversely, if $T^*T = 1 - A^*A$, then all the equations hold.

The preceding paragraphs were intended as motivation. For the proof itself, proceed as follows. Since A is a contraction, $1 - A^*A$ is positive; write $T = \sqrt{1 - A^*A}$, and let $\mathbf{R}$ be the closure of the range of T. Let $\tilde{\mathbf{H}}$ be the direct sum $\mathbf{R} \oplus \mathbf{R} \oplus \mathbf{R} \oplus \cdots$. If $f \in \mathbf{H}$, then $TA^n f \in \mathbf{R}$ for all n, and

$$\sum_{n=0}^{k} \| TA^n f \|^2 = \sum_{n=0}^{k} ((1 - A^*A)A^n f, A^n f)$$

$$= \sum_{n=0}^{k} (\| A^n f \|^2 - \| A^{n+1} f \|^2)$$

$$= \| f \|^2 - \| A^{k+1} f \|^2.$$

Since $\| A^{k+1} f \| \to 0$ by assumption, it follows that if $f \in \mathbf{H}$, and if the mapping V is defined by

$$Vf = \langle Tf, TAf, TA^2 f, \ldots \rangle,$$

then V is an isometric embedding of $\mathbf{H}$ into $\tilde{\mathbf{H}}$. If U is the obvious shift on $\tilde{\mathbf{H}}$ ($U \langle f_0, f_1, f_2, \cdots \rangle = \langle 0, f_0, f_1, \cdots \rangle$), then, clearly, $VAf = U^*Vf$ for all f. Since the V image of $\mathbf{H}$ in $\tilde{\mathbf{H}}$ is invariant under U^*, the proof is complete.

Note that the multiplicity of the shift that the proof gives is equal to the rank of $1 - A^*A$, where "rank" is interpreted to mean the dimension of the closure of the range.

Solution 153. Suppose that A is an operator on a Hilbert space $\mathbf{H}$ such that $r(=r(A)) < 1$. Since $r = \lim_n \| A^n \|^{1/n}$, it follows that the power series $\sum_{n=0}^{\infty} \| A^n \| z^n$ converges in a disc with center 0 and radius $(=1/r)$ greater than 1. This implies that $\sum_{n=0}^{\infty} \| A^n \| < \infty$, and hence, all the more, that $\sum_{n=0}^{\infty} \| A^n \|^2 < \infty$. Let $\mathbf{H}_0$ be the Hilbert space obtained from $\mathbf{H}$ by redefining the inner product; the new inner product is given by

$$(f, g)_0 = \sum_{n=0}^{\infty} (A^n f, A^n g).$$

Since $|(A^n f, A^n g)| \leqq \| A^n f \| \cdot \| A^n g \| \leqq \| A^n \|^2 \cdot \| f \| \cdot \| g \|$, there is no difficulty about convergence. If $\| f \|_0^2 = (f, f)_0$, then

$$\| f \|^2 \leqq \| f \|_0^2 \leqq \left(\sum_{n=0}^{\infty} \| A^n \|^2 \right) \cdot \| f \|^2,$$

and that implies that the identity mapping I from $\mathbf{H}$ to $\mathbf{H}_0$ is an invertible bounded linear transformation. (This, incidentally, is what guarantees that

153

275

H_0 is complete.) If $A_0 = IAI^{-1}$, then A_0 is an operator on H_0, similar to A. If $f \neq 0$, then

$$\frac{\|A_0 f\|_0^2}{\|f\|_0^2} = \frac{\sum_{n=1}^{\infty} \|A^n f\|^2}{\|f\|^2 + \sum_{n=1}^{\infty} \|A^n f\|^2} = \frac{\sum_{n=1}^{\infty} (\|A^n f\|/\|f\|)^2}{1 + \sum_{n=1}^{\infty} (\|A^n f\|/\|f\|)^2}$$

$$\leqq \frac{\sum_{n=1}^{\infty} \|A^n\|^2}{1 + \sum_{n=1}^{\infty} \|A^n\|^2} < 1,$$

so that A_0 is a strict contraction. This implies that the powers of A_0 tend to zero not only strongly, but in the norm.

Corollary 1 is immediate from Problem 152, and Corollary 2 is implied by the proof (given above) that $\|A_0\| < 1$. For Corollary 3: if A is quasi-nilpotent, then

$$r\left(\frac{1}{\varepsilon} A\right) < 1$$

for every positive number ε; if $(1/\varepsilon)A$ is similar to the contraction C, then A is similar to εC.

Corollary 4 requires a little more argument. Clearly $r(A) = r(S^{-1}AS) \leqq \|S^{-1}AS\|$ and therefore $r(A) \leqq \inf_S \|S^{-1}AS\|$. To prove the reverse inequality, let t be a number in the open unit interval and write

$$B = \frac{t}{r(A)} A.$$

(If $r(A) = 0$, apply Corollary 3 instead.) Corollary 2 implies that $\|S^{-1}BS\| < 1$ for some S, so that $t \cdot \|S^{-1}AS\| < r(A)$. Infer that

$$t \cdot \inf_S \|S^{-1}AS\| \leqq r(A),$$

and then let t tend to 1.

154   **Solution 154.** The answer is no, not necessarily.

If an operator A is similar to a contraction, then it is *power bounded*, i.e., the sequence of numbers $\|A^n\|$ is bounded. Reason: if $A = S^{-1}CS$ with $\|C\| \leqq 1$, then $A^n = S^{-1}C^nS$ and $\|A^n\| \leqq \|S^{-1}\| \cdot \|S\|$.

In view of the preceding paragraph the negative answer can be proved by exhibiting an operator A with spectral radius 1 that is not power bounded. Here is one simple example: let X be a non-zero operator with $X^2 = 0$ and put $A = 1 + X$.

To understand power boundedness better it helps to look at the more complicated counterexample of weighted shifts. Under what conditions on a weighted shift A, with weight sequence $\{\alpha_0, \alpha_1, \alpha_2, \cdots\}$, is the spectral radius less than or equal to 1? The answer is known for all weight sequences (Solution 91), and is especially simple in case the α's are strictly positive and monotone decreasing: in that case

$$r(A) = \lim_n (\alpha_0 \cdots \alpha_{n-1})^{1/n}.$$

Under what conditions on the α's (still assumed to be strictly positive and monotone decreasing) is A power bounded? The answer to this question too is known (Solution 91): since $\|A^n\| = \alpha_0 \cdots \alpha_{n-1}$, the condition is that the sequence of partial products $\alpha_0 \cdots \alpha_{n-1}$ be bounded.

Can a bounded sequence of α's be chosen so that $\lim_n (\alpha_0 \cdots \alpha_{n-1})^{1/n} = 1$ and $\alpha_0 \cdots \alpha_{n-1} \to \infty$? That is a question in elementary analysis, and a few experiments will reveal that the answer is yes. The conditions are equivalent to

$$\frac{1}{n} \sum_{j=0}^{n-1} \log \alpha_j \to 0$$

and

$$\sum_{j=0}^{n-1} \log \alpha_j \to \infty$$

respectively. That is: $\log \alpha_n$ must be small, on the average, but the sums of the $\log \alpha_n$'s must be large. Both conditions are satisfied if

$$\log \alpha_n = \log\left(1 + \frac{1}{n+1}\right) = \log\left(\frac{n+2}{n+1}\right) = \log(n+2) - \log(n+1).$$

In that case the (telescoping) sums $\sum_{j=0}^{n-1} \log \alpha_j$ are equal to $\log(n+1)$.

Solution 155. The restriction of U to $\mathbf{M}$ is an isometry. If $\mathbf{N} = \mathbf{M} \cap (U\mathbf{M})^\perp$, then $\mathbf{N}$ is the orthogonal complement of the range of that restriction. Apply the result obtained in Solution 149 to that restriction to obtain   **155**

$$\bigcap_{n=0}^{\infty} U^n \mathbf{M} = \mathbf{M} \cap \bigcap_{n=0}^{\infty} (U^n \mathbf{N})^\perp.$$

Since $\bigcap_{n=0}^{\infty} U^n \mathbf{H}^2 = 0$, it follows that

$$\mathbf{M}^\perp \vee \bigvee_{n=0}^{\infty} U^n \mathbf{N} = \mathbf{H}^2.$$

Since, on the other hand, $U^n \mathbf{N} \subset U^n \mathbf{M} \subset \mathbf{M}$, it follows that

$$\bigvee_{n=0}^{\infty} U^n \mathbf{N} \subset \mathbf{M}.$$

The span of $\mathbf{M}^\perp$ and a proper subspace of $\mathbf{M}$ can never be the whole space. Conclusion:

$$\bigvee_{n=0}^{\infty} U^n \mathbf{N} = \mathbf{M}.$$

It remains to prove that $\dim \mathbf{N} = 1$. For this purpose it is convenient to regard the unilateral shift U as the restriction to $\mathbf{H}^2$ of the bilateral shift W

on the larger space $\mathbf{L}^2$. If f and g are orthogonal unit vectors in $\mathbf{N}$, then the set of all vectors of either of the forms $W^n f$ or $W^m g$ ($n, m = 0, \pm 1, \pm 2, \cdots$) is an orthonormal set in $\mathbf{L}^2$. (This assertion leans on the good behavior of wandering subspaces for unitary operators.) It follows that

$$2 = \|f\|^2 + \|g\|^2 = \sum_n |(f, e_n)|^2 + \sum_m |(g, e_m)|^2$$

$$= \sum_n |(f, W^n e_0)|^2 + \sum_m |(g, W^m e_0)|^2$$

$$= \sum_n |(W^{*n} f, e_0)|^2 + \sum_m |(W^{*m} g, e_0)|^2$$

$$\leqq \|e_0\|^2 = 1.$$

(The inequality is Bessel's.) This absurdity shows that f and g cannot co-exist. The dimension of $\mathbf{N}$ cannot be as great as 2; since it cannot be 0 either, the proof is complete.

The last part of the proof is due to I. Halperin; see [99, p.108]. It is geometric; the original proof in [54] was analytic. See also [116].

156   **Solution 156.** Since $\mathbf{M}_k^\perp(\lambda)$ is spanned by $f_\lambda, \cdots, U^{k-1} f_\lambda$, it is clear that $\dim \mathbf{M}_k^\perp(\lambda) \leqq k$. To prove equality, note first that $\mathbf{M}_k^\perp(\lambda)$ is invariant under U^*. (Indeed $U^* f_\lambda = \lambda f_\lambda$ and, if $j \geqq 1$, $U^* U^j f_\lambda = U^* U U^{j-1} f_\lambda = U^{j-1} f_\lambda$. Note that this proves the invariance of $\mathbf{M}_k(\lambda)$ under U.) If $\dim \mathbf{M}_k^\perp(\lambda) < k$, then $\sum_{i=0}^{k-1} \alpha_i U^i f_\lambda = 0$ for suitable scalars α_i, or, in other words, there exists a polynomial p of degree less than k such that $p(U) f_\lambda = 0$. This implies that $U^n f_\lambda$ is a linear combination of $f_\lambda, \cdots, U^{k-1} f_\lambda$ for all n, and hence that $\mathbf{M}_k^\perp(\lambda)$ is invariant under U also. This is impossible, and therefore $\dim \mathbf{M}_k^\perp(\lambda) = k$.

Since

$$f_\lambda - \lambda U f_\lambda = \sum_{n=0}^{\infty} \lambda^n e_n - \lambda \sum_{n=1}^{\infty} \lambda^{n-1} e_n = e_0,$$

it follows that $e_0 \in \mathbf{M}_k^\perp(\lambda)$ as soon as $k > 1$. This implies that

$$U^j f_\lambda - \lambda U^{j+1} f_\lambda = U^j e_0 = e_j \in \mathbf{M}_k^\perp(\lambda)$$

as soon as $k > j + 1$, and, consequently, $\bigvee_{k=1}^{\infty} \mathbf{M}_k^\perp(\lambda)$ contains all e_j's.

157   **Solution 157.** If $\mathbf{M} = \varphi \cdot \mathbf{H}^2$, then $U\mathbf{M} = e_1 \cdot \mathbf{M} = e_1 \cdot \varphi \cdot \mathbf{H}^2 = \varphi \cdot e_1 \cdot \mathbf{H}^2 \subset \varphi \cdot \mathbf{H}^2 = \mathbf{M}$; this proves the "if". For another proof of the same implication, use the theory of wandering subspaces. If $\mathbf{N}$ is the (one-dimensional) subspace spanned by φ, then $\mathbf{N}$ is wandering; the reason is that $(U^n \varphi, U^m \varphi) = \int e_n e_m^* \, d\mu = \delta_{nm}$. To prove "only if", suppose that $\mathbf{M}$ is invariant under U and use Problem 155 to represent $\mathbf{M}$ in the form $\bigvee_{n=1}^{\infty} U^n \mathbf{N}$, where $\mathbf{N}$ is a wandering subspace for U. Take a unit vector φ in $\mathbf{N}$. Since, by assumption, $(U^n \varphi, \varphi) = 0$ when $n > 0$, or $\int e_n |\varphi|^2 \, d\mu = 0$ when $n > 0$, it follows (by the formation of complex conjugates) that $\int e_n |\varphi|^2 \, d\mu = 0$ when $n < 0$, and hence that $|\varphi|^2$ is a function in $\mathbf{L}^1$ such that all its Fourier coefficients with non-zero index vanish. Conclusion: $|\varphi|$ is constant almost everywhere, and, since $\int |\varphi|^2 \, d\mu = 1$, the constant modulus of φ must be 1. (Note that the preceding argument contains a proof, different in appearance from the one used in Solution 155, that every non-zero wandering subspace of U is one-dimensional.) Since φ, by itself, spans $\mathbf{N}$, the functions $\varphi \cdot e_n$ $(n = 0, 1, 2, \cdots)$ span $\mathbf{M}$. Equivalently, the set of all functions of the form $\phi \cdot p$, where p is a polynomial, spans $\mathbf{M}$. Since multiplication by φ (restricted to $\mathbf{H}^2$) is an isometry, its range is closed; since $\mathbf{M}$ is the span of the image under that isometry of a dense set, it follows that $\mathbf{M}$ is in fact equal to the range of that isometry, and hence that $\mathbf{M} = \varphi \cdot \mathbf{H}^2$.

To prove the first statement of Corollary 1, observe that if $\varphi \cdot \mathbf{H}^2 \subset \psi \cdot \mathbf{H}^2$, then $\varphi = \varphi \cdot e_0 = \psi \cdot f$ for some f in $\mathbf{H}^2$; since $f = \varphi \cdot \psi^*$, it follows that $|f| = 1$, so that f is an inner function. To prove the second statement, it is sufficient to prove that if both θ and θ^* are inner functions, then θ is a constant. To prove that, observe that both $\operatorname{Re} \theta$ and $\operatorname{Im} \theta$ are real functions in $\mathbf{H}^2$, and therefore (Problem 33) both $\operatorname{Re} \theta$ and $\operatorname{Im} \theta$ are constants. As for Corollary 2: if $\mathbf{M} = \varphi \cdot \mathbf{H}^2$ and $\mathbf{N} = \psi \cdot \mathbf{H}^2$, then $\varphi \cdot \psi \in \mathbf{M} \cap \mathbf{N}$.

Solution 158. Given f in $\mathbf{H}^2$, let $\mathbf{M}$ be the least subspace of $\mathbf{H}^2$ that contains f and is invariant under U. By Problem 157, either $f = 0$ or $\mathbf{M}$ contains a function φ such that $|\varphi| = 1$ almost everywhere. Since $p(U)f \in \mathbf{M}$ for every polynomial p, and since the closure of the set of all vectors of the form $p(U)f$ is a subspace of $\mathbf{H}^2$ that contains f and is invariant under U, it follows that φ is the limit in $\mathbf{H}^2$ of a sequence of vectors of the form $p(U)f$. Since every vector of that form vanishes at least when f does, it follows that φ vanishes when f does.

To prove the corollary, observe that if f does not vanish almost everywhere, then, by the F. and M. Riesz theorem, it vanishes almost nowhere, and therefore g must vanish almost everywhere.

Solution 159. Suppose first that $\{\alpha_n\}$ is periodic of period p $(= 1, 2, 3, \cdots)$, and let $\mathbf{M}_j$ $(j = 0, \cdots, p - 1)$ be the span of all those basis vectors e_n for which $n \equiv j \pmod{p}$. Each vector f has a unique representation in the form $f_0 + \cdots + f_{p-1}$ with f_j in $\mathbf{M}_j$. Consider the functional representation of the two-sided shift, and, using it, make the following definition. For each measurable subset E of the circle, let $\mathbf{M}$ $(= \mathbf{M}_E)$ be the set of all those f's for which $f_j(z) = 0$ whenever $j = 0, \cdots, p - 1$ and $z \notin E$. If $f = \sum_{j=0}^{p-1} f_j$ (with f_j in $\mathbf{M}_j$), then

$$Af = \sum_{j=0}^{p-1} \alpha_j W f_j$$

and

$$A^*f = \sum_{j=0}^{p-1} \alpha_{j-1} W^* f_j;$$

279

this proves that $\mathbf{M}$ reduces A. (Note that $W\mathbf{M}_j = \mathbf{M}_{j+1}$ and $W^*\mathbf{M}_j = \mathbf{M}_{j-1}$, where addition and subtraction are interpreted modulo p.)

To show that this construction does not always yield a trivial reducing subspace, let E_0 be a measurable set, with measure strictly between 0 and $1/p$, and let E be its inverse image under the mapping $z \mapsto z^p$. The set E is a measurable set, with measure strictly between 0 and 1. If g is a function that vanishes on the complement of E_0, and if $f_0(z) = g(z^p)$, then f_0 vanishes on the complement of E. If, moreover, $f_j(z) = z^j f_0(z)$, $j = 0, \ldots, p-1$, then the same is true of each f_j. Clearly $f_j \in \mathbf{M}_j$, and $f_0 + \cdots + f_{p-1}$ is a typical non-trivial example of a vector in $\mathbf{M}$. This completes the proof of the sufficiency of the condition.

Necessity is the surprising part. To prove it, suppose first that B is an operator, with matrix $\langle \beta_{ij} \rangle$, that commutes with A. Observe that

$$\beta_{i+1,j+1} = (Be_{j+1}, e_{i+1}) = \left(B\frac{1}{\alpha_j} Ae_j, e_{i+1} \right)$$

$$= \frac{1}{\alpha_j}(Be_j, A^*e_{i+1}) = \frac{\alpha_i}{\alpha_j} \beta_{ij}.$$

Consequence 1: the main diagonal of $\langle \beta_{ij} \rangle$ is constant (put $i = j$). Consequence 2: if $\beta_{ij} = 0$ for some i and j, then $\beta_{i+k,j+k} = 0$ for all k.

If B happens to be Hermitian, then it commutes with A^* also, and hence with A^*A. Since $A^*Ae_n = \alpha_n^2 e_n$, it follows that

$$\beta_{ij} = (Be_j, e_i) = \left(B\frac{1}{\alpha_j^2} A^*Ae_j, e_i \right)$$

$$= \frac{1}{\alpha_j^2}(Be_j, A^*Ae_i) = \frac{\alpha_i^2}{\alpha_j^2} \beta_{ij}.$$

Consequence 3: if $\alpha_i \neq \alpha_j$, then $\beta_{ij} = 0$.

Assume now that the sequence $\{\alpha_n\}$ is not periodic; it is sufficient to prove that every Hermitian B that commutes with A is a scalar. The assumption implies that if m and n are distinct positive integers, then there exist integers i and j such that $\alpha_i \neq \alpha_j$ and $i - j = m - n$. It follows that

$$\begin{aligned} 0 &= \beta_{ij} && \text{(by Consequence 3)} \\ &= \beta_{i-j+n,\,j-j+n} && \text{(by Consequence 2),} \end{aligned}$$

i.e., that $\beta_{mn} = 0$ whenever $m \neq n$. This says that the matrix of B is diagonal; by Consequence 1 it follows that B is a scalar.

CHAPTER 18

Cyclic Vectors

Solution 160. Consider first the simple unilateral shift U. Let $\langle \xi_0, \xi_1, \xi_2, \cdots \rangle$ **160** be a sequence of complex numbers such that

$$\lim_k \frac{1}{|\xi_k|^2} \sum_{n=1}^{\infty} |\xi_{n+k}|^2 = 0.$$

(Concrete example: $\xi_n = 1/n!$.) Assertion: $f = \langle \xi_0, \xi_1, \xi_2, \cdots \rangle$ is a cyclic vector for U^*. For the proof, observe first that $U^{*k}f = \langle \xi_k, \xi_{k+1}, \xi_{k+2}, \cdots \rangle$, and hence that

$$\left\| \frac{1}{\xi_k} U^{*k}f - e_0 \right\|^2 = \left\| \left\langle 1, \frac{\xi_{k+1}}{\xi_k}, \frac{\xi_{k+2}}{\xi_k}, \cdots \right\rangle - \langle 1, 0, 0, \cdots \rangle \right\|^2$$

$$= \left\| \left\langle 0, \frac{\xi_{k+1}}{\xi_k}, \frac{\xi_{k+2}}{\xi_k}, \cdots \right\rangle \right\|^2$$

$$= \frac{1}{|\xi_k|^2} \sum_{n=1}^{\infty} |\xi_{n+k}|^2 \to 0.$$

Consequence: e_0 belongs to the span of $f, U^*f, U^{*2}f, \cdots$. This implies that

$$U^{*k-1}f - \xi_{k-1}e_0 = \langle 0, \xi_k, \xi_{k+1}, \xi_{k+2}, \cdots \rangle$$

belongs to that span ($k = 1, 2, 3, \cdots$). Since

$$\left\| \frac{1}{\xi_k} (U^{*k-1}f - \xi_{k-1}e_0) - e_1 \right\|^2 = \frac{1}{|\xi_k|^2} \sum_{n=1}^{\infty} |\xi_{n+k}|^2 \to 0,$$

it follows that e_1 belongs to the span of $f, U^*f, U^{*2}f, \cdots$. An obvious inductive repetition of this twice-used argument proves that e_n belongs to the span of $f, U^*f, U^{*2}f, \cdots$ for all n ($= 0, 1, 2, \cdots$), and hence that f is cyclic.

281

Once this is settled, the cases of higher multiplicity turn out to be trivial. For multiplicity 2 consider the same sequence $\{\xi_n\}$ and form the vector

$$\langle\langle \xi_0, \xi_2, \xi_4, \cdots \rangle, \langle \xi_1, \xi_3, \xi_5, \cdots \rangle\rangle.$$

For higher finite multiplicities, and even for multiplicity $\aleph_0$, imitate this subsequence formation. Thus, for instance, a cyclic vector for the shift of multiplicity $\aleph_0$ is the vector whose i-th component is the i-th row of the following array:

$$
\begin{array}{cccccccccc}
\xi_0 & \xi_1 & 0 & \xi_3 & 0 & 0 & \xi_6 & 0 & 0 & 0 \\
0 & 0 & \xi_2 & 0 & \xi_4 & 0 & 0 & \xi_7 & 0 & 0 \\
0 & 0 & 0 & 0 & 0 & \xi_5 & 0 & 0 & \xi_8 & 0 \\
0 & 0 & 0 & 0 & 0 & 0 & 0 & 0 & 0 & \xi_9 \\
\end{array}
$$

(Rule: lengthen the diagonals. Each column contains only one non-zero entry; each row is an infinite subsequence of $\{\xi_n\}$.) The point is that each subseries of a series with the property that $\sum_{n=0}^{\infty} |\xi_n|^2$ has (the ratio of terms to tails tends to 0) has the same property.

161 **Solution 161.** If U is the unilateral shift and $V = U \oplus U$, then (Problem 160) V is not cyclic; in fact, no operator in the open ball with center V and radius 1 can be cyclic. Suppose, indeed, that $\| V - A \| < 1$. It follows that

$$\| 1 - V^*A \| = \| V^*V - V^*A \| \leqq \| V - A \| < 1,$$

and hence that V^*A is invertible, and hence that ran $A \cap \ker V^* = 0$. Since ran A is disjoint from the 2-dimensional space ker V^*, the co-dimension of ran A must be at least 2, and hence A cannot be cyclic.

Caution: the preceding reasoning is incomplete. The trouble is that there exist dense linear manifolds disjoint from subspaces of arbitrarily large finite dimension. To avoid error, it is necessary to observe that in the case at hand the linear manifold ran A is closed. Reason: since V^*A is invertible, the operator A is bounded from below.

162 **Solution 162.** The answer is yes.

The usual way to prove that an operator is not cyclic is to prove that it has co-rank 2 or more, or, equivalently, that its adjoint has nullity 2 or more. The multiplicity of the number 0 as an eigenvalue of U^* is 1, and the same is true of λ for all λ in the open unit disc. These observations do not prove but they suggest that if an operator A is near to U, then the multiplicity of λ as an eigenvalue of A^* is likely to be 1 for all λ with $|\lambda| < 1$. The complex numbers λ with $|\lambda| = 1$ are not eigenvalues of U^*, but they are approximate eigenvalues and, as such, they behave as if they had infinite multiplicity; the most promising place to look for operators near to a translate of U^* with nullity greater than 1 is on the unit circle. As far as cyclicity is concerned, translation by a scalar makes no difference; an operator A is just as cyclic or non-cyclic as $A - \lambda$.

So much for the motivation of the proof. The proof itself begins with the observation that 1 is an approximate eigenvalue of U^* in a strong sense: there exists an *orthonormal* infinite sequence $\{f_n\}$ of vectors such that $\|(1 - U^*)f_n\| \to 0$. Indeed, consider U^* as acting on l^2, and, for each positive integer n, consider a vector f_n in l^2 with n consecutive coordinates equal to $1/\sqrt{n}$ and all other coordinates equal to 0. For such a vector, the norm $\|f_n\|$ is 1, and $\|(1 - U^*)f_n\| \leq 2/\sqrt{n}$. Choose the f_n's so that for distinct values of n they have disjoint supports (that is, if $n \neq m$, then the coordinatewise product of f_n and f_m is 0).

If P_n is the projection onto the 2-dimensional space spanned by f_n and f_{n+1}, and if $A_n = (1 - P_n)(1 - U)$, then

$$\|U - (1 - A_n)\| = \|(1 - U) - A_n\| = \|(1 - U)^* - A_n^*\|$$

$$= \|(1 - U^*)P_n\| \leq \frac{2\sqrt{2}}{\sqrt{n}},$$

so that $1 - A_n \to U$. Since

$$\dim \ker A_n^* = \dim \ker(1 - U)^*(1 - P_n) \geq 2,$$

it follows that the co-rank of A_n is at least 2. Conclusion: A_n is not cyclic, and, therefore, neither is $1 - A_n$.

A somewhat more sophisticated version of the same argument proves the general result: if $\dim \mathbf{H} = \aleph_0$, then the set of non-cyclic operators on $\mathbf{H}$ is dense [42].

Solution 163. The answer is no.

Suppose, indeed, that f and g are in l^2, and let f^* and g^* be their co-ordinatewise complex conjugates; then

$$(\langle U^n f, U^{*n}g \rangle, \langle g^*, -f^* \rangle) = (U^n f, g^*) - (U^{*n}g, f^*)$$
$$= (U^n f, g^*) - (g, U^n f^*)$$
$$= (U^n f, g^*) - (U^n f, g^*)$$
$$= 0.$$

Conclusion: no $\langle f, g \rangle$ is a cyclic vector of $U \oplus U^*$.

(This elegant proof is due to N. K. Nikolskii, V. V. Peller, and V. I. Vasunin.)

Solution 164. The answer is yes.

The assertion is that to every vector g and to every positive number ε there corresponds a polynomial q such that $\|g - q(A^*)f\| < \varepsilon$. Since the vectors of the form $A^n f$, $n = 0, 1, 2, \cdots$, span the space, it is sufficient to prove the assertion in case $g = A^n f$.

163

164

283

Given n and ε, find a polynomial p such that

$$\|A^{*n}f - p(A)f\| < \varepsilon;$$

that can be done because f is cyclic for A. If $q(z) = (p(z^*))^*$, then q is a polynomial and

$$\|A^n f - q(A^*)f\| = \|(A^{*n} - p(A))^* f\| = \|(A^{*n} - p(A))f\|;$$

normality is needed for the last step.

165   **Solution 165.** *A vector f is cyclic for the position operator on $\mathbf{L}^2(0, 1)$ if and only if $f(x) \neq 0$ almost everywhere.*

PROOF. The "only if" is obvious: if A is the position operator, then $A^n f$ vanishes at least as much as f does, so that if f vanishes on a set of positive measure, then so do all the $A^n f$'s, and, therefore, so does everything in their span.

The non-trivial part of the problem is to prove that if $f \neq 0$ almost everywhere, and if $\mathbf{K}$ is the closure of the set of all vectors of the form $p(A)f$, where p varies over all polynomials, then $\mathbf{K} = \mathbf{L}^2$.

First step: if φ is a bounded measurable function on $[0, 1]$, then $\varphi\mathbf{K} \subset \mathbf{K}$. To prove this, observe that by Fejér's theorem (see Solution 41) there exists a sequence $\{\varphi_n\}$ of trigonometric polynomials converging boundedly almost everywhere to φ. By the Weierstrass polynomial approximation theorem, each φ_n is uniformly arbitrarily near to a polynomial. It follows that there exists a sequence $\{p_n\}$ of polynomials converging boundedly almost everywhere to φ. If now $h \in \mathbf{K}$, then $\{p_n h\}$ converges dominatedly almost everywhere to φh, and therefore, by Lebesgue, $p_n h \to \varphi h$ in $\mathbf{L}^2$. Since $p\mathbf{K} \subset \mathbf{K}$ for *polynomials* p, so that $p_n h \in \mathbf{K}$, it follows, as asserted, that $\varphi\mathbf{K} \subset \mathbf{K}$.

Next: if g is an arbitrary bounded measurable function on $[0, 1]$, then $g \in \mathbf{K}$. To prove this, write $\varphi_n(x) = 0$ when $|f(x)| < 1/n$ and $\varphi_n(x) = g(x)/f(x)$ otherwise, $n = 1, 2, 3, \cdots$. Each φ_n is a bounded measurable function, and therefore $\varphi_n f \in \mathbf{K}$ by the preceding paragraph. Since $\varphi_n f = 0$ when $|f| < 1/n$ and $\varphi_n f = g$ otherwise, it follows that the sequence $\{\varphi_n f\}$ is uniformly bounded almost everywhere (by $\|g\|_\infty$) and converges to g almost everywhere. Consequence, as asserted: $g \in \mathbf{K}$.

The desired conclusion is now immediate: by the preceding paragraph $\mathbf{K}$ is dense in $\mathbf{L}^2$, and by definition $\mathbf{K}$ is a subspace, so that $\mathbf{K} = \mathbf{L}^2$.

Corollary. *There exists an operator A on a Hilbert space $\mathbf{H}$, and there exists a subspace $\mathbf{K}$ of $\mathbf{H}$ such that both $\mathbf{K}$ and $\mathbf{K}^\perp$ are infinite-dimensional and such that every non-zero vector in either $\mathbf{K}$ or $\mathbf{K}^\perp$ is cyclic for A.*

PROOF. Put $\mathbf{H} = \mathbf{L}^2(0, 1)$, $\mathbf{K} = \mathbf{H}^2$, and let A be the position operator on $\mathbf{H}$.

Another consequence of Solution 165 makes contact with the theory of total sets (cf. Problem 9). It needs only easy analysis to prove that the set of all powers f_n (i.e., $f_n(x) = x^n$, $n = 0, 1, 2, \cdots$) is total in $\mathbf{L}^2(0, 1)$; what takes a little more is that that set remains total after the omission of each finite subset. Indeed: to prove that $\{f_n, f_{n+1}, f_{n+2}, \cdots\}$ is total for each n, just observe that f_n is cyclic for the position operator, and

$$f_{n+k} = A^k f_n, \quad k = 0, 1, 2, \cdots.$$

That all cofinite sets of powers are total is a very small fragment of the whole truth. The fact is that a set of powers is total if and only if the reciprocals of the exponents form a divergent series; this surprising and satisfying statement is the Müntz–Szász theorem [139].

Solution 166. *The span of the set of all cyclic vectors of an operator on a Hilbert space* **H** *is either* 0 *or* **H**.   **166**

The main step in the proof is to show that if f is cyclic for A, then so is $f_p = (1 - \alpha A)^p f$, $p = 0, 1, 2, \cdots$, provided only that α is sufficiently small.

If $\alpha = 0$, the conclusion is trivial. Suppose now that $0 < \|\alpha A\| < 1$. It follows that the series

$$\sum_{n=0}^{\infty} (\alpha A)^n (1 - \alpha A)$$

converges (in the norm); its sum is $(1 - \alpha A)^{-1}(1 - \alpha A) = 1$. This implies that

$$f_p = (1 - \alpha A)^{-1} f_{p+1} = \sum_{n=0}^{\infty} \alpha^n A^n f_{p+1},$$

and hence that

$$A^m f_p = \sum_{n=0}^{\infty} \alpha^n A^{n+m} f_{p+1}, \quad m = 0, 1, 2, \cdots.$$

Consequence: if f_p is cyclic, then so is f_{p+1}, and, therefore, by induction, every f_p is cyclic.

The proof of the principal assertion is to show that the f_p's span **H**. For that purpose, note that

$$(\alpha A)^n = (1 - (1 - \alpha A))^n = \sum_{p=0}^{n} (-1)^p \binom{n}{p}(1 - \alpha A)^p,$$

and that, therefore,

$$A^n f = \frac{1}{\alpha^n} \sum_{p=0}^{n} (-1)^p \binom{n}{p} f_p,$$

for $n = 0, 1, 2, \cdots$. Since the vectors $A^n f$ span **H**, it follows that the vectors $f_0, f_1, f_2, \cdots$ span **H**.

Both the result and the proof are due to L. Gehér [46]; note that they are valid for arbitrary Banach spaces.

Solution 167. (a) If A is an operator whose matrix with respect to a basis $\{e_0, e_1, e_2, \cdots\}$ is triangular $+1$ and has no zero entries in the diagonal just   **167**

below the main one, then e_0 is a cyclic vector of A. The reasoning goes as follows. The span of $\{e_0, Ae_0, A^2e_0, \cdots\}$ contains e_1, because Ae_0 is a linear combination of e_0 and e_1, with the coefficient of e_1 different from zero. (That coefficient is the first entry in the diagonal just below the main one.) Next: the span of $\{e_0, Ae_0, A^2e_0, \cdots\}$ contains e_2, because A^2e_0 is a linear combination of e_0, e_1, and e_2, with the coefficient of e_2 different from zero. These are the first steps of an induction; the rest are just like them.

(b) Yes, the converse is true. Assume, for convenience, that the underlying Hilbert space is of dimension $\aleph_0$. (The result (a) is true for finite-dimensional spaces, and the proof in the preceding paragraph applies. The converse (b) is also true for finite-dimensional spaces; its proof is only a slight modification of what follows.) If f is a cyclic vector of A, consider the total sequence $\{f, Af, A^2f, \cdots\}$ and note that it is linearly independent. (This is where infinite-dimensionality is used.) If $\{e_0, e_1, e_2, \cdots\}$ is the basis obtained by orthonormalization, then the matrix of A with respect to that basis is triangular $+1$. The reason none of the entries just below the main diagonal is zero is that if e_{n+1} belonged to the span of $\{e_0, \cdots, e_n\}$, then $A^{n+1}f$ would belong to the span of $\{f, \cdots, A^nf\}$, and it does not.

168 **Solution 168.** Let $\mathbf{H}$ be l^2, let U be the unilateral shift, and write $A = 2U^*$. (There is nothing magic about 2; any number greater than 1 would do just as well.)

Consider now a sequence $\langle f_1, f_2, f_3, \cdots \rangle$ of finite sequences of complex numbers (i.e., vectors in finite-dimensional Euclidean spaces) such that if each one is converted into an element of l^2 by tacking on an infinite tail of 0's, then the resulting sequence of vectors is dense in l^2.

The vector f to be constructed depends on the f_n's and on two sequences $\langle k_1, k_2, k_3, \cdots \rangle$ and $\langle p_1, p_2, p_3, \cdots \rangle$ of positive integers, and it looks like this: a sequence of 0's, of length k_1, followed by $f_1/2^{p_1}$ (i.e., followed by the coordinates of $f_1/2^{p_1}$, in order), followed by a sequence of 0's, of length k_2, followed by $f_2/2^{p_2}$, and so on ad infinitum. A simple (but unnecessarily generous) way to determine the k's and the p's is as follows: choose k_n so that $\|f_n\|/2^{k_n} < 1/2^n$, and let p_n be the (unique) exponent such that $A^{p_n}f$ begins with f_n. (That is: the first coordinate of $f_n/2^{p_n}$ in f has exactly p_n predecessors. Alternatively: p_n is the sum of $k_1, \cdots, k_n$ and the lengths of $f_1, \cdots, f_{n-1}$.)

Since $\sum_n \|f_n\|/2^{p_n} < \sum_n \|f_n\|/2^{k_n} < \infty$, it is clear that $f \in l^2$ (in fact, $f \in l^1$). The density of the $A^{p_n}f$'s follows from the facts that $A^{p_n}f$ begins with f_n and the norm of the tail of $A^{p_n}f$ is the square root of

$$\sum_j \left(\frac{\|f_{n+j}\|}{2^{p_{n+j}}} \right)^2 \leq \sum_j \left(\frac{\|f_{n+j}\|}{2^{k_{n+j}}} \right)^2$$

$$\leq \sum_j \left(\frac{1}{2^{n+j}} \right)^2 < \frac{1}{4^n}.$$

Indeed: to approximate an arbitrary h in l^2 by $A^{p_n}f$ within ε, choose n_0 so large that every tail beyond n_0 has norm below $\varepsilon/2$, and then approximate h within $\varepsilon/2$ by a vector of the form $\langle f_n, 0, 0, 0, \cdots \rangle$, with $n > n_0$.

The result here described is due to Rolewicz [117]. A related result (with scalar multiples allowed but sums still not) was discussed later, in a different analytic context, in [73].

Here is a pertinent question: does U have a dense orbit? The answer is no. Reason: every orbit is bounded. A more complicated proof is worthwhile. If $\{f, Uf, U^2f, \cdots\}$ is dense, then the initial coordinate of f cannot be 0; assume, with no loss, that it is 1. If the next non-zero coordinate is the one with index k so that

$$f = \langle 1, 0, \cdots, 0, \xi_k, \xi_{k+1}, \xi_{k+2}, \cdots \rangle$$

with $\xi_k \neq 0$, then consider the vector

$$e_0 = \langle 1, 0, \cdots, 0, 0, 0, 0, \cdots \rangle.$$

If $\varepsilon < \min(1, |\xi_k|)$, then no vector of the form $U^n f$ can be within ε of e_0. Reason: if $n > 0$, then $\|e_0 - U^n f\| \geq 1$ (look at the coordinate with index 0); if $n = 0$, then $\|e_0 - U^n f\| \geq |\xi_k|$ (look at the coordinate with index k). The merit of this technique is that a slight refinement of it can be used to show that even the set of scalar multiples of the vectors in the orbit of f is not dense.

CHAPTER 19

Properties of Compactness

169   **Solution 169.** If A is (s → s) continuous, and if $\{f_j\}$ is a net w-convergent to f, then $(Af_j, g) = (f_j, A^*g) \to (f, A^*g) = (Af, g)$ for all g, so that $Af_j \to Af$ (w). This proves that A is (w → w) continuous. Note that the assumption of (s → s) continuity was tacitly, but heavily, used via the existence of the adjoint A^*.

If A is (w → w) continuous, and if $\{f_j\}$ is a net s-convergent to f, then, a fortiori, $f_j \to f$ (w), and the assumption implies that $Af_j \to Af$ (w). This proves that A is (s → w) continuous.

To prove that if A is (s → w) continuous, then A is bounded, assume the opposite. That implies the existence of a sequence $\{f_n\}$ of unit vectors such that $\|Af_n\| \geqq n^2$. Since

$$\frac{1}{n} f_n \to 0 \quad \text{(s)},$$

the assumption implies that

$$\frac{1}{n} Af_n \to 0 \quad \text{(w)},$$

and hence that

$$\left\{ \frac{1}{n} Af_n \right\}$$

is a bounded sequence; this is contradicted by

$$\left\| \frac{1}{n} Af_n \right\| \geqq n.$$

Suppose, finally, that A is (w → s) continuous. It follows that the inverse image under A of the open unit ball is a weak open set, and hence that it includes a basic weak neighborhood of 0. In other words, there exist vectors $f_1, \cdots, f_k$ and there exists a positive number ε such that if $|(f, f_i)| < \varepsilon$, $i = 1, \cdots, k,$ then $\|Af\| < 1$. If f is in the orthogonal complement of the span of $\{f_1, \cdots, f_k\}$, then certainly $|(f, f_i)| < \varepsilon$, $i = 1, \cdots, k$, and therefore $\|Af\| < 1$. Since this conclusion applies to all scalar multiples of f too, it follows that Af must be 0. This proves that A annihilates a subspace of finite co-dimension, and this is equivalent to the statement that A has finite rank. (If there is an infinite-dimensional subspace on which A is one-to-one, then the range of A is infinite-dimensional; to prove the converse, note that the range of A is equal to the image under A of the orthogonal complement of the kernel.)

To prove the corollary, use the result that an operator (i.e., a linear transformation that is continuous (s → s)) is continuous (w → w). Since the closed unit ball is weakly compact, it follows that its image is weakly compact, therefore weakly closed, and therefore strongly closed.

Solution 170. The proof that **K** is an ideal is elementary. The proof that **K** is self-adjoint is easy via the polar decomposition. Indeed, if $A \in \mathbf{K}$ and $A = UP$, then $P = U^*A$ (see Corollary 1, Problem 134), so that $P \in \mathbf{K}$; since $A^* = PU^*$, it follows that $A^* \in \mathbf{K}$.

Suppose now that $A_n \in \mathbf{K}$ and $\|A_n - A\| \to 0$; it is to be proved that $Af_j \to Af$ whenever $\{f_j\}$ is a bounded net converging weakly to f. Note that

$$\|Af_j - Af\| \leqq \|Af_j - A_n f_j\| + \|A_n f_j - A_n f\| + \|A_n f - Af\|.$$

The first term on the right is dominated by $\|A - A_n\| \cdot \|f_j\|$; since $\{\|f_j\|\}$ is bounded, it follows that the first term is small for all large n, uniformly in j. The last term is dominated by $\|A_n - A\| \cdot \|f\|$, and, consequently, it too is small for large n. Fix some large n; the compactness of A_n implies that the middle term is small for "large" j. This completes the proof that **K** is closed.

Solution 171. Let A be an operator with diagonal $\{\alpha_n\}$, and, for each positive integer n, consider the diagonal operator A_n with diagonal

$$\{\alpha_0, \cdots, \alpha_{n-1}, 0, 0, 0, \cdots\}.$$

Since $A - A_n$ is a diagonal operator with diagonal

$$\{0, \cdots, 0, \alpha_n, \alpha_{n+1}, \cdots\},$$

so that $\|A - A_n\| = \sup_k |\alpha_{n+k}|$, it is clear that the assumption $\alpha_n \to 0$ implies the conclusion $\|A - A_n\| \to 0$. Since the limit (in the norm) of compact operators is compact, it follows that if $\alpha_n \to 0$, then A is compact.

To prove the converse, consider the orthonormal basis $\{e_n\}$ that makes A diagonal. If A is compact, then $Ae_n \to 0$ strongly (because $e_n \to 0$ weakly; cf. Solution 19). In other words, if A is compact, then $\|\alpha_n e_n\| \to 0$, and this says exactly that $\alpha_n \to 0$.

170

171

289

If $Se_n = e_{n+1}$, then each of A and SA is a multiple of the other (recall that $S*S = 1$), which implies that A and SA are simultaneously compact or not compact. This remark proves the corollary.

172   **Solution 172.** With a sufficiently powerful tool (the spectral theorem) the proof becomes easy. Begin with the observation that a compact operator on an infinite-dimensional Hilbert space cannot be invertible. (Proof: the image of the unit ball under an invertible operator is strongly compact if and only if the unit ball itself is strongly compact.) Since the restriction of a compact operator to an invariant subspace is compact, it follows that if the restriction of a compact operator to an invariant subspace is invertible, then the subspace is finite-dimensional.

Suppose now that A is a compact normal operator; by the spectral theorem there is no loss of generality in assuming that A is a multiplication operator induced by a bounded measurable function φ on some measure space. For each positive number ε, let M_ε be the set $\{x : |\varphi(x)| > \varepsilon\}$, and let $\mathbf{M}_\varepsilon$ be the subspace of $\mathbf{L}^2$ consisting of the functions that vanish outside M_ε. Clearly each $\mathbf{M}_\varepsilon$ reduces A, and the restriction of A to $\mathbf{M}_\varepsilon$ is bounded from below; it follows that $\mathbf{M}_\varepsilon$ is finite-dimensional.

The spectrum of A is the essential range of φ. The preceding paragraph implies that, for each positive integer n, the part of the spectrum that lies outside the disc $\{\lambda : |\lambda| \leq 1/n\}$ can contain nothing but a finite number of eigenvalues each of finite multiplicity; from this everything follows.

173   **Solution 173.** Recall that a simple function is a measurable function with a finite range; equivalently, a simple function is a finite linear combination of characteristic functions of measurable sets. A simple function belongs to $\mathbf{L}^2$ if and only if the inverse image of the complement of the origin has finite measure; an equivalent condition is that it is a finite linear combination of characteristic functions of measurable sets of finite measure. The simple functions in $\mathbf{L}^2(\mu)$ are dense in $\mathbf{L}^2(\mu)$. It follows that the finite linear combinations of characteristic functions of measurable rectangles of finite measure are dense in $\mathbf{L}^2(\mu \times \mu)$. In view of these remarks it is sufficient to prove that if A is an integral operator with kernel K, where

$$K(x, y) = \sum_{i=1}^{n} g_i(x)h_i(y),$$

and where each g_i and each h_i is a scalar multiple of a characteristic function of a measurable set of finite measure, then A is compact. It is just as easy to prove something much stronger: as long as each g_i and each h_i belongs to $\mathbf{L}^2(\mu)$, the operator A has finite rank. In fact the range of A is included in the span of the g's. The proof is immediate: if $f \in \mathbf{L}^2(\mu)$, then

$$(Af)(x) = \sum_{i=1}^{n} g_i(x) \int h_i(y)f(y)d\mu(y).$$

Solution 174. *If A is a Hilbert–Schmidt operator, then the sum of the eigenvalues of A^*A is finite.* **174**

PROOF. To say that A is a Hilbert-Schmidt operator means, of course, that A is an integral operator on, say, $L^2(\mu)$, induced by a kernel K in $L^2(\mu \times \mu)$. Since A^*A is a compact normal operator, there exists an orthonormal basis $\{f_j\}$ consisting of eigenvectors of A^*A (Problem 172); write

$$A^*Af_j = \lambda_j f_j.$$

The useful way to put the preceding two statements together is to introduce a suitable basis for $L^2(\mu \times \mu)$ and, by Parseval's equality, express the $L^2(\mu \times \mu)$ norm of K (which is finite, of course) in terms of that basis. There is only one sensible looking basis in sight, the one consisting of the functions g_{ij}, where $g_{ij}(x, y) = f_i(x)f_j(y)$. It turns out, however, that a slightly less sensible looking basis is algebraically slightly more convenient; it consists of the functions g_{ij} defined by $g_{ij}(x, y) = f_i(x)f_j(y)^*$.

The rest is simple computation:

$$\|K\|^2 = \sum_i \sum_j |(K, g_{ij})|^2 \text{ (by Parseval)}$$

$$= \sum_i \sum_j \left| \iint K(x, y)f_i(x)^*f_j(y)d\mu(x)d\mu(y) \right|^2$$

$$= \sum_j \sum_i \left| \int \left(\int K(x, y)f_j(y)d\mu(y) \right)f_i(x)^* \, d\mu(x) \right|^2$$

$$= \sum_j \sum_i \left| \int (Af_j)(x)f_i(x)^* \, d\mu(x) \right|^2$$

$$= \sum_j \sum_i |(Af_j, f_i)|^2 = \sum_j \|Af_j\|^2 \quad \text{(by Parseval)}$$

$$= \sum_j (Af_j, Af_j) = \sum_j (A^*Af_j, f_j) = \sum_j \lambda_j.$$

The proof is over. The construction of a concrete compact operator that does not satisfy the Hilbert–Schmidt condition is now easy. Consider an infinite matrix (i.e., a "kernel" on l^2). By definition, if the sum of the squares of the moduli of the entries is finite, the matrix defines a Hilbert–Schmidt operator. This is true, in particular, if the matrix is diagonal. The theorem just proved implies that in that case the finiteness condition is not only sufficient but also necessary for the result to be a Hilbert–Schmidt operator. Thus, in the diagonal case, the difference between compact and Hilbert–Schmidt is the difference between a sequence that tends to 0 and a sequence that is square-summable.

175 **Solution 175.** If A is compact and UP is its polar decomposition, then P $(= U^*A)$ is compact. By Problem 172, P is the direct sum of 0 and a diagonal operator on a separable space, and the sequence of diagonal terms of the diagonal operator tends to 0. This implies that P is the limit (in the norm) of a sequence $\{P_n\}$ of operators of finite rank, and hence that $A = U \cdot \lim_n P_n = \lim_n UP_n$. Since UP_n has finite rank for each n, the proof is complete.

176 **Solution 176.** Suppose that $\mathbf{I}$ is a non-zero closed ideal of operators. The first step is to show that $\mathbf{I}$ contains every operator of rank 1. To prove this, observe that if u and v are non-zero vectors, then the operator A defined by $Af = (f, u)v$ has rank 1, and every operator of rank 1 has this form. To show that each such operator belongs to $\mathbf{I}$, take a non-zero operator A_0 in $\mathbf{I}$, and let u_0 and v_0 be non-zero vectors such that $A_0 u_0 = v_0$. Let B be the operator defined by $Bf = (f, u)u_0$, and let C be an arbitrary operator such that $Cv_0 = v$. It follows that

$$CA_0 Bf = CA_0(f, u)u_0 = (f, u)Cv_0 = (f, u)v = Af,$$

i.e., that $CA_0 B = A$. Since $\mathbf{I}$ is an ideal, it follows that $A \in \mathbf{I}$, as promised.

Since $\mathbf{I}$ contains all operators of rank 1, it contains also all operators of finite rank, and, since $\mathbf{I}$ is closed, it follows that $\mathbf{I}$ contains every compact operator. (Note that separability was not needed yet.)

The final step is to show that if $\mathbf{I}$ contains an operator A that is not compact, then $\mathbf{I}$ contains every operator. If UP is the polar decomposition of A, then $P \in \mathbf{I}$ (because $P = U^*A$), and P is not compact (because $A = UP$). Since P is Hermitian, there exists an infinite-dimensional subspace $\mathbf{M}$, invariant under P, on which P is bounded from below, by ε say. (If not, P would be compact.) Let V be an isometry from $\mathbf{H}$ onto $\mathbf{M}$. (Here is where the separability of $\mathbf{H}$ comes in.) Since $P\mathbf{M} = \mathbf{M}$, it follows that $V^*PV\mathbf{H} = V^*P\mathbf{M} = V^*\mathbf{M} = \mathbf{H}$. Since, moreover, $Vf \in \mathbf{M}$ for all f, it follows that

$$\|V^*PVf\| = \|PVf\| \geqq \varepsilon\|Vf\| = \varepsilon\|f\|.$$

These two assertions imply that V^*PV is invertible. Since $V^*PV \in \mathbf{I}$, the proof is complete; an ideal that contains an invertible element contains everything.

177 **Solution 177.** The answer to the "some" question is no; there are many counterexamples.

A simple one is given by the classically important Cesàro matrix

$$C = \begin{pmatrix} 1 & 0 & 0 & 0 \\ \frac{1}{2} & \frac{1}{2} & 0 & 0 \\ \frac{1}{3} & \frac{1}{3} & \frac{1}{3} & 0 \\ \frac{1}{4} & \frac{1}{4} & \frac{1}{4} & \frac{1}{4} \\ & & & & \ddots \end{pmatrix}.$$

To prove that it works it is, of course, necessary to prove that C is bounded but not compact. Both those statements are true [21] but their proofs would be a digression that is not worth making here. An alternative possibility is

$$B = \begin{pmatrix} 1 & 0 & 0 & 0 & 0 & 0 \\ 0 & \dfrac{1}{\sqrt{2}} & \dfrac{1}{\sqrt{2}} & 0 & 0 & 0 \\ 0 & 0 & 0 & \dfrac{1}{\sqrt{3}} & \dfrac{1}{\sqrt{3}} & \dfrac{1}{\sqrt{3}} \\ 0 & 0 & 0 & 0 & 0 & 0 \\ 0 & 0 & 0 & 0 & 0 & 0 \\ 0 & 0 & 0 & 0 & 0 & 0 \\ & & & & & & \ddots \end{pmatrix}.$$

In this case it is easy to prove directly that B is bounded but not compact, but for the present purpose even easy proofs are unnecessary effort; there is an almost trivial construction that settles everything. Note that $BB^* = 1$, so that B is a co-isometry; the most efficient answer to Problem 177 is the projection $A = B^*B$,

$$A = \begin{pmatrix} 1 & 0 & 0 & 0 & 0 & 0 \\ 0 & \frac{1}{2} & \frac{1}{2} & 0 & 0 & 0 \\ 0 & \frac{1}{2} & \frac{1}{2} & 0 & 0 & 0 \\ 0 & 0 & 0 & \frac{1}{3} & \frac{1}{3} & \frac{1}{3} \\ 0 & 0 & 0 & \frac{1}{3} & \frac{1}{3} & \frac{1}{3} \\ 0 & 0 & 0 & \frac{1}{3} & \frac{1}{3} & \frac{1}{3} \\ & & & & & & \ddots \end{pmatrix}.$$

Since the matrix A is the direct sum of an infinite sequence of projections, it is obviously bounded but not compact; since $n/n^2 \to 0$ as $n \to \infty$, the operator A obviously maps the natural coordinate basis onto a strong null sequence.

The answer to the "every" question is yes. Consider the collection $\mathbf{J}$ of all the operators that map every orthonormal basis onto a strong null sequence. The collection $\mathbf{J}$ is a norm-closed vector space, and, moreover, if $A \in \mathbf{J}$, then $BA \in \mathbf{J}$ for every operator B: in other words, $\mathbf{J}$ is a closed left ideal. More is true: $\mathbf{J}$ is, in fact, a two-sided ideal. The reason is that (a) if $A \in \mathbf{J}$ and U is unitary, then $AU \in \mathbf{J}$, and (b) every operator is a linear combination of four unitary ones. To prove (a), note that if $\{e_n\}$ is an orthonormal basis, $A \in \mathbf{J}$, and U is unitary, then $\{Ue_n\}$ is an orthonormal basis, so that $\|AUe_n\| \to 0$, and therefore $AU \in \mathbf{J}$. To prove (b) note that if A is an arbitrary Hermitian contraction, then both $U = A + i\sqrt{1 - A^2}$ and $V = A - i\sqrt{1 - A^2}$ are unitary, and $\frac{1}{2}(U + V) = A$; write an arbitrary operator as a scalar multiple of a contraction, and then express its real and imaginary parts as linear combinations of unitary operators.

The preceding paragraph proves that J is a closed ideal of operators. Clearly J has non-zero elements (namely, all compact operators), and J does not contain every operator (for example, $1 \notin J$). Conclusion (via Problem 138): an operator is in J if and only if it is compact.

The answer is finished, but it is worthwhile to call attention to an interesting and useful corollary. What is now known is that if A is not compact, then there exists an orthonormal basis $\{e_n\}$ such that $\{Ae_n\}$ does not converge strongly to 0. That implies that there exists a positive number ε such that $\|Ae_n\| \geqq \varepsilon$ for infinitely many values of n. The corollary is the following strengthened statement: *if A is not compact, then there exists an orthonormal basis $\{e_n\}$, and there exists a positive number ε, such that $\|Ae_n\| \geqq \varepsilon$ for every value of n.* The idea of the proof is to start with a basis $\{e_n\}$ such that $\|Ae_n\| \geqq \varepsilon$ for many n and then change both the basis and the number ε. Consider the indices k for which $\|Ae_k\| < \varepsilon/2$. Pair each such k with some n such that $\|Ae_n\| \geqq \varepsilon$, and replace the pair $\{e_k, e_n\}$ by $\{(e_k + e_n)/\sqrt{2}, (e_k - e_n)/\sqrt{2}\}$; the unpaired e_n's with $\|Ae_n\| \geqq \varepsilon$, the e_n's with $\varepsilon/2 < \|Ae_n\| < \varepsilon$, and the replacements together form an orthonormal basis; that basis and the number $\varepsilon/4$ do what is required.

178

Solution 178. *If A is normal and if A^n is compact for some positive integer n, then A is compact.*

PROOF. Represent A as a multiplication operator, induced by a bounded measurable function φ on a suitable measure space, and note that this automatically represents A^n as the multiplication operator induced by φ^n. Since, by Problem 172, A^n is the direct sum of 0 and a diagonal operator with diagonal terms converging to 0, it follows that the essential range of φ^n is a countable set that can cluster at 0 alone. This implies that A is the direct sum of 0 and a diagonal operator with diagonal terms converging to 0, and hence that A is compact.

179

Solution 179. With the approximate point spectrum $\Pi(C)$ in place of spec C the proof is easy. Indeed: if $\lambda \neq 0$ and $Cf_n - \lambda f_n \to 0$, with $\|f_n\| = 1$, then use compactness to drop down to a suitable subsequence so as to justify the assumption that the sequence $\{Cf_n\}$ converges, to f say. It follows that $\lambda f_n \to f$, so that $\|f_n\| \to (1/|\lambda|)\|f\|$, and therefore $f \neq 0$. Since $C(\lambda f_n) \to Cf$ and $\lambda(Cf_n) \to \lambda f$, it follows that $Cf = \lambda f$.

The next useful step is to observe that $\Pi_0(C)$ is countable. To prove that, it is sufficient to prove that if $\lambda_1, \lambda_2, \lambda_3, \cdots$ are distinct eigenvalues of C, then $\lambda_n \to 0$. If $Cf_n = \lambda_n f_n$, with $f_n \neq 0$, then the sequence $\{f_1, f_2, f_3, \cdots\}$ is linearly independent. Let $\{g_1, g_2, g_3, \cdots\}$ be an orthonormal sequence such that $g_n \in \bigvee \{f_1, \cdots, f_n\}$ for each n. Expand Cg_n in terms of $f_1, \cdots, f_n$ and observe the coefficient of f_n in the expansion; it is the same as the co-efficient of f_n in the expansion of $\lambda_n\, g_n$. Consequence:

$$Cg_n - \lambda_n g_n \in \bigvee \{f_1, \cdots, f_{n-1}\}$$

whenever $n > 1$, and, therefore, $\lambda_n = (Cg_n, g_n)$; the asserted convergence follows from the compactness of C.

The proof of the full Fredholm alternative is now easy to complete. By Problem 78, the boundary of spec C is included in $\Pi(C)$, and therefore, by the preceding two paragraphs, the boundary of spec C is countable. The only way a bounded set in the plane can have a countable boundary is to be included in its own boundary.

This proof was shown to me by Peter Rosenthal.

Solution 180. Suppose that A is compact and suppose that **M** is a subspace such that $\mathbf{M} \subset \operatorname{ran} A$. The inverse image of **M** under A is a subspace, say **N**, and so is the intersection $\mathbf{N} \cap \ker^\perp A$. The restriction of A to that intersection is a one-to-one bounded linear transformation from that intersection onto **M**, and therefore (Problem 52) it is invertible. The image of the closed unit ball of **N** (i.e., of the intersection of the closed unit ball with **N**) is a strongly compact subset of **M**, and, by invertibility, it includes a closed ball in **M** (i.e., it includes the intersection of a closed ball with **M**). This implies that **M** is finite-dimensional; the proof is complete.

180

Solution 181. (1) implies (2). It is always true that A maps $\ker^\perp A$ one-to-one onto ran A, and hence that the inverse mapping maps ran A one-to-one onto $\ker^\perp A$. In the present case ran A is closed, and therefore, by the closed graph theorem, the inverse mapping is bounded. Let B be the operator that is equal to that inverse on ran A and equal to 0 on $\operatorname{ran}^\perp A$. Note that both P and Q have ker A, and let Q be the projection on $\operatorname{ran}^\perp A$. Note that both P and Q have finite rank. Since $BA = 1 - P$ on both $\ker^\perp A$ and ker A, and since $AB = 1 - Q$ on both ran A and $\operatorname{ran}^\perp A$, it follows that both $1 - BA$ and $1 - AB$ have finite rank.

181

(2) implies (3). Trivial: an operator of finite rank is compact.

(3) implies (1). If $C = 1 - AB$ and $D = 1 - BA$, with C and D compact, then both ker B^*A^* and ker BA are finite-dimensional. It follows that both ker A^* and ker A are finite-dimensional, and hence that both $\operatorname{ran}^\perp A$ and ker A are finite-dimensional. To prove that ran A is closed, note first that BA is bounded from below on $\ker^\perp BA$. (See Solution 179.) Since $\|BAf\| \leq \|B\| \cdot \|Af\|$ for all f, it follows that A is bounded from below on $\ker^\perp BA$, and hence that the image under A of $\ker^\perp BA$ is closed. Since ker BA is finite-dimensional, the image under A of ker BA is finite-dimensional and hence closed, and ran A is the sum $A(\ker BA) + A \ker^\perp BA$. (Recall that the sum of two subspaces, of which one is finite-dimensional, is always a subspace; see Problem 13.)

182   **Solution 182.** Translate by λ and reduce the assertion to this: if A is not invertible but ker $A = 0$, then B is not invertible. Contrapositively: if B is invertible, then either ker $A \neq 0$ or A is invertible. For the proof, assume that B is invertible and write

$$A = B + (A - B) = B(1 + B^{-1}(A - B)).$$

The operator $B^{-1}(A - B)$ is compact along with $A - B$. It follows that either -1 is an eigenvalue of $B^{-1}(A - B)$ (in which case ker $A \neq 0$), or $1 + B^{-1}(A - B)$ is invertible (in which case A is invertible).

183   **Solution 183.** The bilateral shift is an example. Suppose that

$$\{e_n : n = 0, \pm 1, \pm 2, \cdots\}$$

is the basis that is being shifted ($We_n = e_{n+1}$), and let C be the operator defined by $Cf = (f, e_{-1})e_0$. The operator C has rank 1 (its range is the span of e_0), and it is therefore compact. What is the operator $W - C$? Since $\mathbf{H}^2$ (the span of the e_n's with $n \geq 0$) is invariant under both W and C, it is invariant under $W - C$ also. The orthogonal complement of $\mathbf{H}^2$ (the span of the e_n's with $n < 0$) is invariant under neither W nor C (since $We_{-1} = Ce_{-1} = e_0$), but it is invariant under $W - C$. (Reason: if $n < 0$, then $W - C$ maps e_n onto e_{n+1} or 0, according as $n < -1$ or $n = -1$.) Conclusion: $\mathbf{H}^2$ reduces $W - C$. This conclusion makes it easy to describe $W - C$; it agrees with the unilateral shift on $\mathbf{H}^2$ and it agrees with the adjoint of the unilateral shift on the orthogonal complement of $\mathbf{H}^2$. In other words, $W - C$ is the direct sum $U^* \oplus U$, and, consequently, its spectrum is the union of the spectra of U^* and U.

It helps to look at all this via matrices. The matrix of W (with respect to the shifted basis) has 1's on the diagonal just below the main one and 0's elsewhere; the effect of subtracting C is to replace one of the 1's, the one in row 0 and column -1, by 0.

184   **Solution 184.** *No perturbation makes the unilateral shift normal.*

PROOF. The technique is to examine the spectrum and to use the relative stability of the spectrum under perturbation.

If $U = B - C$, with B normal and C compact, then

$$U^*U = B^*B - D,$$

where

$$D = C^*B + B^*C - C^*C,$$

so that D is compact. Since $U^*U = 1$, and since spec $1 = \{1\}$, it follows (Problem 182) that every number in spec B^*B, except possibly 1, must in fact be an eigenvalue of B^*B. (Alternatively, use the Fredholm alternative.) Since a Hermitian operator on a separable Hilbert space can have only countably many eigenvalues, it follows that the spectrum of B^*B must be

296

countable. Since spec U is the closed unit disc, and since U has no eigenvalues, another consequence of Problem 182 is that the spectrum of B can differ from the disc by the set of eigenvalues of B only. A normal operator on a separable Hilbert space can have only countably many eigenvalues. Conclusion: modulo countable sets, spec B is the unit disc, and therefore (Problem 123), modulo countable sets, spec B^*B is the interval $[0, 1]$. This contradicts the countability of spec B^*B.

Solution 185. If U is the unilateral shift, then $\|U - 0\| = 1$, and therefore the distance from U to the set of all compact operators is not more than 1. The distance is, in fact, equal to 1. For the proof, suppose that C is compact, and observe that $(U - C)^*(U - C) = 1 - C'$ with C' compact. Consequence: $1 \in \text{spec}((U - C)^*(U - C))$, whence $r((U - C)^*(U - C)) \geqq 1$. Conclusion: $\|(U - C)^*(U - C)\| \geqq 1$, and therefore $\|U - C\| \geqq 1$.

185

    Use of a somewhat more powerful tool yields a somewhat simpler proof; Problem 182 implies that spec$(U - C)$ includes the unit disc.

CHAPTER 20

Examples of Compactness

186

Solution 186. If H and K are Volterra kernels, then so is their "product" (matrix composition). Reason:

$$(HK)(x, y) = \int_0^1 H(x, z)K(z, y)dz,$$

and if $x < y$, then, for all z, either $x < z$ (in which case $H(x, z) = 0$), or $z < y$ (in which case $K(z, y) = 0$). In other words $(HK)(x, y) = 0$ if $x < y$; if $x \geq y$, then

$$(HK)(x, y) = \int_y^x H(x, z)K(z, y)dz,$$

because unless z is between y and x one of $H(x, z)$ and $K(z, y)$ must vanish. It follows that if K is a bounded Volterra kernel, with, say, $|K(x, y)| \leq c$, and if $x \geq y$, then

$$|K^2(x, y)| = \left| \int_y^x K(x, z)K(z, y)dz \right| \leq c^2 \cdot (x - y).$$

(In this context symbols such as K^2, K^3, etc., refer to the "matrix products" KK, KKK, etc.) From this in turn it follows that if $x \geq y$, then

$$|K^3(x, y)| = \left| \int_y^x K(x, z)K^2(z, y)dz \right| \leq c^3 \int_y^x (z - y)dz = \frac{c^3}{2}(x - y)^2.$$

These are the first two steps of an obvious inductive procedure; the general result is that if $n \geq 1$ and $x \geq y$, then

$$|K^n(x, y)| \leq \frac{c^n}{(n - 1)!}(x - y)^{n - 1}.$$

This implies, a fortiori, that

$$|K^n(x, y)| \leq \frac{c^n}{(n-1)!},$$

and hence that if A is the induced integral operator, then

$$\|A^n\| \leq \|K^n\| \leq \frac{c^n}{(n-1)!}.$$

Since

$$\left(\frac{1}{(n-1)!}\right)^{1/n} \to 0 \quad \text{as } n \to \infty$$

(recall that the radius of convergence of the exponential series is ∞), the proof that A is quasinilpotent is complete.

Solution 187. *A Volterra operator has no eigenvalue different from 0.* **187**

PROOF. Suppose that K is a Volterra kernel with associated Volterra operator V, and suppose that λ is an eigenvalue of V with associated (non-zero) eigenfunction f, so that $Vf = \lambda f$. If

$$g(x) = \int_0^x |f(y)|^2 \, dy,$$

then g is a monotone differentiable function and

$$g'(x) = |f(x)|^2$$

almost everywhere. Let a be the infimum of the support of g, i.e., $g(a) = 0$ and $g(x) > 0$ whenever $a < x \leq 1$; note that $0 < g(1) < \infty$. Since

$$\lambda f(x) = \int_0^x K(x, y) f(y) dy$$

almost everywhere, so that

$$|\lambda|^2 |f(x)|^2 \leq \int_0^x |K(x, y)|^2 \, dy \cdot \int_0^x |f(y)|^2 \, dy$$

almost everywhere, it follows that

$$|\lambda|^2 \frac{g'(x)}{g(x)} \leq \int_0^x |K(x, y)|^2 \, dy$$

almost everywhere in $(a, 1)$. By integration the last inequality becomes

$$|\lambda|^2 \log g(x) \Big|_a^1 = \int_a^1 |\lambda|^2 \frac{g'(x)}{g(x)} \, dx \leq \int_0^1 \int_0^x |K(x, y)|^2 \, dy \, dx = \|K\|^2.$$

Since $\log g(1) - \log 0 = \infty$ and $\|K\| < \infty$, this can happen only if $\lambda = 0$; the proof is complete.

Corollary. *Every Volterra operator is quasinilpotent.*

PROOF. Since V is compact (Problem 173), the Fredholm alternative (Problem 179) implies that 0 is the only number that can possibly be in the spectrum of V.

This elegant proof is due to Frode D. Poulsen.

188 **Solution 188.** $\|V\| = 2/\pi$.

PROOF. A direct attack on the problem seems not to lead anywhere. Here is a not unnatural indirect attack: evaluate $\|V^*V\|$ and take its square root. The reason this is promising is that V^*V is not only compact (as is V), but is also Hermitian. It follows that V^*V is diagonal; the obvious way to find its norm is to find its largest eigenvalue. (Note that V^*V is positive, so that its eigenvalues are positive.)

Since V^* is given by

$$(V^*f)(x) = \int_x^1 f(y)dy,$$

it is easy to find the integral kernel that induces V^*V. A simple computation shows that that kernel, K say, is given by

$$K(x, y) = 1 - \max(x, y) = \begin{cases} 1 - x & \text{if } 0 \leqq y \leqq x \leqq 1, \\ 1 - y & \text{if } 0 \leqq x < y \leqq 1. \end{cases}$$

It follows that

$$(V^*Vf)(x) = \int_0^1 f(y)dy - x\int_0^x f(y)dy - \int_x^1 yf(y)dy$$

for almost every x, whenever $f \in \mathbf{L}^2(0, 1)$. This suggests that the eigenvalues of V^*V can be explicitly determined by setting $V^*Vf = \lambda f$, differentiating (twice, to get rid of all integrals), and solving the resulting differential equation. There is no conceptual difficulty in filling in the steps. The outcome is that if

$$c_k(x) = \frac{1}{\sqrt{2}}(e^{i\pi(k + \frac{1}{2})x} + e^{-i\pi(k + \frac{1}{2})x}),$$

for $k = 0, 1, 2, \cdots$, then the c_k's form an orthonormal basis for $\mathbf{L}^2$, and each c_k is an eigenvector of V^*V, with corresponding eigenvalue $1/(k + \frac{1}{2})^2\pi^2$. The largest of these eigenvalues is the one with $k = 0$.

The outline above shows how the eigenvalues and eigenvectors can be *discovered*. If all that is wanted is an answer to the question (how much is $\|V\|$?), it is enough to *verify* that the c_k's are eigenvectors of V^*V, with the eigenvalues as described above, and that the c_k's form an orthonormal basis for $\mathbf{L}^2$. The first step is routine computation. The second step is necessary in order to guarantee that V^*V has no other eigenvalues, possibly larger than any of the ones that go with the c_k's.

Here is a way to prove that the c_k's form a basis. For each f in $\mathbf{L}^2$ write

$$(Uf)(x) = \frac{1}{\sqrt{2}}(f(x)e^{i\pi x/2} + f(1 - x)e^{-i\pi x/2}).$$

It is easy to verify that U is a unitary operator. If

$$e_n(x) = e^{2\pi i n x}, \qquad n = 0, \pm 1, \pm 2, \cdots,$$

then

$$Ue_n = c_{2n} \quad \text{for} \quad n = 0, 1, 2, \cdots,$$

and

$$Ue_n = c_{-(2n+1)} \quad \text{for } n = -1, -2, -3, \cdots.$$

Solution 189. spec $V_0 = \{0\}$, $\|V_0\| = 4/\pi$. **189**

PROOF. The most illuminating remark about V_0 is that its range is included in the set of all odd functions in $\mathbf{L}^2(-1, +1)$. (Recall that f is even if $f(x) = f(-x)$, and f is odd if $f(x) = -f(-x)$.) The second most illuminating remark (suggested by the first) is that if f is odd, then $V_0 f = 0$. These two remarks imply that V_0 is nilpotent of index 2, and hence that the spectrum of V_0 consists of 0 only.

One way to try to find the norm of V_0 is to identify $\mathbf{L}^2(-1, +1)$ with $\mathbf{L}^2(0, 1) \oplus \mathbf{L}^2(0, 1)$, determine the two-by-two operator matrix of V_0 corresponding to such an identification, and hope that the entries in the matrix are simple and familiar enough to make the evaluation of the norm feasible. One natural way to identify $\mathbf{L}^2(-1, +1)$ with $\mathbf{L}^2(0, 1) \oplus \mathbf{L}^2(0, 1)$ is to map f onto $\langle g, h \rangle$, where $g(x) = f(x)$ and $h(x) = f(-x)$ whenever $x \in (0, 1)$. This gives something, but it is not the best thing to do. For present purposes another identification of $\mathbf{L}^2(-1, +1)$ with $\mathbf{L}^2(0, 1) \oplus \mathbf{L}^2(0, 1)$ is more pertinent; it is the one that maps f onto $\langle g, h \rangle$, where

$$g(x) = \tfrac{1}{2}(f(x) - f(-x)) \quad \text{and} \quad h(x) = \tfrac{1}{2}(f(x) + f(-x))$$

whenever $x \in (0, 1)$. The inverse map sends $\langle g, h \rangle$ onto f, where

$$f(x) = h(x) + g(x) \quad \text{and} \quad f(-x) = h(x) - g(x)$$

whenever $x \in (0, 1)$. Since

$$(V_0 f)(x) = 2 \int_0^x \tfrac{1}{2}(f(y) + f(-y))dy,$$

it follows that if $x \in (0, 1)$, then

$$(V_0 f)(x) = 2(Vh)(x) \quad \text{and} \quad (V_0 f)(-x) = -2(Vh)(x).$$

The conclusion can be expressed in the form

$$V_0 \langle g, h \rangle = \langle 2Vh, 0 \rangle.$$

From this form the matrix of V_0 can be read off; it is

$$\begin{pmatrix} 0 & 0 \\ 2V & 0 \end{pmatrix}.$$

This proves, again, that $V_0^2 = 0$, and it shows, moreover, that $\|V_0\| = 2\|V\|$.

190
    Solution 190. *If V is the Volterra integration operator, and if $A =$ $(1 + V)^{-1}$, then* spec $A = \{1\}$ *and* $\|A\| = 1$.

PROOF. The example is simple, but it is the sort that takes either inspiration or experience to produce; reason alone does not seem to be enough. To prove that the example works, begin by recalling that spec $V = \{0\}$ (cf. Problems 186 and 88 or else Problem 187); it follows that spec$(1 + V) = \{1\}$, so that $1 + V$ is invertible, and hence that the definition of A makes sense. Since spec$(1 + V) = \{1\}$, it follows that spec $A = \{1\}$. Since $r(A) = 1$, it follows that $\|A\| \geqq 1$. Clearly $A \neq 1$. This settles all properties except one; everything except the inequality $\|A\| \leqq 1$ is obvious.

One way to prove that $\|Af\| \leqq \|f\|$ for all f, i.e., that A is bounded from above by 1, is to prove that A^{-1} is bounded from below by 1. Since

$$\|A^{-1}f\|^2 = \|(1 + V)f\|^2 = (f + Vf, f + Vf)$$
$$= \|f\|^2 + (Vf, f) + (V^*f, f) + \|Vf\|^2,$$

it is sufficient to prove that $((V + V^*)f, f) \geqq 0$ (i.e., that the real part of V is positive). This is true and already known; the operator $V + V^*$ is, in fact, the projection onto the (one-dimensional) space of constant functions (see Problem 188).

191
    Solution 191. Let $\{\alpha_n\}$ be the weight sequence, so that $\alpha_0 \geqq \alpha_1 \geqq \alpha_2 \geqq \cdots$, $\alpha_n \neq 0$, and $\sum_{n=0}^{\infty} \alpha_n^2 < \infty$; the operator A is given by $Ae_n = \alpha_{n-1}e_{n-1}$ when $n > 0$ and $Ae_0 = 0$.

Each non-zero vector f in the given Hilbert space **H** has a "degree", namely the largest index n (or ∞ if there is no largest) such that the Fourier coefficient (f, e_n) is not zero. Suppose that $f \in \mathbf{M}$ and that deg $f = n < \infty$. It is easy to see that the vectors $f, \cdots, A^n f$ are linearly independent; the point is that the non-vanishing of the α's implies that

$$\deg A^i f = n - i, \qquad i = 0, \cdots, n.$$

Since $A^i f \in \mathbf{M}_n$, $i = 0, \cdots, n$, it follows that the span of $\{f, \cdots, A^n f\}$ is $\mathbf{M}_n$, and hence that $\mathbf{M}_n \subset \mathbf{M}$.

The degrees of the non-zero vectors in **M** are either bounded or not. If they are, and if their maximum is n, then $\mathbf{M} \subset \mathbf{M}_n$, and the preceding paragraph implies that $\mathbf{M} = \mathbf{M}_n$. It remains to show that if **M** is a subspace invariant under A and if the degrees of the non-zero vectors in **M** are not bounded, then $\mathbf{M} = \mathbf{H}$. If **M** contains vectors of arbitrarily large finite degree, then, by the preceding paragraph, $\mathbf{M}_n \subset \mathbf{M}$ for infinitely many n, and hence $\mathbf{M} = \mathbf{H}$. The only remaining case is the one in which **M** contains a vector of infinite degree.

Consider the following lemma: if **M** is a subspace invariant under A, and if **M** contains a vector of infinite degree, then **M** contains e_0. Assertion: the lemma implies the theorem. To prove this, it is sufficient to prove that $\mathbf{M}_k \subset \mathbf{M}$ for all k. The idea of the proof is that nothing changes if the first few terms of the basis are omitted. In precise language, the proof is induction on k.

The initial step is the lemma itself. Suppose now that $\mathbf{M}_k \subset \mathbf{M}$, let P_k be the projection onto $\mathbf{M}_k^\perp$, and let A_k be the operator on $\mathbf{M}_k^\perp$ defined by $A_k f = P_k A f$ for each f in $\mathbf{M}_k^\perp$. The induction hypothesis implies that $P_k f \in \mathbf{M}$ for all f, and hence that $\mathbf{M} \cap \mathbf{M}_k^\perp$ is invariant under A_k. Since A_k is a weighted shift on $\mathbf{M}_k^\perp$ (with respect to the orthonormal basis $\{e_k, e_{k+1}, \cdots\}$), satisfying exactly the same conditions as A on $\mathbf{H}$ and since the image under P_k of a vector of infinite degree in $\mathbf{H}$ has infinite degree in $\mathbf{M}_k^\perp$ (with respect to the basis $\{e_k, e_{k+1}, \cdots\}$), the lemma is applicable. The conclusion is that $\mathbf{M} \cap \mathbf{M}_k^\perp$ contains e_k (so that, in particular, $e_k \in \mathbf{M}$, whence $\mathbf{M}_k \subset \mathbf{M}$), and the derivation of the theorem from the lemma is complete.

Turn now to the proof of the lemma. Suppose that $f \in \mathbf{M}$ and $\deg f = \infty$. If $f = \sum_{i=0}^\infty \xi_i e_i$, then

$$A^n f = \sum_{i=n}^\infty \xi_i \alpha_{i-1} \cdots \alpha_{i-n} e_{i-n}.$$

If n is such that $\xi_n \neq 0$, then

$$\frac{1}{\xi_n \alpha_{n-1} \cdots \alpha_0} A^n f = e_0 + f_n,$$

where

$$f_n = \sum_{i=n+1}^\infty \frac{\xi_i}{\xi_n} \cdot \frac{\alpha_{i-1} \cdots \alpha_{i-n}}{\alpha_{n-1} \cdots \alpha_0} e_{i-n}.$$

It is sufficient to prove that for each positive number ε the integer n can be chosen so that $\|f_n\| < \varepsilon$. To do this, first choose k so that

$$\sum_{i=k}^\infty \alpha_i^2 < \varepsilon^2 \alpha_0^2,$$

and then choose n so that $n \geq k$ and so that

$$|\xi_n| \geq \max\{|\xi_i| : i \geq k\}.$$

With this choice, $\xi_n \neq 0$, and, if $i \geq n$, then $|\xi_i/\xi_n| \leq 1$. Note also that if $i \geq n+1$, then $\alpha_{i-2} \leq \alpha_{n-1}, \cdots, \alpha_{i-n} \leq \alpha_1$ (here is where monotoneness is used). Conclusion:

$$\begin{aligned}
\|f_n\|^2 &= \sum_{i=n+1}^\infty \left| \frac{\xi_i}{\xi_n} \right|^2 \left(\frac{\alpha_{i-1} \cdots \alpha_{i-n}}{\alpha_{n-1} \cdots \alpha_0} \right)^2 \\
&\leq \sum_{i=n+1}^\infty \left(\frac{\alpha_{i-1}}{\alpha_0} \right)^2 \\
&= \sum_{i=n}^\infty \left(\frac{\alpha_i}{\alpha_0} \right)^2 < \varepsilon^2.
\end{aligned}$$

CHAPTER 21

Subnormal Operators

192   Solution 192. Every known proof of Fuglede's theorem can be modified so as to yield this generalized conclusion. Alternatively, there is a neat derivation, via operator matrices, of the statement for two normal operators from the statement for one. Write

$$\hat{A} = \begin{pmatrix} A_1 & 0 \\ 0 & A_2 \end{pmatrix} \quad \text{and} \quad \hat{B} = \begin{pmatrix} 0 & B \\ 0 & 0 \end{pmatrix}.$$

The operator $\hat{A}$ is normal, and a straightforward verification proves that $\hat{B}$ commutes with it. The Fuglede theorem implies that $\hat{B}$ commutes with $\hat{A}^*$ also, and (multiply the matrices $\hat{A}^*$ and $\hat{B}$ in both orders and compare corresponding entries) this implies the desired conclusion.

The corollary takes a little more work. If B is invertible, and if UP is its polar decomposition, then U is unitary and P is, as always, the positive square root of B^*B. If A_1 and A_2 are normal and $A_1B = BA_2$, then

$$A_2(B^*B) = (A_2B^*)B = (B^*A_1)B = B^*(A_1B) = B^*(BA_2) = (B^*B)A_2,$$

so that

$$A_2P^2 = P^2A_2;$$

it follows that

$$A_2P = PA_2.$$

(Compare Solution 137.) Since $A_1UP = UPA_2$ (by assumption) $= UA_2P$ (by what was just proved), it follows that $A_1U = UA_2$, and the proof of the corollary is complete.

There is a breathtakingly elegant and simple proof of the Putnam–Fuglede theorem in [118]. The original proof is in [45]; a variant is in [50] or

304

[56]; the two-operator generalization first appeared in [109]. The ingenious matrix derivation of Putnam from Fuglede is due to Berberian [9].

Solution 193. The answer is surprising: if a collection of normal operators **193** is a vector space, then each pair of operators in the collection is commutative. The assumption that the collection is an algebra (is closed under multiplication) is irrelevant.

The proof is based on a computational trick and Fuglede's theorem. If **M** is a vector space of normal operators, and if A and B are in **M**, then $A + B$ and $A + iB$ are in **M**, and therefore both $A + B$ and $A + iB$ are normal. In other words, if

$$C = (A + B)^*(A + B) - (A + B)(A + B)^*$$

and

$$D = (A + iB)^*(A + iB) - (A + iB)(A + iB)^*,$$

then $C = D = 0$, and therefore $C + iD = 0$. Multiply everything out and infer that $2(B^*A - AB^*) = 0$, i.e., that B^* commutes with A. By Fuglede's theorem, B commutes with A.

Reference: [110].

Solution 194. If $\chi_{\varphi^{-1}(D)} f = f$, then $\{x: f(x) \neq 0\} \subset \varphi^{-1}(D)$, and therefore **194**

$$\|A^n f\|^2 = \int |\varphi^n f|^2 \, d\mu = \int_{\varphi^{-1}(D)} |\varphi^n f|^2 \, d\mu$$

$$\leq \int_{\varphi^{-1}(D)} |f|^2 \, d\mu = \|f\|^2.$$

If, conversely, $\|A^n f\| \leq \|f\|$ for all n, and if $M_r = \{x: |\varphi(x)| \geq r > 1\}$, then

$$\|f\|^2 \geq \int |\varphi^n f|^2 \, d\mu \geq \int_{M_r} r^{2n} |f|^2 \, d\mu.$$

Unless f vanishes on M_r, the last written integral becomes infinite with n. Conclusion: f vanishes on M_r, for every r, and therefore $\{x: f(x) \neq 0\} \subset \varphi^{-1}(D)$.

Solution 195. It is convenient to begin with the observation that if A is **195** quasinormal, then ker A reduces A. Reason: ker $A = $ ker A^*A for every operator A; since quasinormality implies that A^* commutes with A^*A, it follows that ker A^*A is invariant under A^*.

In view of the preceding paragraph every quasinormal operator is the direct sum of 0 and an operator with trivial kernel. Since the direct summands can be treated separately, there is no loss of generality in assuming that ker $A = 0$ in the first place. If, in that case, UP is the polar decomposition of A, then U is an isometry, and (by Problem 137) $UP = PU$ and

305

$U*P = PU*$. The isometric character of U implies that if E is the projection $UU*$, then $(1 - E)U = U*(1 - E) = 0$. In view of these algebraic relations, A can be shown to be subnormal by explicitly constructing a normal extension for it. If A acts on $\mathbf{H}$, then a normal extension B can be constructed that acts on $\mathbf{H} \oplus \mathbf{H}$. (If $\mathbf{H}$ is identified with $\mathbf{H} \oplus 0$, then $\mathbf{H}$ is a subspace of $\mathbf{H} \oplus \mathbf{H}$.) An operator on $\mathbf{H} \oplus \mathbf{H}$ is given by a two-by-two matrix whose entries are operators on $\mathbf{H}$. If, in particular,

$$V = \begin{pmatrix} U & 1 - E \\ 0 & U* \end{pmatrix} \quad \text{and} \quad Q = \begin{pmatrix} P & 0 \\ 0 & P \end{pmatrix},$$

then V is unitary, Q is positive, V and Q commute, and therefore

$$B = VQ = \begin{pmatrix} UP & (1 - E)P \\ 0 & U*P \end{pmatrix}$$

is a normal extension of A.

196 **Solution 196.** If A is quasinormal and $B = A*A$, then B is Hermitian and $AB = BA$. If E is the spectral measure associated with B, then $AE(M) = E(M)A$ for every Borel subset M of the real line: this is the trivial, Hermitian, special case of Fuglede's theorem. If B is not a scalar, then there exists an M with $E(M) \neq 0$ and $E(M) \neq 1$; in that case the proof is over. If B is equal to the (positive) scalar β, then $A/\sqrt{\beta}$ is an isometry, and the desired conclusion is obviously true; see Problem 149.

197 **Solution 197.** Let $\mathbf{M}_1$ be the set of all finite sums of the form $\sum_j B_1*^j f_j$, where $f_j \in \mathbf{H}$ for all $j\, (= 0, 1, 2, \cdots)$. The set $\mathbf{M}_1$ is a linear manifold; since $B_1(\sum_j B_1*^j f_j) = \sum_j B_1*^j(B_1 f_j)$ and $B_1*(\sum_j B_1*^j f_j) = \sum_j B_1*^{j+1} f_j$, the closure of $\mathbf{M}_1$ reduces B_1. Since $\mathbf{H}$ itself is included in $\mathbf{M}_1$, the minimality of B_1 implies that $\mathbf{K}_1 = \overline{\mathbf{M}}_1$. Similarly, of course, the set $\mathbf{M}_2$ of all finite sums of the form $\sum_j B_2*^j f_j$, where each f_j is in $\mathbf{H}$, is dense in $\mathbf{K}_2$.

It is tempting to try to complete the proof by setting $U(\sum_j B_1*^j f_j) = \sum_j B_2*^j f_j$. This works, but it takes a little care. First: does this equation really define anything? That is: if $\sum_j B_1*^j f_j = \sum_j B_1*^j g_j$ (with f_j and g_j in $\mathbf{H}$), does it follow that $\sum_j B_2*^j f_j = \sum_j B_2*^j g_j$? Equivalently (subtract): if $\sum_j B_1*^j f_j = 0$, does it follow that $\sum_j B_2*^j f_j = 0$? The answer is yes; the reason is contained in the following computation:

$$\left\| \sum_j B_1*^j f_j \right\|^2 = \left(\sum_j B_1*^j f_j, \sum_k B_1*^k f_k \right)$$
$$= \sum_j \sum_k (B_1^k f_j, B_1^j f_k) = \sum_j \sum_k (A^k f_j, A^j f_k).$$

This computation accomplishes much more than the proof that U is unambiguously defined; it implies that U is an isometry (from $\mathbf{M}_1$ onto $\mathbf{M}_2$), that therefore U has a unique isometric extension that maps $\mathbf{K}_1$ onto $\mathbf{K}_2$, and

306

that U is the identity on $\mathbf{H}$. The proof that $UB_1 = B_2 U$ is another computation. It suffices to verify that UB_1 agrees with $B_2 U$ on $\mathbf{M}_1$, and this is implied by

$$UB_1\left(\sum_j B_1{}^{*j}f_j\right) = U\left(\sum_j B_1{}^{*j}B_1 f_j\right) = \sum_j B_2{}^{*j}Af_j$$

$$= \sum_j B_2{}^{*j}B_2 f_j = B_2\sum_j B_2{}^{*j}f_j = B_2 U\left(\sum_j B_1{}^{*j}f_j\right).$$

Solution 198. The natural guess is $p(W)$, where W is the bilateral shift, and that is right. In this context it is most convenient to consider W as the multiplication operator on $\mathbf{L}^2$ of the unit circle, defined by $Wf(z) = zf(z)$, and U as its restriction to $\mathbf{H}^2$ (see Problems 82 and 84). Clearly $p(W)$ is a normal extension of $p(U)$; it is to be proved that if $\mathbf{M}$ is a subspace of $\mathbf{L}^2$ such that $\mathbf{H}^2 \subset \mathbf{M}$ and $\mathbf{M}$ reduces $p(W)$, then $\mathbf{M} = \mathbf{L}^2$.

The operators $p(W)$ and $p(W)^*$ are multiplication operators induced by the function p and its complex conjugate p^*. The invariance of $\mathbf{M}$ under $p(W)^*$ and the presence in $\mathbf{M}$ of e_0 imply therefore the presence in $\mathbf{M}$ of the complex conjugates of polynomials of arbitrarily high degrees.

Suppose now that a polynomial q of degree n $(\geqq 1)$ is a power of p, so that $q^* \in \mathbf{M}$; assume (with no loss of generality) that q is monic, so that $q = e_n + \sum_{j=0}^{n-1} \alpha_j e_j$. Since $e_{n-1} \in \mathbf{M}$, it follows that $q^*e_{n-1} \in \mathbf{M}$; since, however, $\mathbf{H}^2 \subset \mathbf{M}$, so that every polynomial is in $\mathbf{M}$, it follows that $e_{-1} \in \mathbf{M}$. An inductive repetition of the argument shows that $e_{-n} \in \mathbf{M}$ for $n = 1, 2, 3, \cdots$, and hence that $\mathbf{M} = \mathbf{L}^2$.

The result is a special case of a general theory of minimal normal extensions; see [104].

Solution 199. *There exist two subnormal operators that are similar but not unitarily equivalent.*

PROOF. Consider the measure space consisting of the unit circle together with its center, with measure v defined so as to be normalized Lebesgue measure in the circle and a unit mass at the center. Let B be the position operator on $\mathbf{L}^2(v)$ (that is, $(Bf)(z) = zf(z)$), and let A be its restriction to the closure $\mathbf{H}^2(v)$ of all polynomials. Clearly B is normal and A is subnormal.

An orthonormal basis for $\mathbf{H}^2(v)$ consists of the functions

$$e_n \qquad (n = 1, 2, 3, \cdots),$$

defined by $e_n(z) = z^n$, together with the function e_0, defined by

$$e_0(z) = 1/\sqrt{2}.$$

The action of A on this basis is easy to describe: $Ae_0 = (1/\sqrt{2})e_1$ and $Ae_n = e_{n+1}$ for $n = 1, 2, 3, \cdots$. In other words, A is a unilateral weighted shift, with weight sequence $\{1/\sqrt{2}, 1, 1, 1, \cdots\}$. It follows from Problem 90

198

199

(but it is just as easy to verify directly) that A is similar to the ordinary unweighted unilateral shift U. There are several ways of proving that U and A are not unitarily equivalent. One way is to recall that two unilateral weighted shifts are unitarily equivalent only if corresponding weights have equal moduli (cf. Problem 90); the simplest way, however, is to observe that U is an isometry and A is not.

It is worth noting that B is the minimal normal extension of A (see Problem 197). This is not obvious at a glance, but it is quite easy to prove. From this it follows again that U and A are not unitarily equivalent. Reason: their minimal normal extensions are not.

This example is due to D. E. Sarason.

200 **Solution 200.** It is to be proved that if λ is a complex number such that $B - \lambda$ is not invertible, then neither is $A - \lambda$. By simple geometry (translate) and equally simple logic (form the contrapositive), the assertion reduces to this: if A is invertible, then so is B. Suppose therefore that A is invertible; without loss of generality normalize so that $\|A^{-1}\| = 1$. Let ε be an arbitrary number in the open interval $(0, 1)$, fixed from now on, and write $\mathbf{E} = \{f : \|B^n f\| \leqq \varepsilon^n \|f\|, n = 1, 2, 3, \cdots\}$. If $\mathbf{I}$ and $\mathbf{K}$ are the domains of A and B, and if $f \in \mathbf{E}$ and $g \in \mathbf{H}$, then

$$|(f, g)| = |(f, A^n A^{-n} g)| = |(f, B^n A^{-n} g)|$$

$$= |(B^{*n} f, A^{-n} g)| \leqq \|B^{*n} f\| \cdot \|A^{-n} g\|$$

$$= \|B^n f\| \cdot \|A^{-n} g\| \leqq \varepsilon^n \cdot \|f\| \cdot \|g\|$$

for all n, and, consequently, $(f, g) = 0$. In other words, $\mathbf{E} \perp \mathbf{H}$, and therefore $\mathbf{H} \subset \mathbf{E}^{\perp}$. Since (Problem 194) $\mathbf{E}$ is a reducing subspace for B, it follows that $\mathbf{E}^{\perp} = \mathbf{K}$, so that $\mathbf{E} = 0$; from this in turn (see Problem 194) it follows that B is invertible.

201 **Solution 201.** The proof depends on the trivial spectral inclusion

$$\Pi(A) \subset \Pi(B)$$

only. The conclusion holds for a pair of operators A and B whenever their spectra and approximate point spectra are so related; no deeper or more special properties of subnormal and normal operators are needed.

Consider the sets $\Delta^- = \Delta - \operatorname{spec} A$ and $\Delta^+ = \Delta \cap \operatorname{spec} A$. Since Δ is open and $\operatorname{spec} A$ is closed, the set Δ^- is open. Assertion: Δ^+ is also open. To prove this, consider an arbitrary point λ in Δ^+. Since $\lambda \in \Delta$ and Δ is a hole of $\operatorname{spec} B$, the point λ cannot belong to $\operatorname{spec} B$. This implies, of course, that λ is not in $\Pi(B)$, hence that λ is not in $\Pi(A)$, and hence that λ is not on the boundary of $\operatorname{spec} A$ (see Problem 78). Since, however, $\lambda \in \Delta^+$ and $\Delta^+ \subset \operatorname{spec} A$, it follows that the only place λ can be is in the interior of $\operatorname{spec} A$. This argument proves that Δ^+ is, in fact, the intersection of Δ with the interior of spec A, and it follows, as asserted, that Δ^+ is open.

Since Δ is the union of the disjoint open sets Δ^- and Δ^+, the connectedness of Δ implies that one of them is empty.

The result is due to Bram [16]; for a generalization see [77]. The simple proof above is due to S. K. Parrott.

Solution 202. *Every finite-dimensional subspace invariant under a normal operator B reduces B.*   **202**

PROOF. Since on a finite-dimensional space every operator has an eigenvalue, it is sufficient to prove that each one-dimensional invariant subspace of B reduces B. This is easy: in fact each eigenvector of B is an eigenvector of B^* too. (If $Bf = \lambda f$, then, by normality,

$$0 = \|(B - \lambda)f\| = \|(B^* - \lambda^*)f\|.)$$

Corollary. *On finite-dimensional spaces every subnormal operator is normal.*

PROOF. The restriction of a normal operator to a reducing subspace is normal.

From the result thus proved it follows that the answer to the dimension question is no. Reason: if B is a normal extension of A to $\mathbf{K}$, then $\mathbf{K} \cap \mathbf{H}^\perp$ is invariant under the normal operator B^*, and, therefore, if $\dim(\mathbf{K} \cap \mathbf{H}^\perp)$ is finite, $\mathbf{H}$ reduces B. Since A was assumed to be non-normal, this is impossible.

Solution 203. The difficulty is to prove that something is *not* subnormal. Since   **203**
subnormality was defined by requiring the existence of something, what is wanted here is a non-existence theorem. The best way to prove such a theorem (the only way?) is to assume existence, derive a usable "constructive" necessary condition from it (with luck it will be sufficient as well), and then look for something that violates the condition.

If B (on $\mathbf{K}$) is a normal extension of A (on $\mathbf{H}$), and if $f_0, \cdots, f_n$ are vectors in $\mathbf{H}$, then $\|\sum_j B^{*j}f_j\| \geq 0$. This triviality can be rewritten in a non-trivial way, as follows:

$$0 \leq \left(\sum_j B^{*j}f_j, \sum_i B^{*i}f_i \right) = \sum_j \sum_i (B^{*j}f_j, B^{*i}f_i)$$

$$= \sum_j \sum_i (B^i B^{*j}f_j, f_i)$$

$$= \sum_j \sum_i (B^{*j}B^i f_j, f_i) \qquad \text{(because B is normal)}$$

$$= \sum_j \sum_i (B^i f_j, B^j f_i) = \sum_j \sum_i (A^i f_j, A^j f_i).$$

Replace each f_j by some scalar multiple $\xi_j f_j$, and conclude that

$$\sum_j \sum_i (A^i f_j, A^j f_i)\xi_j \xi_i^* \geqq 0,$$

i.e., that the finite matrix $\langle (A^i f_j, A^j f_i) \rangle$ is positive definite. This is a "constructive" intrinsic necessary condition that follows from subnormality; it will be used to exhibit a hyponormal operator that is not subnormal. First, however, it is pertinent to comment that the condition is sufficient, as well as necessary, for subnormality. To put it precisely: if, for each finite set of vectors $f_0, \cdots, f_n$, the corresponding matrix $\langle (A^i f_j, A^j f_i) \rangle$ is positive definite, then the operator A is subnormal. The proof is somewhat involved; the fact will not be used in the sequel.

The desired counterexample can be found among weighted shifts. When is a weighted shift S, with weights $\{\alpha_0, \alpha_1, \alpha_2, \cdots\}$, hyponormal? Since both S^*S and SS^* are diagonal, there is an easy answer in terms of the α's. The diagonal of S^*S is $\{|\alpha_0|^2, |\alpha_1|^2 \cdot |\alpha_2|^2, \cdots\}$, and the diagonal of SS^* is $\{0, |\alpha_0|^2, |\alpha_1|^2, |\alpha_2|^2, \cdots\}$; it follows that S is hyponormal if and only if the sequence $\{|\alpha_n|\}$ is monotone increasing.

With this much information available, the construction of a counterexample along these lines (if it is possible at all) should be easy. A finite amount of experimentation might lead to the weighted shift S with weights $\{\alpha, \beta, 1, 1, 1, \cdots\}$, where $0 < \alpha < \beta < 1$. The preceding paragraph implies that S is hyponormal. To prove that S is not subnormal, examine the matrix $\langle (S^i e_j, S^j e_i) \rangle$, where $\{e_0, e_1, e_2, \cdots\}$ is the orthonormal basis that S shifts, and where i and j take the values 0, 1, 2. Written explicitly, the matrix is

$$\begin{pmatrix} 1 & \alpha & \alpha\beta \\ \alpha & \beta^2 & \beta \\ \alpha\beta & \beta & 1 \end{pmatrix}.$$

Its determinant is $-\alpha^2(1 - \beta^2)^2$, which is negative.
Examples of this type have been studied by J. G. Stampfli.

204 **Solution 204.** The "if" for normal partial isometries is trivial and for subnormal ones is a consequence of Problem 149. (The point is that the typical non-unitary isometry, the unilateral shift, is subnormal.) To prove "only if", suppose that U is a partial isometry, so that U^*U is the projection on the initial space (the orthogonal complement of the kernel), and UU^* is the projection on the final space (the range). If U is subnormal, then it is hyponormal, and consequently the initial space includes the range. This implies that the initial space is invariant under U, and hence that it reduces U; clearly the restriction of U to the initial space is an isometry. If, moreover, U is normal, then the initial space is equal to the range, and therefore the restriction of U to the initial space is unitary.

It is interesting to note (as a consequence of the proof) that a partial isometry is subnormal if and only if it is hyponormal.

Solution 205. For $n = 1$, the equality is trivial; proceed by induction. Since **205**

$$\|A^n f\|^2 = (A^n f, A^n f) = (A^* A^n f, A^{n-1} f)$$
$$\leqq \|A^* A^n f\| \cdot \|A^{n-1} f\| \leqq \|A^{n+1} f\| \cdot \|A^{n-1} f\|$$
$$\leqq \|A^{n+1}\| \cdot \|A^{n-1}\| \cdot \|f\|^2$$

for every vector f, it follows that

$$\|A^n\|^2 \leqq \|A^{n+1}\| \cdot \|A^{n-1}\|.$$

In view of the induction hypothesis ($\|A^k\| = \|A\|^k$ whenever $1 \leqq k \leqq n$), this can be rewritten as

$$\|A\|^{2n} \leqq \|A^{n+1}\| \cdot \|A\|^{n-1},$$

from which it follows that

$$\|A\|^{n+1} \leqq \|A^{n+1}\|.$$

Since the reverse inequality is universal, the induction step is accomplished.

Reference: [4, 136]. The proof above is a slight simplification of the simple proof in [136].

Solution 206. Suppose that A is hyponormal. The program is to prove that **206** the span of the eigenvectors of A reduces A; compactness does not enter here. In the presence of compactness the orthogonal complement of that span becomes amenable; an application of Problem 205 will yield the conclusion.

(1) For each complex number λ,

$$\{f: Af = \lambda f\} \subset \{f: A^* f = \lambda^* f\}.$$

The reason is that $A - \lambda$ is just as hyponormal as A, and that, on general grounds that have nothing to do with hyponormality, a necessary and sufficient condition that $(A^* - \lambda^*)f = 0$ is that

$$(A - \lambda)(A^* - \lambda^*)f = 0.$$

(2) For each complex number λ, the subspace $\{f: Af = \lambda f\}$ reduces A. Indeed: invariance under A is trivial, and invariance under A^* follows from (1).

(3) If $\lambda_1 \neq \lambda_2$, then

$$\{f: Af = \lambda_1 f\} \perp \{f: Af = \lambda_2 f\}.$$

A straightforward and often-used argument: if $Af_1 = \lambda_1 f_1$ and $Af_2 = \lambda_2 f_2$, then

$$\lambda_1(f_1, f_2) = (Af_1, f_2) = (f_1, A^* f_2) = \lambda_2(f_1, f_2).$$

(4) The span of all the eigenvectors of A reduces A and the restriction of A to that span is normal. Proof: use (2) and observe that, by (3), the restriction of A to each eigenspace is normal (in fact equal to a scalar).

311

(5)   Now assume that A is compact, and consider the restriction of A to the orthogonal complement of the span of all the eigenvectors. The resulting operator is still hyponormal (by the reduction assertion of (4)), and still compact. Since the point spectrum of this compact operator is empty, it is quasinilpotent (Problem 179); an application of Problem 205 implies that it must be 0. If the orthogonal complement on which all this action is taking place is not 0, then there is a contradiction: the non-zero vectors in it both must be and cannot be eigenvectors of eigenvalue 0.

207   **Solution 207.** *If A is hyponormal, $A = B + iC$, with B and C Hermitian and C compact, then A must be normal.*

Consider an eigenvalue γ of C with corresponding eigenvector f, and observe that

$$\|Af\| = \|(B + i\gamma)f\| = \|(B + i\gamma)^* f\|,$$

because $B + i\gamma$ is normal. It follows that

$$\|Af\| = \|(B - i\gamma)f\| = \|A^* f\|.$$

If, in other words, **M** is the set of all those vectors f for which

$$\|Af\| = \|A^* f\|,$$

then **M** contains every eigenvector of C.

To say that $\|Af\| = \|A^* f\|$ is the same as to say $(A^*Af, f) = (AA^*f)$, or, in other words, it is the same as to say $(Df, f) = 0$, where D is the positive operator $A^*A - AA^*$. For a positive operator $(Df, f) = 0$ is equivalent to $Df = 0$. Consequence: **M** = ker D, so that, in particular, **M** is a subspace.

Since the eigenvectors of C span the whole space (Problem 172), **M** = **H**. Conclusion: A is normal.

Reference: [153].

208   **Solution 208.** *Every hyponormal idempotent is a projection* (and, therefore, the same is true for quasinormal and subnormal idempotents).

For the proof, observe first that if $P^2 = P$, then ran $P = \{f : Pf = f\}$, and, consequently, ran P is closed. Express P as an operator matrix with respect to the decomposition ran $P \oplus \text{ran}^\perp P$; the result is of the form

$$P = \begin{pmatrix} 1 & A \\ 0 & 0 \end{pmatrix}.$$

It follows that

$$PP^* = \begin{pmatrix} 1 & A \\ 0 & 0 \end{pmatrix}\begin{pmatrix} 1 & 0 \\ A^* & 0 \end{pmatrix} = \begin{pmatrix} 1 + AA^* & 0 \\ 0 & 0 \end{pmatrix},$$

$$P^*P = \begin{pmatrix} 1 & 0 \\ A^* & 0 \end{pmatrix}\begin{pmatrix} 1 & A \\ 0 & 0 \end{pmatrix} = \begin{pmatrix} 1 & A \\ A^* & A^*A \end{pmatrix}.$$

If now P is assumed to be hyponormal, then $1 + AA^* \leqq 1$. That implies $A = 0$, and the proof is complete.

Solution 209. If U is the unilateral shift, and

$$A = U^* + 2U,$$

then A is hyponormal but A^2 is not.

The proof that A is hyponormal can be done in (at least) two ways, each of which is illuminating. Algebraically:

$$A^*A = (U + 2U^*)(U^* + 2U) = UU^* + 2U^{*2} + 2U^2 + 4,$$

$$AA^* = (U^* + 2U)(U + 2U^*) = 1 + 2U^2 + 2U^{*2} + 4UU^*,$$

and therefore

$$A^*A - AA^* = 3 - 3UU^* = 3(1 - UU^*) \geqq 0.$$

Numerically: since

$$A = \begin{pmatrix} 0 & 1 & 0 & 0 \\ 2 & 0 & 1 & 0 \\ 0 & 2 & 0 & 1 \\ 0 & 0 & 2 & 0 \\ & & & & \ddots \end{pmatrix} \quad \text{and} \quad A^* = \begin{pmatrix} 0 & 2 & 0 & 0 \\ 1 & 0 & 2 & 0 \\ 0 & 1 & 0 & 2 \\ 0 & 0 & 1 & 0 \\ & & & & \ddots \end{pmatrix},$$

it follows that if $f = \langle \xi_0, \xi_1, \xi_2, \cdots \rangle$, then

$$Af = \langle \xi_1, 2\xi_0 + \xi_2, 2\xi_1 + \xi_3, 2\xi_2 + \xi_4, \cdots \rangle$$

and

$$A^*f = \langle 2\xi_1, \xi_0 + 2\xi_2, \xi_1 + 2\xi_3, \xi_2 + 2\xi_4, \cdots \rangle.$$

Each of $\xi_1, 2\xi_1, \xi_2, 2\xi_2, \xi_3, 2\xi_3, \cdots$ occurs in both Af and A^*f exactly once; their contributions to the sum of the squares of the moduli exactly balance. What does not balance is $4|\xi_0|^2$ in $\|Af\|^2$ and $|\xi_0|^2$ in $\|A^*f\|^2$; the former always dominates.

The proof that A^2 is not hyponormal is less pleasant. The quickest way is to exhibit a vector f such that $\|A^2f\| < \|A^{*2}f\|$. One such vector f is $e_0 - 2e_2$. Once that is said, then nothing remains except numerical computation, which does not shed any light at all. The answer is that

$$\|A^2f\|^2 = 80 \quad \text{and} \quad \|A^{*2}f\|^2 = 89.$$

This example is due to Ito and Wong [78]; it is much simpler than the original one [49].

CHAPTER 22

Numerical Range

210

Solution 210. To prove that the numerical range W of an operator A is convex, it is sufficient to prove that the intersection of W with every straight line in the complex plane is connected. (The ingenious idea of basing the proof on this observation is due to N. P. Dekker [35].) Consider a line with the (real) equation $px + qy + r = 0$ (where the pair $\langle x, y \rangle$ is identified with the complex number $x + iy$). If $A = B + iC$, with B and C Hermitian, then the intersection in question consists of all those complex numbers $x + iy$ for which $px + qy + r = 0$ and for which there exists a unit vector f such that $x = (Bf, f)$, $y = (Cf, f)$. In other words, the intersection is the set of all $(Bf, f) + i(Cf, f)$ with $\|f\| = 1$ and $((pB + qC + r)f, f) = 0$. There is still another way of describing the intersection: it is the image under the (continuous) mapping $f \mapsto (Bf, f) + i(Cf, f)$ of the set $\mathbf{N}$ of all those unit vectors f for which $((pB + qC + r)f, f) = 0$. An efficient way to conclude the proof is to show that $\mathbf{N}$ itself is connected.

With superfluous notation discarded, the desired conclusion is just this: if L is a Hermitian operator, then the set of all unit vectors f such that $(Lf, f) = 0$ is connected. Although this general statement is true, and is easy to prove directly, a slightly indirect two-step process gives more geometric insight. Step 1: prove it for the special case when the dimension of the underlying Hilbert space is 2. Step 2: reduce the general Toeplitz–Hausdorff theorem to its 2-dimensional special case.

Step 1. Assume (with no loss) that L is defined on $\mathbf{C}^2$ by a diagonal matrix $\begin{pmatrix} \alpha & 0 \\ 0 & \beta \end{pmatrix}$. The desired conclusion now is that the set $\mathbf{N}$ of all $\langle \xi, \eta \rangle$ with $|\xi|^2 + |\eta|^2 = 1$ and $\alpha|\xi|^2 + \beta|\eta|^2 = 0$ is connected. As $|\xi|$ goes from 0 to 1, $\alpha|\xi|^2 + \beta|\eta|^2$ goes from β to α and can be equal to 0 at most once.

314

To get the most general element of **N**, multiply the two coordinates of the unique positive element by any two complex numbers of absolute value 1. From this point of view **N** is a torus, the Cartesian product of two copies of the unit circle, and therefore clearly connected.

Step 2. Suppose that A is an operator on a Hilbert space **H** and f and g are unit vectors in **H**. Let P be the projection from **H** onto the 2-dimensional space **K** spanned by f and g. (If the span of f and g is 1-dimensional, then $f = \lambda g$ with $|\lambda| = 1$, whence $(Af, f) = (Ag, g)$, and the desired conclusion degenerates to a triviality.) Apply the Toeplitz–Hausdorff theorem (2-dimensional case) to the operator PAP acting on the space **K** to infer that to every complex number z on the segment joining (Af, f) $(=(PAPf, f))$ to (Ag, g) $(=(PAPg, g))$ there corresponds a unit vector h (in **K**) such that $z = (PAPh, h)$ $(=(Ah, h))$. That is what was wanted—plus the extra information that h is in **K**.

For an alternative proof see [30].

Solution 211. *For every operator A and for every positive integer k, the k-numerical range $W_k(A)$ is convex.*   **211**

PROOF. Suppose to begin with that **M** and **N** are k-dimensional Hilbert spaces and that T is a linear transformation from **M** into **N**. There is a useful sense in which T and T^* (from **N** into **M**) can be simultaneously diagonalized. The assertion is that there exist orthonormal bases $\{f_1, \cdots, f_k\}$ for **M** and $\{g_1, \cdots, g_k\}$ for **N**, and there exist positive $(\geqq 0)$ scalars $\alpha_1, \cdots, \alpha_k$ such that $Tf_i = \alpha_i g_i$ and $T^*g_i = \alpha_i f_i$, $i = 1, \cdots, k$. To prove this, let UP be the polar decomposition of T, and diagonalize P. That is: find an orthonormal basis $\{f_1, \cdots, f_k\}$ for **M** and find positive scalars $\alpha_1, \cdots, \alpha_k$ such that $Pf_i = \alpha_i f_i$. If the partial isometry U is not an isometry from **M** onto **N**, it can be replaced by one (since dim **M** = dim **N** = k); assume that that has been done. Then put $g_i = Uf_i$, $i = 1, \cdots, k$, and reap the consequences:

$$Tf_i = UPf_i = U(\alpha_i f_i) = \alpha_i g_i,$$

and

$$T^*g_i = PU^*g_i = Pf_i = \alpha_i f_i, i = 1, \cdots, k.$$

That is a lemma; now for the theorem. Suppose that P and Q are projections of rank k, with respective ranges **M** and **N**. If T is the restriction of QP to **M**, then the preceding lemma is applicable. For each i $(=1, \cdots, k)$, let L_i be the span of f_i and g_i. Assertion: the subspaces L_i are pairwise orthogonal. Suppose, indeed, that $i \neq j$; since $f_i \perp f_j$ and $g_i \perp g_j$, it is sufficient to prove that $f_i \perp g_j$ (for then $f_j \perp g_i$ follows by symmetry). The proof is easy:

$$(f_i, g_j) = (Pf_i, Qg_j) = (QPf_i, g_j) = (\alpha_i g_i, g_j).$$

315

The desired convexity proof is now near at hand. If $0 \leq t \leq 1$, use the classical Toeplitz–Hausdorff theorem k times to obtain a unit vector h_i in $\mathbf{L}_i$ so that

$$(Ah_i, h_i) = t(Af_i, f_i) + (1 - t)(Ag_i, g_i).$$

Since $\{h_1, \cdots, h_k\}$ is an orthonormal set, the projection R onto its span has rank k, and

$$t \cdot \operatorname{tr} PAP + (1 - t) \cdot \operatorname{tr} QAQ = t \cdot \sum_i (Af_i, f_i) + (1 - t) \cdot \sum_i (Ag_i, g_i)$$

$$= \sum_i (Ah_i, h_i) = \operatorname{tr} RAR.$$

The proof of the theorem is complete.

The problem was raised in [57]. The first solution, somewhat more complicated than the one above, is due to C. A. Berger.

212

Solution 212. *If A is an operator and λ is a complex number such that $|\lambda| = \|A\|$ and $\lambda \in W(A)$, then λ is an eigenvalue of A.*

PROOF. If $\lambda = (Af, f)$ with $\|f\| = 1$, then

$$\|A\| = |\lambda| = |(Af, f)| \leq \|Af\| \cdot \|f\| \leq \|A\|,$$

so that equality holds everywhere. The known facts about when the Schwarz inequality becomes an equation imply that $Af = \lambda_0 f$ for some λ_0, and this in turn implies that

$$\lambda_0 = \lambda_0(f, f) = (\lambda_0 f, f) = (Af, f) = \lambda,$$

so that λ is an eigenvalue of A.

It follows from this theorem that if λ is a number in $\overline{W(A)}$ such that $|\lambda| = \|A\|$ and λ is *not* an eigenvalue of A (and, in particular, if A has no eigenvalues), then λ does not belong to $W(A)$. In view of this comment it is easy to construct examples of operators whose numerical range is not closed.

(1) Observe that the eigenvalues of every operator A belong to $W(A)$. (Proof: if $Af = \lambda f$ with $\|f\| = 1$, then $(Af, f) = \lambda$.) If A is normal, then $\|A\| = \sup\{|\lambda| : \lambda \in W(A)\}$, so that there always exists a λ in $\overline{W(A)}$ such that $|\lambda| = \|A\|$. It follows that if a normal operator has sufficiently many eigenvalues to approximate its norm, but does not have one whose modulus is as large as the norm, then its numerical range will not be closed. Concrete example: a diagonal operator such that the modulus of the diagonal terms does not attain its supremum. Another example, along slightly different lines: take A to be the diagonal operator with diagonal $\{1, \frac{1}{2}, \frac{1}{3}, \cdots\}$. Since $A \geq 0$ and $\ker A = 0$, it follows that $0 \notin W(A)$; in fact $W(A) = (0, 1]$. This shows, by the way, that the numerical range may fail to be closed even for compact operators.

316

(2) Take A to be the unilateral shift. Since every number in the open unit disc is an eigenvalue of A^*, it follows that the open unit disc is included in $W(A^*)$. Since $W(A^*)$ is always $(W(A))^*$ (proof: $(A^*f, f) = (Af, f)^*$), it follows that the open unit disc is included in $W(A)$. Since, finally, A has no eigenvalues, the theorem proved above implies that $W(A)$ cannot contain any number of modulus 1, so that $W(A)$ is equal to the open unit disc.

Solution 213. *If A is compact and $0 \in W(A)$, then $W(A)$ is closed*   **213**

PROOF. Observe to begin with that if $0 \in W(A)$, then $(Af, f) \in W(A)$ for every vector f in the unit ball (and not only for the unit vectors). Reason: if $\|f\| = 1$ and $0 \leq t \leq 1$, then

$$(A(tf), tf) = t^2(Af, f) = t^2(Af, f) + (1 - t^2) \cdot 0 \in W(A),$$

by convexity.

The argument of the preceding paragraph did not need compactness; it is valid for every operator. The next step is to show that in the presence of compactness the quadratic form $f \mapsto (Af, f)$ is weakly continuous on bounded sets. Indeed, if $\{f_n\}$ is a bounded net weakly convergent to f, then

$$|(Af_n, f_n) - (Af, f)| \leq |(Af_n, f_n) - (Af, f_n)| + |(Af, f_n) - (Af, f)|;$$

the first summand tends to 0 because (by compactness) $\{Af_n\}$ is strongly convergent, and the second summand tends to 0 because $f_n \to f$ weakly.

If both hypotheses are satisfied ($0 \in W(A)$ and A is compact), then $W(A) = \{(Af, f): \|f\| \leq 1\}$ (by the first paragraph above), so that $W(A)$ is the image of a weakly compact set (the unit ball) under a mapping that is weakly continuous on that set (by the preceding paragraph).

Reference: [31].

Solution 214. If λ is in the compression spectrum of A, then λ^* is an eigen-   **214**
value of A^*, so that $\lambda^* \in W(A^*)$, and therefore $\lambda \in W(A)$. Conclusion: the numerical range includes the compression spectrum.

If λ is in the approximate point spectrum of A, then there exist unit vectors f_n such that $(A - \lambda)f_n \to 0$. Since

$$|(Af_n, f_n) - \lambda| = |((A - \lambda)f_n, f_n)|$$

$$\leq \|(A - \lambda)f_n\|,$$

it follows that $(Af_n, f_n) \to \lambda$. Conclusion: the closure of the numerical range includes the approximate point spectrum.

These two paragraphs complete the proof. A slightly different proof can be obtained by combining the fact just proved for the approximate point spectrum with two other facts: the boundary of the spectrum is included in the approximate point spectrum, and the numerical range is convex.

215   **Solution 215.** *If V is the Volterra integration operator and if*

$$A = 1 - (1 + V)^{-1} (= V(1 + V)^{-1}),$$

then A is quasinilpotent but $W(A)$ does not contain 0.

PROOF. Since the quasinilpotence of A is obvious (Problem 186), it is sufficient to prove that if f is a vector such that $(Af, f) = 0$, then $f = 0$. If, indeed, $(Af, f) = 0$, then

$$\|f\|^2 = ((1 + V)^{-1}f, f) \leqq \|(1 + V)^{-1}\| \cdot \|f\|^2 = \|f\|^2.$$

(See Solution 190. The trick of considering $(1 + V)^{-1}$ has more than one application.) It follows (by what is known about when the Schwarz inequality degenerates) that f must be an eigenvalue of $(1 + V)^{-1}$. Since $\mathrm{spec}(1 + V)^{-1} = \{1\}$, it follows that $(1 + V)^{-1}f = f$, or $f = (1 + V)f$, or $Vf = 0$. This implies that $f = 0$ (see Problem 188); the proof is complete. Observe that the operator A is compact.

216   **Solution 216.** Suppose that A is a normal operator. Since $\overline{W(A)}$ is convex and $\mathrm{spec}\,A \subset \overline{W(A)}$ (Problems 210 and 214), it follows that conv $\mathrm{spec}\,A \subset \overline{W(A)}$. It remains to prove the reverse inclusion. In view of the characterization of convex hulls in terms of half planes, the desired result can be formulated this way: if a closed half plane includes $\mathrm{spec}\,A$, then it includes $W(A)$. If A is replaced by $\alpha A + \beta$ (where α and β are complex numbers), then spec and W are replaced by $\alpha\,\mathrm{spec} + \beta$ and $\alpha W + \beta$. This remark makes it possible to "normalize" the problem. Its effect is to reduce the problem to the study of any one particular half plane, for instance the right half plane. The desired result now is this: if every number in the spectrum of A has a positive ($\geqq 0$) real part, then the same is true of the numerical range of A. (Observe that the reduction to this point did not use normality; that assumption enters in the proof of the reduced statement.)

Use the spectral theorem to justify the assumption that A is a multiplication, induced by a bounded measurable function φ on a measure space with measure μ. If $f \in \mathbf{L}^2(\mu)$, then $(Af, f) = \int \varphi |f|^2 \, d\mu$. In these terms, the reduced statement says that if $0 \leqq \mathrm{Re}\,\varphi$ almost everywhere (this says that the essential range of φ is included in the right half plane), then

$$0 \leqq \mathrm{Re} \int \varphi |f|^2 \, d\mu = \int (\mathrm{Re}\,\varphi)|f|^2 \, d\mu.$$

This, finally, is obvious; if $dv = |f|^2 \, d\mu$, then v is a positive measure, and the assertion is just that the integral of a positive function with respect to a positive measure is positive.

217   **Solution 217.** *The closure of the numerical range of a subnormal operator is the convex hull of its spectrum.*

318

PROOF. If A is subnormal and B is its minimal normal extension (see Problem 197), then spec $B \subset$ spec A (Problem 200), and, trivially, $W(A) \subset W(B)$. It follows that

$$\overline{W(B)} = \text{conv spec } B \qquad \text{(Problem 216)}$$
$$\subset \text{conv spec } A$$
$$\subset \overline{W(A)} \qquad \text{(Problems 210 and 214)}$$
$$\subset \overline{W(B)},$$

and hence that all the sets that enter are the same.

Note, as a corollary of the proof, that the closure of the numerical range of a subnormal operator is the same as the closure of the numerical range of its minimal normal extension.

Solution 218. (a) If A is not invertible, then $0 \in$ spec A, so that $1 \in$ spec$(1 - A)$; it follows that $1 \le r(1 - A) \le w(1 - A)$. (b) Assume, with no loss of generality, that $\|A\| = 1$. (Multiply by a suitable positive constant.) The hypothesis $w(A) = \|A\|$ then guarantees the existence of a sequence $\{f_n\}$ of unit vectors such that $|(Af_n, f_n)| \to 1$; assume with no loss of generality that $(Af_n, f_n) \to 1$. (Multiply by a suitable constant of modulus 1.) Since $|(Af_n, f_n)| \le \|Af_n\| \le 1$ and $(Af_n, f_n) \to 1$, it follows that $\|Af_n\| \to 1$. This implies that

$$\|Af_n - f_n\|^2 = \|Af_n\|^2 - 2 \operatorname{Re}(Af_n, f_n) + 1 \to 0,$$

so that 1 is an approximate eigenvalue of A, and therefore $r(A)$ must be equal to 1.

218

Solution 219. *There exist convexoid operators that are not normaloid and vice versa.*

219

PROOF. Write

$$M = \begin{pmatrix} 0 & 0 \\ 1 & 0 \end{pmatrix},$$

and let N be a normal operator whose spectrum is the closed disc D with center 0 and radius $\frac{1}{2}$. If

$$A = \begin{pmatrix} M & 0 \\ 0 & N \end{pmatrix},$$

then spec $A = \{0\} \cup D = D$, and $W(A) = \text{conv}(W(M) \cup W(N)) = D$. This shows that A is convexoid. Since $\|A\| = 1$ (in fact $\|M\| = 1$), A is not normaloid.

Next write

$$A = \begin{pmatrix} M & 0 \\ 0 & 1 \end{pmatrix}.$$

319

Since $\|A\| = 1$, and $W(A) = \mathrm{conv}(D \cup \{1\})$, it follows that $w(A) = 1$ and hence that A is normaloid. Since, however, spec $A = \{0\} \cup \{1\}$, so that conv spec A is the closed unit interval, A is not convexoid.

Many of these concepts were first studied by Wintner [159]. The paper contains a small error; it asserts that every normaloid operator is convexoid.

220   **Solution 220.** *The function $\overline{W}$ is continuous with respect to the uniform (norm) topology; if the underlying Hilbert space is infinite-dimensional, then the function w is discontinuous with respect to the strong topology (and hence with respect to the weak).*

PROOF. If $\|A - B\| < \varepsilon$, and if f is a unit vector, then

$$|((A - B)f, f)| < \varepsilon,$$

and therefore

$$(Af, f) = (Bf, f) + ((A - Bf, f) \in W(B) + (\varepsilon).$$

It follows that $W(A) \subset W(B) + (\varepsilon)$; symmetrically, $W(B) \subset W(A) + (\varepsilon)$. This proves the first assertion. (The proof is due to A. Brown.)

As for the second assertion, consider the unilateral shift U. The sequence $\{U^{*n}\}$ tends to 0 in the strong topology (more and more Fourier coefficients get lost as n increases), but $w(U^{*n}) = 1$ for all n.

221   **Solution 221.** If a is a complex number with $|a| \leq 1$ and if $|z| < 1$, then $\mathrm{Re}(1 - za) = 1 - \mathrm{Re}(za) \geq 1 - |z| > 0$. If conversely the complex number a is such that $\mathrm{Re}(1 - za) \geq 0$ for each z with $|z| < 1$, then this is true, in particular, if $za = t|a|, 0 < t < 1$; since, therefore,

$$1 - t|a| = \mathrm{Re}(1 - t|a|) \geq 0,$$

it follows (let t tend to 1) that $|a| \leq 1$.

The operator fact corresponding to (and implied by) this numerical fact is that $w(A) \leq 1$ if and only if $\mathrm{Re}(1 - zA) \geq 0$. Indeed, the following assertions about A are pairwise equivalent:

$$w(A) \leq 1,$$

$$|(Af, f)| \leq 1 \quad \text{whenever } \|f\| = 1,$$

$$(\mathrm{Re}(1 - zA)f, f) \geq 0 \quad \text{whenever } \|f\| = 1 \text{ and } |z| < 1.$$

If $w(A) \leq 1$, then $r(A) \leq 1$, and therefore $1 - zA$ is invertible whenever $|z| < 1$. Since an invertible operator has positive real part if and only if its inverse has positive real part (if B is invertible, then

$$(B^{-1}f, f) = (B^{-1}f, BB^{-1}f) = (B(B^{-1}f), (B^{-1}f))^*),$$

it follows that $w(A) \leq 1$ if and only if $\mathrm{Re}(1 - zA)^{-1} \geq 0$ in the unit disc.

320

Observe next that if n is a positive integer and if ω is a primitive n-th root of unity (i.e., n is the smallest positive integer such that $\omega^n = 1$), then

$$\frac{1}{1 - z^n} = \frac{1}{n} \sum_{k=0}^{n-1} \frac{1}{1 - \omega^k z}$$

for all z other than the powers of ω. This identity is, in fact, the partial fraction expansion of the left side. For a direct verification, multiply through by $1 - z^n$, observe that the right side becomes a polynomial of degree $n - 1$ at most that is invariant under each of the n substitutions

$$z \mapsto \omega^k z \qquad (k = 0, \cdots, n - 1)$$

and is therefore constant, and then evaluate the constant by setting z equal to 0.

The identity of the preceding paragraph implies that if $w(A) \leqq 1$, then

$$(1 - z^n A^n)^{-1} = \frac{1}{n} \sum_{k=0}^{n-1} (1 - \omega^k z A)^{-1}$$

whenever $|z| < 1$. Since each summand on the right side has positive real part (because $w(\omega^k A) \leqq 1$), it follows that the left side has positive real part, and that implies that $w(A^n) \leqq 1$.

One step in the proof might be unfamiliar enough to deserve a second look. To prove an identity between operators by substitution into an identity between rational functions is to make use of the functional calculus for rational functions (cf. Problem 123). Explicitly: if φ_1 and φ_2 are rational functions whose poles are not in the spectrum of A, so that $\varphi_1(A)$ and $\varphi_2(A)$ make sense, then the same is true of each polynomial p in φ_1 and φ_2; if $\varphi(\lambda) = p(\varphi_1(\lambda), \varphi_2(\lambda))$, then $\varphi(A) = p(\varphi_1(A), \varphi_2(A))$. The proof is obvious.

The equivalence of $w(A) \leqq 1$ and $\mathrm{Re}(1 - zA)^{-1} \geqq 0$ for $|z| < 1$ is elementary, but basic for the argument; it was Berger's main new idea. That idea is visible in some form in all subsequent proofs. The proof given above is a simplification of a simplification discovered by Pearcy [107].

321

CHAPTER 23

Unitary Dilations

222 Solution 222. (a) As a heuristic guide to the proof, consider the very special case in which the given Hilbert space **H** is one-dimensional real Euclidean space and the dilation space **K** is the plane. In that case the given contraction A is a scalar α (with $|\alpha| \leqq 1$), and, in geometric terms, the assertion is that multiplication (on the line) by α can be achieved by a suitable rotation (in the plane), followed by projection (back to the line). A picture makes all this crystal clear; simple analytic geometry shows that the matrix of the rotation is

$$\begin{pmatrix} \alpha & \sqrt{1 - \alpha^2} \\ \sqrt{1 - \alpha^2} & -\alpha \end{pmatrix}.$$

The proof itself is the most direct possible imitation of the technique that worked for the plane. A few experiments are needed, to see whether the role of α^2 should be played by A^2, or AA^*, or A^*A, or sometimes one and sometimes another. The result can be described as follows. Given **H**, write **K** = **H** $\oplus$ **H** and identify **H** with the first summand; then each operator on **K** is a two-rowed matrix of operators on **H**, and, in particular,

$$P = \begin{pmatrix} 1 & 0 \\ 0 & 0 \end{pmatrix}.$$

Given A, write

$$S = \sqrt{1 - AA^*} \quad \text{and} \quad T = \sqrt{1 - A^*A},$$

where the positive square roots are meant, of course; note that since $\|A\| \leqq 1$, it follows that $1 - AA^*$ and $1 - A^*A$ are positive. The desired dilation B can be defined by

$$B = \begin{pmatrix} A & S \\ T & -A^* \end{pmatrix}.$$

322

That B is a dilation of A is clear. Since

$$B^* = \begin{pmatrix} A^* & T \\ S & -A \end{pmatrix},$$

it follows by direct computation that

$$B^*B = \begin{pmatrix} A^*A + T^2 & A^*S - TA^* \\ SA - AT & S^2 + AA^* \end{pmatrix},$$

$$BB^* = \begin{pmatrix} AA^* + S^2 & AT - SA \\ TA^* - A^*S & T^2 + A^*A \end{pmatrix}.$$

It remains only to prove that $AT = SA$. Trivially $AT^2 = S^2A$, and it follows, by induction, that $AT^{2n} = S^{2n}A$ for $n = 0, 1, 2, \cdots$. This implies that $Ap(T^2) = p(S^2)A$ for every polynomial p (cf. Solution 137), and hence that $AT = SA$, as desired.

(b) The proof is similar to that of (a), and simpler. Given A, with $0 \leqq A \leqq 1$, let R be the positive square root of $A(1 - A)$, and write

$$B = \begin{pmatrix} A & R \\ R & 1 - A \end{pmatrix}.$$

The verification that B is a projection is painless. (The result (b) is due to E. A. Michael; see [49].)

Solution 223. The answer is: not necessarily. A good way to get a counter-example is to consider a suitable operator with non-closed range and make a projection out of it. A standard example of an operator A with non-closed range is given by the matrix

$$\mathrm{diag}\langle \tfrac{1}{2}, \tfrac{1}{3}, \tfrac{1}{4}, \cdots \rangle.$$

Since $0 \leqq A \leqq 1$, the operator matrix

$$P = \begin{pmatrix} A & \sqrt{A(1 - A)} \\ \sqrt{A(1 - A)} & 1 - A \end{pmatrix}$$

is a projection. (The use of P is the reason this discussion is here; see Solution 222.) Assertion: the image under P of the "x-axis", i.e., of the set of vectors of the form $\langle f, 0 \rangle$, is not closed.

The image in question is the set of all vectors of the form

$$\langle Af, \sqrt{A(1 - A)}f \rangle.$$

It is therefore pertinent to observe that a vector $f = \langle \xi_0, \xi_1, \xi_2, \cdots \rangle$ in l^2 belongs to ran A if and only if

$$\sum_{n=0}^{\infty} (n + 2)^2 |\xi_n|^2 < \infty;$$

it belongs to $\mathrm{ran}\sqrt{A(1 - A)}$ if and only if

$$\sum_{n=0}^{\infty} \frac{(n + 2)^2}{n + 1} |\xi_n|^2 < \infty.$$

In view of these observations it is an easy exercise in elementary analysis to produce a sequence $\{f_k\}$ of vectors such that the limit of the image sequence $\{P\langle f_k, 0\rangle\}$ is not in ran P. One possibility is

$$f_k = \left\langle \frac{1}{\sqrt{2}}, \frac{1}{\sqrt{3}}, \cdots, \frac{1}{\sqrt{k + 1}}, 0, 0, 0, \cdots \right\rangle.$$

224   **Solution 224.** It follows from Problem 222(b) and the projection characterization of the weak topology that the weak closure of **P** includes the "interval" **I** of all positive contractions (i.e., the set of all operators A such that $0 \leqq A \leqq 1$). Since $A \in \mathbf{I}$ if and only if $0 \leqq (Af, f) \leqq \|f\|^2$ for each vector f, and since $A \mapsto (Af, f)$ is weakly continuous, if follows that **I** is weakly closed.

A similar argument (based on Problem 222(a)) implies that the weak closure of **U** is the set of all contractions. Caution: is that set weakly closed?

Since $\mathbf{U} \subset \mathbf{N}$, it follows that the weak closure of **N** contains every contraction, and, therefore, that the weak closure of **N** is the set of all operators. (Similar use of the monotoneness of the closure operation yields, trivially, the weak closures of many other sets, such as the sets of isometries, co-isometries, partial isometries, subnormal operators, and hyponormal operators.) There is a longer way to get the same result, which also has some merit, as follows. For every operator A_0 the operator matrix

$$B_0 = \begin{pmatrix} A_0 & A_0^* \\ A_0^* & A_0 \end{pmatrix}$$ is normal. If, in particular, A_0 is a compression of an arbitrary operator A on **H** to a finite-dimensional subspace $\mathbf{H}_0$, then extend the embedding of $\mathbf{H}_0$ into **H** to an embedding of $\mathbf{H}_0 \oplus \mathbf{H}_0$ into **H**, extend the operator B_0 to a normal operator B on **H** (for instance by defining B to be 0 on $(\mathbf{H}_0 \oplus \mathbf{H}_0)^{\perp}$), and thus infer that every compression of A to a finite-dimensional subspace has a dilation in **N**.

225   **Solution 225.** Every isometry has a unitary extension to a larger space (Problem 149). It follows that the restriction of an isometry to a finite-dimensional subspace of an infinite-dimensional space **H** always has a unitary extension to **H**. Consequence: every isometry is in the strong closure of **U**. Since A is an isometry if and only if $\|Af\| = \|f\|$ for each vector f, and since $A \mapsto \|Af\|$ is strongly continuous, it follows that the set of isometries is strongly closed.

Since the set **P** of projections can be characterized as the set of all idempotent contractions, it follows that **P** is strongly closed.

As for co-isometries: recall that if A is a strict contraction, then A has a co-isometric extension (Problem 152). Since the set of all strict contractions is strongly dense in the set of all contractions, it follows that the strong closure of the set of all co-isometries is the set of all contractions. A general principle is at work here: in most statements about the strong operator topology, the behavior of the adjoint is what makes the result more interesting and the proof less trivial than for similar statements about the weak operator topology or the norm topology. The effect of the misbehavior of the adjoint is strikingly visible in the difference between the strong closure of the set of isometries and the strong closure of their adjoints.

Solution 226. *The set of hyponormal operators is strongly closed.* **226**

PROOF. A typical (basic) strong neighborhood of an operator B is the set

$$\{A: \|AF - BF\| < \varepsilon\},$$

where $\varepsilon > 0$ and F is a projection of finite rank. The assertion is that if every such neighborhood of B contains a hyponormal operator, then

$$\|B^*g\| \leq \|Bg\|$$

for every vector g. Given g, let F be the projection onto the span of g and B^*g. To get an idea about how to use the neighborhood hypothesis, assume for a moment a much stronger hypothesis, namely that there exists a hyponormal operator A such that

$$AF = BF.$$

In that case, of course,

$$FA^* = FB^*,$$

and it follows that

$$\|B^*g\| = \|FB^*g\| = \|FA^*g\| \leq \|A^*g\|$$
$$\leq \|Ag\| = \|AFg\| = \|BFg\| = \|Bg\|.$$

The proof proper does all this again, carrying an ε along, as follows. The assumption is that for every $\varepsilon > 0$ there exists a hyponormal operator A ($= A(\varepsilon, F)$) such that

$$\|AF - BF\| < \varepsilon.$$

In that case, of course,

$$\|FA^* - FB^*\| < \varepsilon,$$

and it follows that

$$\|B^*g\| = \|FB^*g\| \leq \|FB^*g - FA^*g\| + \|FA^*g\| \leq \varepsilon\|g\| + \|A^*g\|$$
$$\leq \varepsilon\|g\| + \|Ag\| = \varepsilon\|g\| + \|AFg\| \leq \varepsilon\|g\| + \|AFg - BFg\| + \|BFg\|$$
$$\leq 2\varepsilon\|g\| + \|BFg\| = 2\varepsilon\|g\| + \|Bg\|.$$

Comparison of the first and last terms of this chain of relations (true for every $\varepsilon > 0$) implies that $\|B^*g\| \leq \|Bg\|$, as desired.

227   **Solution 227.** The proof is constructive. Given **H**, let **K** be the direct sum of countably infinitely many copies of **H**, indexed by all integers (positive, negative, zero); then each operator on **K** is an infinite operator matrix, and, in particular, the projection P from **K** to **H** is given by

$$P = \begin{pmatrix} \ddots & & & \\ & 0 & 0 & 0 & \\ & 0 & (1) & 0 & \\ & 0 & 0 & 0 & \\ & & & & \ddots \end{pmatrix}.$$

(The parentheses indicate the entry in position $\langle 0, 0 \rangle$.) Given A, put

$$B = \begin{pmatrix} \ddots & & & & & & & \\ 0 & 0 & 0 & 0 & 0 & 0 & 0 \\ 1 & 0 & 0 & 0 & 0 & 0 & 0 \\ 0 & 1 & 0 & 0 & 0 & 0 & 0 \\ 0 & 0 & S & (A) & 0 & 0 & 0 \\ 0 & 0 & -A^* & T & 0 & 0 & 0 \\ 0 & 0 & 0 & 0 & 1 & 0 & 0 \\ 0 & 0 & 0 & 0 & 0 & 1 & 0 \\ & & & & & & & \ddots \end{pmatrix},$$

where S and T are as in Solution 222. Since B is triangular, its powers are triangular, and the diagonal entries of the powers are the corresponding powers of the diagonal entries of B. This makes it obvious that B is a power dilation of A. The proof that B is unitary is an obvious computation (which uses the results of Solution 222).

Although this may not be the most revealing proof of the theorem, it is certainly the shortest; it is due to Schäffer [125].

228   **Solution 228.** The theorem can be proved directly, but the proof via unitary operators and dilation theory has an elegance that is hard to surpass. As for the theorem for unitary operators, it can be proved by relatively elementary and widely generalizable geometric methods (cf. [60, p. 185]), but the parochial Hilbert space proof via the spectral theorem is more transparent.

If U is a unitary operator on **H**, then the spectral theorem justifies the assumption that $\mathbf{H} = \mathbf{L}^2(\mu)$ for some measure μ, on some suitable measure space, in such a way that U is the multiplication induced by a measurable function φ of constant modulus 1 almost everywhere. If $f \in \mathbf{H}\ (= \mathbf{L}^2(\mu))$, then

$$\frac{1}{n} \sum_{j=0}^{n-1} U^j f = \left(\frac{1}{n} \sum_{j=0}^{n-1} \varphi^j \right) f.$$

Since $|\varphi| = 1$ almost everywhere, it follows that

$$\left| \frac{1}{n} \sum_{j=0}^{n-1} \varphi^j \right| \leq 1$$

326

almost everywhere. Since, moreover, the assumption that $|\varphi| = 1$ almost everywhere implies that the averages

$$\frac{1}{n}\sum_{j=0}^{n-1}\varphi^j$$

form a convergent sequence almost everywhere (whose limit is the characteristic function of the set where $\varphi = 1$), it follows that the Lebesgue dominated convergence theorem (not necessarily the bounded convergence theorem) is applicable to the sequence

$$\left\{\left(\frac{1}{n}\sum_{j=0}^{n-1}\varphi^j\right)f\right\}.$$

This completes the proof of convergence; a quick second glance will even reveal what the limit is.

If A is an arbitrary contraction on H, then let U be a unitary power dilation of it on a Hilbert space K, say, and let P be the projection from K to H. This means that if $f \in H$, then $A^n f = PU^n f$, $n = 0, 1, 2, \cdots$, and it follows that

$$\frac{1}{n}\sum_{j=0}^{n-1}A^j f = P\left(\frac{1}{n}\sum_{j=0}^{n-1}U^j f\right).$$

Since

$$\frac{1}{n}\sum_{j=0}^{n-1}U^j f$$

has a limit as $n \to \infty$ for each f, and since P is continuous (i.e., bounded), it follows that

$$\frac{1}{n}\sum_{j=0}^{n-1}A^j f$$

has a limit as $n \to \infty$ for each f.

The mean ergodic theorem for unitary operators was first proved by von Neumann [149]. The extension to contractions is due to Riesz-Nagy [113]; the proof via dilation theory is due to Nagy [96]. A good recent reference to ergodic theory in general is [152].

Solution 229. Relatively hard analytic proofs can be given; with dilation theory all becomes simple [96]. Given A on H, let U on K be a unitary power dilation of it, and let P be the projection from K to H. If p is a polynomial and if f is in H, then

$$\|p(A)f\| = \|Pp(U)f\| \qquad \text{(by the definition of power dilation)}$$
$$\leqq \|p(U)\| \cdot \|f\| \qquad \text{(because } \|P\| = 1)$$
$$\leqq \|p\|_D \cdot \|f\| \qquad \text{(because } U \text{ is normal)},$$

and it follows, as stated, that $\|p(A)\| \leqq \|p\|_D$.

That's all there is to the proof, but the success of this "one-variable" theory made it tempting to look for "several-variable" extensions. Is it true that if $\|A_j\| \leqq 1$, $j = 1, \cdots, k$, and if D is the closed unit disc, then

$$\|p(A_1, \cdots, A_k)\| \leqq \|p\|_D$$

for every polynomial p in k variables?

What does the question mean? A sensible definition of $\|p\|_D$ is near the surface:

$$\|p\|_D = \sup\{|p(\lambda_1, \cdots, \lambda_k)|: \lambda_j \in D, j = 1, \cdots, k\}.$$

A sensible definition of $p(A_1, \cdots, A_k)$ probably cannot be formulated in general; if, however, the A_j's are pairwise commutative, then $p(A_1, \cdots, A_k)$ makes unambiguous sense. Very well then, assume that the A_j's commute; now is the k-variable von Neumann inequality true?

A natural way to try to find the answer is first to look for a k-variable generalization of dilation theory. Is it true that if $A_1, \cdots, A_k$ are pairwise commutative contractions, then there exist pairwise commutative unitary operators $U_1, \cdots, U_k$ on a suitable large space such that $A_1^{m_1} \cdots A_k^{m_k}$ is the compression to the original space of $U_1^{m_1} \cdots U_k^{m_k}$, for every finite sequence $\{m_1, \cdots, m_k\}$ of exponents? Each time that the answer is yes, the argument for $k = 1$ can be applied and yields the desired inequality. If ever the answer is no, the question about the inequality must be asked again; a natural attempt at a proof failed, but the conclusion could be true just the same.

The growth of the subject went through three periods of suspense. First: is the commutative dilation theorem true for $k = 2$? Answer by Ando [3] (eight years after Nagy's dilation proof for $k = 1$): yes. (Consequence: the 2-variable von Neumann inequality is true.) Next: is the commutative dilation theorem true for $k = 3$? Answer by Parrott [105] (seven years after Ando's proof for $k = 2$): no. Finally: is the k-variable von Neumann inequality for $k \geqq 3$ true nevertheless? Answer by Varopoulos [145] and Crabb–Davie [28]: no; counterexamples exist with $k = 3$ on spaces of dimension 5.

CHAPTER 24

Commutators

Solution 230. Wintner's proof. If $PQ - QP = \alpha$, replace P by $P + \lambda$, where **230** λ is an arbitrary scalar, and observe that the new P satisfies the same commutation relation. There is, consequently, no loss of generality in assuming that P is invertible. Since, in that case, $QP = P^{-1}(PQ)P$, and therefore spec QP = spec PQ, the relation $PQ = QP + \alpha$ implies that

$$\text{spec}(PQ) = \text{spec}(QP + \alpha) = \text{spec}(QP) + \alpha = \text{spec}(PQ) + \alpha.$$

The only translation that can leave a non-empty compact subset (such as spec PQ) of the complex plane invariant is the trivial translation (i.e., no translation at all); in other words, α must be 0.

Wielandt's proof. If $PQ - QP = \alpha$, then

$$P^2Q - QP^2 = P^2Q - PQP + PQP - QP^2$$
$$= P(PQ - QP) + (PQ - QP)P = 2P\alpha,$$

and more generally (induction)

$$P^nQ - QP^n = nP^{n-1}\alpha, \qquad n = 1, 2, 3, \cdots.$$

If P is nilpotent, of index n, say, then $nP^{n-1}\alpha = 0$, and therefore $\alpha = 0$. If P is not nilpotent, then the inequality

$$n\|P^{n-1}\| \cdot |\alpha| \leqq 2\|P\| \cdot \|Q\| \cdot \|P^{n-1}\|,$$

true for $n = 1, 2, 3, \cdots$, implies that

$$n|\alpha| \leqq 2\|P\| \cdot \|Q\|,$$

and hence that, again, $\alpha = 0$.

231   **Solution 231.** Given the Hilbert space **H**, let **M** be the normed vector space of all bounded sequences $f = \langle f_1, f_2, f_3, \cdots \rangle$ of vectors in **H** (coordinatewise vector operations, supremum norm), and let **N** be the subspace of all null sequences in **M** (i.e., sequences f with $\lim_n \|f_n\| = 0$). The quotient space $\hat{\mathbf{M}} = \mathbf{M}/\mathbf{N}$ is a normed vector space. Each bounded sequence $A = \langle A_1, A_2, A_3, \cdots \rangle$ of operators on **H** induces an operator on **M**; the image of $\langle f_1, f_2, f_3, \cdots \rangle$ under $\langle A_1, A_2, A_3, \cdots \rangle$ is $\langle A_1 f_1, A_2 f_2, A_3 f_3, \cdots \rangle$. Since the subspace **N** is invariant under each such induced operator, the sequence A also induces, in a natural manner, an operator on $\hat{\mathbf{M}}$; call it $\hat{A}$. Bounded sequences of operators on **H** form a normed algebra (coordinatewise operations, supremum norm). The correspondence $A \mapsto \hat{A}$ from such bounded sequences to operators on $\hat{\mathbf{M}}$ is a norm-decreasing homomorphism. If $P = \langle P_1, P_2, P_3, \cdots \rangle$ and $Q = \langle Q_1, Q_2, Q_3, \cdots \rangle$ are such that $P_n Q_n - Q_n P_n \to C$, then $\hat{P}\hat{Q} - \hat{Q}\hat{P}$ is a commutator on $\hat{\mathbf{M}}$; since that commutator cannot be equal to $\hat{1}$ (= the identity operator on $\hat{\mathbf{M}}$), the proof is complete.

232   **Solution 232.** Fix P and consider $C = \Delta Q = PQ - QP$ as a function of Q. The operation Δ is obviously a linear transformation on the vector space of operators; since

$$\|\Delta Q\| = \|PQ - QP\| \leqq 2\|P\| \cdot \|Q\|,$$

that linear transformation is bounded (on the Banach space of operators), and

$$\|\Delta\| \leqq 2\|P\|.$$

Mappings such as Δ often play an important algebraic role. The most important property of Δ is that it is a *derivation* in the sense that

$$\Delta(QR) = \Delta Q \cdot R + Q \cdot \Delta R.$$

Proof: $PQR - QRP = (PQR - QPR) + (QPR - QRP)$.

Derivations have many of the algebraic properties of differentiation, but, as is visible in the definition itself, they have them in a non-commutative way. First among those properties is the validity of the Leibniz formula for "differentiating" products with several factors. The assertion is that $\Delta(Q_1 \cdots Q_n)$ is the sum of n terms; to obtain the j-th term, replace Q_j by ΔQ_j in the product $Q_1 \cdots Q_n$. The proof is an obvious induction. For $n = 1$, there is nothing to do; for the step from n to $n + 1$, write $Q_1 \cdots Q_{n+1}$ as $(Q_1 \cdots Q_n)Q_{n+1}$ and use the given (two-factor) product formula. The result is, of course, applicable to the special case in which all the Q_j's coincide, but it does not become much more pleasant to contemplate.

A special property of the derivation Δ is that $\Delta^2 Q = 0$. Here Δ^2 is, of course, the composition of Δ with itself, so that

$$\Delta^2 Q = \Delta(\Delta Q) = P \cdot \Delta Q - \Delta Q \cdot P;$$

the vanishing of $\Delta^2 Q$ expresses exactly that ΔQ commutes with P.

The Leibniz formula and the vanishing of $\Delta^2 Q$ make it easy to evaluate higher order derivatives of higher powers of Q. The process begins with ΔQ^n: it is equal to the sum of the n possible products each of which has one factor equal to ΔQ and $n - 1$ factors equal to Q. When Δ is applied to one of these summands, the result is the sum of only $n - 1$ products. (Reason: when Δ is applied to ΔQ, the result is 0.) Each of the $n - 1$ products so obtained has two factors equal to ΔQ and $n - 2$ factors equal to Q. Consequence: $\Delta^2 Q^n$ is equal to the sum of the $n(n - 1)$ possible products of that kind. The argument continues from here on with no surprises and yields a description of $\Delta^k Q^n$. With $k = n$, the result is that $\Delta^n Q^n$ is the sum of $n!$ terms, each of which is $(\Delta Q)^n$; in other words

$$\Delta^n Q^n = n!(\Delta Q)^n.$$

The last equation is the crucial point of the proof; the desired result is a trivial consequence of it. Indeed, since

$$\|(\Delta Q)^n\| = \frac{1}{n!} \|\Delta^n Q^n\| \leq \frac{1}{n!} \|\Delta^n\| \cdot \|Q^n\| \leq \frac{1}{n!} \|\Delta\|^n \cdot \|Q\|^n,$$

it follows that

$$\|(\Delta Q)^n\|^{1/n} \leq \left(\frac{1}{n!}\right)^{1/n} \cdot \|\Delta\| \cdot \|Q\|,$$

and hence that ΔQ is quasinilpotent.

As a dividend, the equation for $\Delta^n Q^n$ yields a proof of Jacobson's original algebraic result. Statement: if an element Q of an algebra over a field of characteristic greater than $n!$ satisfies a polynomial equation of degree n, and if Δ is a derivation of that algebra such that $\Delta^2 Q = 0$, then ΔQ is nilpotent of index n. Proof: from $\Delta(\Delta Q) = 0$ infer that $\Delta(\Delta Q)^k = 0$ for every positive integer k, and hence, from the equation for $\Delta^n Q^n$ infer that $\Delta^{n+1} Q^n = 0$. Consequence: $\Delta^n Q^i = 0$ whenever $n > i$. If $Q^n = \sum_{i=0}^{n-1} \alpha_i Q^i$ is the polynomial equation satisfied by Q, then it follows that $\Delta^n Q^n = 0$, and hence it follows, again from the equation for $\Delta^n Q^n$, that $n!(\Delta Q)^n = 0$. The conclusion follows from the assumption about the characteristic.

Solution 233. (a) The trick is to generalize the formula for the "derivative" of a power to the non-commutative case; cf. Solutions 230 and 232. The generalization that is notationally most convenient here says that

$$P^n Q - QP^n = \sum_{i=0}^{n-1} P^{n-i-1} CP^i;$$

the proof is a straightforward induction. It follows that

$$P^n Q - QP^n = nP^{n-1} - \sum_{i=0}^{n-1} P^{n-i-1}(1 - C)P^i,$$

and hence that

$$n\|P^{n-1}\| \leq 2\|P\| \cdot \|Q\| \cdot \|P^{n-1}\| + \|1 - C\| \cdot \sum_{i=0}^{n-1} \|P^{n-i-1}\| \cdot \|P^i\|.$$

233

Up to now P could have been arbitrary. Since P was assumed hyponormal, the last written sum is equal to $n\|P^{n-1}\|$ (see Problem 162). Divide through by $n\|P^{n-1}\|$. (If $P = 0$, everything is trivial, and if $P \neq 0$, then $P^{n-1} \neq 0$.) The result is that

$$1 \leqq \frac{2}{n} \|P\| \cdot \|Q\| + \|1 - C\|,$$

and the conclusion follows.

(b) If $\|1 - C\| < 1$, then C is invertible; since, by the Kleinecke–Shirokov theorem, C is quasinilpotent, that is impossible.

234

Solution 234. If A has a large kernel, then that kernel is the direct sum of $\aleph_0$ subspaces all of the same dimension. The orthogonal complement of the kernel may or may not be large. If, however, one of the direct summands of the kernel is adjoined to that orthogonal complement, the result is a representation of the underlying Hilbert space in the form of an infinite direct sum $\mathbf{H} \oplus \mathbf{H} \oplus \mathbf{H} \oplus \cdots$ in such a way that the direct sum of all the summands beginning with the second one is annihilated by A. If corresponding to this representation of the space the operator A is represented as a matrix, it will have the form

$$A = \begin{pmatrix} A_0 & 0 & 0 & 0 \\ A_1 & 0 & 0 & 0 \\ A_2 & 0 & 0 & 0 \\ A_3 & 0 & 0 & 0 \\ & & & & \ddots \end{pmatrix}$$

where each A_n (and each 0) is an operator on $\mathbf{H}$. Write

$$P = \begin{pmatrix} 0 & 0 & 0 & 0 \\ 1 & 0 & 0 & 0 \\ 0 & 1 & 0 & 0 \\ 0 & 0 & 1 & 0 \\ & & & & \ddots \end{pmatrix}$$

and

$$Q = \begin{pmatrix} A_1 & -A_0 & 0 & 0 \\ A_2 & 0 & -A_0 & 0 \\ A_3 & 0 & 0 & -A_0 \\ A_4 & 0 & 0 & 0 \\ & & & & \ddots \end{pmatrix};$$

then P and Q are operators and (straightforward computation) $PQ - QP = A$. The proof of the main assertion is complete.

To prove Corollary 1, suppose that $\{f_1, \cdots, f_n\}$ is a finite set of vectors in an infinite-dimensional Hilbert space $\mathbf{H}$, and let $\mathbf{M}$ be their span. If A is an operator on **H**, let C be the operator that is A on **M** and 0 on **M**$^\perp$; by what was just proved, C is a commutator, and C agrees with A on each f_i, $i = 1, \cdots, n$. This implies that every basic strong neighborhood of A contains commutators.

The proof of Corollary 2 is similar. Given **H**, let **M** be a large subspace with a large orthogonal complement; given A, define C as in the preceding paragraph, and write $B = A - C$. Since $A = B + C$ and both B and C are commutators, the proof is complete.

Solution 235. Since A is not a scalar, there exists a vector f such that f and Af are linearly independent. Let T be an invertible operator such that $Tf = f$ and $TAf = -Af$. Since this implies that

$$(A + T^{-1}AT)f = Af - Af = 0,$$

it follows that the direct sum

$$S = (A + T^{-1}AT) \oplus (A + T^{-1}AT) \oplus (A + T^{-1}AT) \oplus \cdots$$

has a large kernel. (What really follows is that the kernel is infinite-dimensional; since the whole space is separable, this implies that the kernel is large.) By Problem 234, the direct sum S is a commutator. If

$$B = A \oplus A \oplus A \oplus \cdots$$

and

$$C = T^{-1}AT \oplus T^{-1}AT \oplus T^{-1}AT \oplus \cdots,$$

then

$$S = B + C.$$

The next step is the following somewhat surprising lemma: whenever B and C are operators such that $B + C$ is a commutator, then $B \oplus C$ is a commutator. The proof is an inspired bit of elementary algebra. If $B + C = PQ - QP$, then write $R = C + QP = PQ - B$, and compute the commutator of

$$\begin{pmatrix} 0 & P \\ 1 & 0 \end{pmatrix} \quad \text{and} \quad \begin{pmatrix} 0 & R \\ Q & 0 \end{pmatrix}:$$

$$\begin{pmatrix} 0 & P \\ 1 & 0 \end{pmatrix}\begin{pmatrix} 0 & R \\ Q & 0 \end{pmatrix} - \begin{pmatrix} 0 & R \\ Q & 0 \end{pmatrix}\begin{pmatrix} 0 & P \\ 1 & 0 \end{pmatrix}$$

$$= \begin{pmatrix} PQ & 0 \\ 0 & R \end{pmatrix} - \begin{pmatrix} R & 0 \\ 0 & QP \end{pmatrix} = \begin{pmatrix} B & 0 \\ 0 & C \end{pmatrix}.$$

Consequence: $B \oplus C$ is a commutator. Since, however, C is clearly similar to B, it follows that $B \oplus B$ is a commutator; since, finally, $B \oplus B$ is unitarily equivalent to B, the proof is complete.

236   **Solution 236.** Suppose that $C = A^*A - AA^* \geqq 0$. The problem is to show that $0 \in \operatorname{spec} C$. It is sufficient to show that there exists a sequence $\{f_n\}$ of unit vectors such that $Cf_n \to 0$ (i.e., that $0 \in \Pi(C)$). For this purpose, take a complex number λ in the approximate point spectrum of A, and, corresponding to λ, find a sequence $\{f_n\}$ of unit vectors such that $(A - \lambda)f_n \to 0$. Since the self-commutator of $A - \lambda$ is equal to C, and since $C \geqq 0$, it follows that

$$(A - \lambda)^*(A - \lambda) \geqq (A - \lambda)(A - \lambda)^*.$$

Since $(A - \lambda)f_n \to 0$, it follows that $(A - \lambda)^*(A - \lambda)f_n \to 0$. The last two remarks imply that $(A - \lambda)(A - \lambda)^*f_n \to 0$. Since, therefore, C is the difference of two operators, each of which annihilates $\{f_n\}$, the operator C does so too.

237   **Solution 237.** (a) The program is to show that (1) A is quasinormal, (2) $\ker(1 - A^*A)$ reduces A, and (3) the orthogonal complement of $\ker(1 - A^*A)$ is included in $\ker(A^*A - AA^*)$.

     (1)   Write $P = A^*A - AA^*$. Since, for all f,

$$\|f\|^2 \geqq \|Af\|^2 = (A^*Af, f) = (AA^*f, f) + (Pf, f)$$
$$= \|A^*f\|^2 + \|Pf\|^2,$$

it follows that if $Pf = f$, then $A^*P = 0$. (The norm condition was used at the first step.) This implies that $A^*P = 0$, and hence that $PA = 0$, or, equivalently, that $(A^*A)A = A(A^*A)$.

     (2)   Write $\mathbf{M} = \ker(1 - A^*A)$. If $f \in \mathbf{M}$, then $f - A^*Af = 0$. It follows that $Af - (A^*A)Af = Af - A(A^*A)f = A(f - A^*Af) = 0$, so that $\mathbf{M}$ is invariant under A. Similarly (instead of replacing f by Af, replace f by A^*f) $\mathbf{M}$ is invariant under A^*. (Cf. Solution 195.)

     (3)   Since P is idempotent, it follows that

$$A^*A - AA^* = A^*AA^*A - AA^*A^*A - A^*AAA^* + AA^*AA^*.$$

Since A^*A commutes with both A and A^*, this can be rewritten as

$$A^*A - AA^* = A^*A(A^*A - AA^*).$$

In other words,

$$P = A^*AP,$$

or

$$(1 - A^*A)P = 0.$$

It follows that

$$\operatorname{ran} P \subset \mathbf{M},$$

or

$$\mathbf{M}^\perp \subset \ker P.$$

Now use the assumption that A is abnormal: it says exactly that ker P includes no non-zero subspace that reduces A. Conclusion: $\mathbf{M}^\perp = 0$, and this means that A is an isometry.

(b) *If A is the bilateral shift with weights $\{\alpha_n\}$ such that $\alpha_n = 1$ or $\sqrt{2}$ according as $n \leq 0$ or $n > 0$, then A is abnormal and the self-commutator of A is a projection.*

PROOF. The self-commutator of A is easy to compute; it turns out to be the projection of rank 1 whose range is spanned by e_1. The abnormality of A follows from Problem 159: according to that result, A is irreducible, and hence as abnormal as can be.

Solution 238. An infinite-dimensional Hilbert space is the direct sum of Hilbert spaces of dimension $\aleph_0$. To prove that every scalar of modulus 1 is a commutator, it is therefore sufficient to prove that on a Hilbert space of dimension $\aleph_0$ every scalar of modulus 1 is the commutator of two *unitary* operators. (The unitary character of the factors guarantees that the possibly uncountable direct sum is bounded.) In a Hilbert space of dimension $\aleph_0$ there always exists an orthonormal basis $\{e_n : n = 0, \pm 1, \pm 2, \cdots\}$. Given α with $|\alpha| = 1$, let P be the diagonal operator defined by $Pe_n = \alpha^n e_n$, and let Q be the bilateral shift, $Qe_n = e_{n+1}$. Both P and Q are unitary; a straightforward computation shows that $PQP^{-1}Q^{-1} = \alpha$.

The proof that if $\alpha = PQP^{-1}Q^{-1}$, then $|\alpha| = 1$ is an adaptation of the Wintner argument (Solution 230). Since $PQ = \alpha QP$, it follows that spec $PQ = \alpha$ spec QP; since, however, PQ is similar to QP, it follows that spec $QP = $ spec QP, and hence that spec $QP = \alpha$ spec QP. Since spec QP is a non-empty compact set different from $\{0\}$ (remember that QP is invertible), and since the only homothety that can leave such a set fixed is a rotation, the result follows.

238

Solution 239. The proof is an adaptation of Solution 238. The first step is to use Problem 142 to represent the given space as the direct sum of $\aleph_0$ subspaces, all of the same dimension, each of which reduces the given unitary operator U. The direct sum decomposition serves to represent U as a diagonal operator matrix whose n-th diagonal entry is U_n, say, for $n = 0, \pm 1, \pm 2, \cdots$.

Solution 238 suggests that the multiplicative commutator of a diagonal operator and a bilateral shift may work here too. To avoid writing down large matrices, it is convenient to introduce some more notation. Think of the given Hilbert space as the set of all sequences $f = \{f_n : n = 0, \pm 1, \pm 2, \cdots\}$ of vectors in some fixed Hilbert space (subject of course to the usual condition $\sum_n \|f_n\|^2 < \infty$). A typical diagonal operator matrix P is defined by

$$(Pf)_n = V_n f_n,$$

and the bilateral shift Q is defined by

$$(Qf)_n = f_{n-1}.$$

The commutator is easy to compute; the result is that

$$(PQP^{-1}Q^{-1}f)_n = V_n V_{n-1}^{-1} f_n.$$

239

335

The equations

$$U_n = V_n V_{n-1}{}^{-1}$$

can be solved for the V's in terms of the U's. If, for instance, V_0 is set equal to 1, then

$$V_n = U_n \cdots U_1 \quad \text{for } n \geq 1,$$

and

$$V_{-(n+1)} = U_{-n}{}^{-1} \cdots U_0{}^{-1} \quad \text{for } n \geq 0.$$

The unitary character of the U's implies that the transformation P given by these V's is a unitary operator, and all is well.

240   **Solution 240.** *On an infinite-dimensional Hilbert space, the commutator subgroup of the full linear group is the full linear group itself.*

PROOF. The assertion is that every invertible operator is the product of a finite (but not necessarily bounded) number of multiplicative commutators. The fact is that every invertible operator is the product of two commutators [23]. The proof of that fact takes more work than the present purpose is worth; it is sufficient to prove that every invertible operator is the product of three commutators, and that is much easier.

Given an arbitrary invertible operator A on an arbitrary infinite-dimensional Hilbert space, consider the infinite operator matrices

$$P = \begin{pmatrix} \ddots & & & & & & & \\ 0 & 1 & 0 & 0 & 0 & 0 & 0 \\ 0 & 0 & 1 & 0 & 0 & 0 & 0 \\ 0 & 0 & 0 & 1 & 0 & 0 & 0 \\ 0 & 0 & 0 & (0) & A & 0 & 0 \\ 0 & 0 & 0 & 0 & 0 & A & 0 \\ 0 & 0 & 0 & 0 & 0 & 0 & A \\ 0 & 0 & 0 & 0 & 0 & 0 & 0 \\ & & & & & & & \ddots \end{pmatrix}$$

and

$$Q = \begin{pmatrix} \ddots & & & & & & & \\ 0 & 0 & 0 & 0 & 0 & 0 & 0 \\ 1 & 0 & 0 & 0 & 0 & 0 & 0 \\ 0 & 1 & 0 & 0 & 0 & 0 & 0 \\ 0 & 0 & 1 & (0) & 0 & 0 & 0 \\ 0 & 0 & 0 & 1 & 0 & 0 & 0 \\ 0 & 0 & 0 & 0 & 1 & 0 & 0 \\ 0 & 0 & 0 & 0 & 0 & 1 & 0 \\ & & & & & & & \ddots \end{pmatrix}.$$

Routine computation proves that

$$PQP^{-1}Q^{-1} = \begin{pmatrix} 1 & 0 & 0 & 0 & 0 & 0 & 0 \\ 0 & 1 & 0 & 0 & 0 & 0 & 0 \\ 0 & 0 & 1 & 0 & 0 & 0 & 0 \\ 0 & 0 & 0 & (A) & 0 & 0 & 0 \\ 0 & 0 & 0 & 0 & 1 & 0 & 0 \\ 0 & 0 & 0 & 0 & 0 & 1 & 0 \\ 0 & 0 & 0 & 0 & 0 & 0 & 1 \end{pmatrix}.$$

Regard the direct sum on which these matrices act as the direct sum of the summand with index 0 and the others, and identify the direct sum of the others with one of them. With an obvious change of notation, the result of the above computations becomes this: every operator matrix of either of the forms

$$\begin{pmatrix} A & 0 \\ 0 & 1 \end{pmatrix} \quad \text{or} \quad \begin{pmatrix} 1 & 0 \\ 0 & A \end{pmatrix}$$

is a multiplicative commutator (provided that the matrix entries operate on an infinite-dimensional space, and that A is invertible).

Every invertible normal operator on an infinite-dimensional Hilbert space has large reducing subspaces with large orthogonal complements (Problem 142), and is therefore representable as the product of two matrices of the indicated forms. Consequence: every invertible normal operator is the product of two commutators. Since every invertible operator is the product of a unitary operator and an invertible positive operator (polar decomposition), it follows, as stated, that every invertible operator is the product of three multiplicative commutators. (Apply Problem 239 to dispose of the unitary factor.)

CHAPTER 25

Toeplitz Operators

241   **Solution 241.** If $\varphi = \sum_n \alpha_n e_n$, then the matrix entries of L_φ are given by

$$\lambda_{ij} = (L_\varphi e_j, e_i) = \left(\sum_n \alpha_n e_{n+j}, e_i\right)$$

$$= \sum_n \alpha_n \delta_{n+j, i} = \alpha_{i-j}.$$

If, conversely, A is an operator on $\mathbf{L}^2$ such that

$$(Ae_{j+1}, e_{i+1}) = (Ae_j, e_i)$$

for all i and j, and if W is the bilateral shift (multiplication by e_1), then

$$(AWe_j, e_i) = (Ae_{j+1}, e_i) = (Ae_j, e_{i-1})$$
$$= (Ae_j, W^*e_i) = (WAe_j, e_i).$$

This implies that A commutes with W, and hence (Problem 146) that A is a multiplication.

242   **Solution 242.** The proof of necessity is a simple computation: if $i, j = 0, 1, 2, \cdots$, then

$$(T_\varphi e_j, e_i) = (PL_\varphi e_j, e_i) = (L_\varphi e_j, e_i) = (L_\varphi e_{j+1}, e_{i+1})$$
$$= (PL_\varphi e_{j+1}, e_{i+1}) = (T_\varphi e_{j+1}, e_{i+1}).$$

To prove sufficiency, assume that A is an operator on $\mathbf{H}^2$ such that

$$(Ae_{j+1}, e_{i+1}) = (Ae_j, e_i) \qquad (i, j = 0, 1, 2, \cdots);$$

it is to be proved that A is a Toeplitz operator. Consider for each nonnegative integer n the operator on $\mathbf{L}^2$ given by

$$A_n = W^{*n}APW^n.$$

338

(where W is, as before, the bilateral shift). If $i, j \geq 0$, then

$$(A_n e_j, e_i) = (A_0 e_{j+n}, e_{i+n}) = (A e_j, e_i).$$

Something like this is true even for negative indices. Indeed, for n sufficiently large both $j + n$ and $i + n$ are positive, and from then on $(A_0 e_{j+n}, e_{i+n})$ is independent of n. Consequence: if p and q are trigonometric polynomials (finite linear combinations of the e_i's, $i = 0, \pm 1, \pm 2, \cdots$), then the sequence $\{(A_n p, q)\}$ is convergent. Since

$$\|A_n\| \leq \|A_0\| = \|A\|,$$

it follows on easy general grounds that the sequence $\{A_n\}$ of operators on $\mathbf{L}^2$ is weakly convergent to an operator A_∞ on $\mathbf{L}^2$.

Since, for all i and j,

$$(A_\infty e_j, e_i) = \lim_n (W^{*n} A P W^n e_j, e_i)$$

$$= \lim_n (W^{*n+1} A P W^{n+1} e_j, e_i)$$

$$= \lim_n (W^{*n} A P W^n e_{j+1}, e_{i+1})$$

$$= (A_\infty e_{j+1}, e_{i+1}),$$

it follows that the operator A_∞ has a Laurent matrix and hence that it is a Laurent operator (Problem 241). If f and g are in $\mathbf{H}^2$, then

$$(P A_\infty f, g) = (A_\infty f, g) = \lim_n (W^{*n} A P W^n f, g) = (A f, g),$$

so that $P A_\infty f = A f$ for each f in $\mathbf{H}^2$. Conclusion: A is the compression to $\mathbf{H}^2$ of a Laurent operator, and hence, by definition, A is a Toeplitz operator.

How can the function φ that induces A be recaptured from the matrix of A? If $A = T_\varphi$, then $A_\infty = L_\varphi$, and therefore the Fourier coefficients of φ are the entries in the 0 column of the matrix of A_∞. This is an answer, but not a satisfying one; it is natural to wish for an answer expressed in terms of A instead of A_∞. That turns out to be easy. If $i, j \geq 0$, then

$$(A e_j, e_i) = (A_\infty e_j, e_i) = (\varphi, e_{i-j});$$

this implies that

$$(\varphi, e_i) = (A e_0, e_i) \quad \text{for } i \geq 0,$$

and

$$(\varphi, e_{-j}) = (A e_j, e_0) \quad \text{for } j \geq 0.$$

339

Conclusion: φ is the function whose forward Fourier coefficients (the ones with positive index) are the terms of the 0 column of the matrix of A and whose backward Fourier coefficients are the terms of the 0 row of that matrix.

To prove Corollary 1, observe that

$$(Ae_{j+1}, e_{i+1}) = (AUe_j, Ue_i) = (U^*AUe_j, e_i).$$

To prove Corollary 2, observe that if φ is a bounded measurable function, and if both n and $n + k$ are non-negative integers, then

$$(\varphi, e_k) = (T_\varphi e_n, e_{n+k}).$$

If T_φ is compact, then $\|T_\varphi e_n\| \to 0$ (since $e_n \to 0$ weakly); it follows that $(\varphi, e_k) = 0$ for all k (positive, negative, or zero), and hence that $\varphi = 0$.

243 **Solution 243.** Write $C = T_\varphi T_\psi$ and let $\langle \gamma_{ij} \rangle$ be the (not necessarily Toeplitz) matrix of C. If the Fourier expansions of φ and ψ are $\varphi = \sum_i \alpha_i e_i$ and $\psi = \sum_j \beta_j e_j$, so that the matrices of T_φ and T_ψ are $\langle \alpha_{i-j} \rangle$ and $\langle \beta_{i-j} \rangle$, respectively, then

$$\gamma_{i+1, j+1} = \gamma_{ij} + \alpha_{i+1}\beta_{-j-1}$$

whenever $i, j \geq 0$. The proof is straightforward. Since

$$\gamma_{ij} = \sum_{k=0}^{\infty} \alpha_{i-k}\beta_{k-j},$$

it follows that

$$\gamma_{i+1, j+1} = \sum_{k=0}^{\infty} \alpha_{i+1-k}\beta_{k-j-1}$$

$$= \alpha_{i+1}\beta_{-j-1} + \sum_{k=1}^{\infty} \alpha_{i+1-k}\beta_{k-j-1}$$

$$= \alpha_{i+1}\beta_{-j-1} + \sum_{k=0}^{\infty} \alpha_{i-k}\beta_{k-j}$$

$$= \alpha_{i+1}\beta_{-j-1} + \gamma_{ij}.$$

If now ψ is analytic, then

$$T_\varphi T_\psi f = T_\varphi(\psi \cdot f) = P(\varphi \cdot \psi \cdot f) = T_{\varphi\psi} f$$

for all f in $\mathbf{H}^2$, so that $T_\varphi T_\psi = T_{\varphi\psi}$; if φ^* is analytic, then

$$T_\varphi T_\psi = (T_{\psi^*} T_{\varphi^*})^* = T_{\varphi\psi}.$$

This proves the sufficiency of the condition and the last assertion of the problem. If, conversely, the product $T_\varphi T_\psi$ is a Toeplitz operator, then its matrix is a Toeplitz matrix (Problem 242); the equation for $\gamma_{i+1,j+1}$ then implies that $\alpha_{i+1}\beta_{-j-1} = 0$ whenever $i, j \geqq 0$. From this, in turn, it follows that either $\alpha_{i+1} = 0$ for all $i \geqq 0$ or else $\beta_{-j-1} = 0$ for all $j \geqq 0$, which is equivalent to the desired conclusion.

As for the corollary, sufficiency is trivial. If, conversely, $T_\varphi T_\psi = 0$, then, since 0 is a Toeplitz operator, it follows from Problem 243 that either φ^* or ψ is analytic and that $\varphi\psi = 0$. The F. and M. Riesz theorem applies (Problem 158) and proves that if φ^* is analytic (and not zero), then $\psi = 0$, and if ψ is analytic (and not zero), then $\varphi = 0$.

Solution 244. If φ is the characteristic function of the closed upper semicircle and ψ is the characteristic function of the open lower semicircle, then $T_\varphi T_\psi - T_{\varphi\psi}$ is not compact.

244

For computational ease, it's best to identify the interval $[0, 1)$ with the unit circle, via the mapping $x \mapsto e^{2\pi ix}$. The function φ becomes then the characteristic function of $[0, \frac{1}{2}]$. The Fourier coefficients of φ are easy to determine:

$$(\varphi, e_n) = \begin{cases} \dfrac{1}{2} & \text{if } n = 0, \\[2mm] \dfrac{1}{\pi in} & \text{if } n \text{ is odd}, \\[2mm] 0 & \text{if } n \text{ is even}, n \neq 0. \end{cases}$$

Consequence: the matrix of T_φ is

$$\frac{1}{\pi i}\begin{pmatrix} \dfrac{\pi i}{2} & -1 & 0 & -\dfrac{1}{3} & 0 & -\dfrac{1}{5} \\[2mm] 1 & \dfrac{\pi i}{2} & -1 & 0 & -\dfrac{1}{3} & 0 \\[2mm] 0 & 1 & \dfrac{\pi i}{2} & -1 & 0 & -\dfrac{1}{3} \\[2mm] \dfrac{1}{3} & 0 & 1 & \dfrac{\pi i}{2} & -1 & 0 \\[2mm] 0 & \dfrac{1}{3} & 0 & 1 & \dfrac{\pi i}{2} & -1 \\[2mm] \dfrac{1}{5} & 0 & \dfrac{1}{3} & 0 & 1 & \dfrac{\pi i}{2} \end{pmatrix}.$$

The pattern suggests that the basis be split into two parts according to the parity of the subscripts; if all the even ones are written first, followed by all the odd ones, the matrix of T_φ takes the form

$$\frac{1}{\pi i}\begin{pmatrix} \dfrac{\pi i}{2} & 0 & 0 & -1 & -\dfrac{1}{3} & -\dfrac{1}{5} \\[2mm] 0 & \dfrac{\pi i}{2} & 0 & 1 & -1 & -\dfrac{1}{3} \\[2mm] 0 & 0 & \dfrac{\pi i}{2} & \dfrac{1}{3} & 1 & -1 \\[2mm] 1 & -1 & -\dfrac{1}{3} & \dfrac{\pi i}{2} & 0 & 0 \\[2mm] \dfrac{1}{3} & 1 & -1 & 0 & \dfrac{\pi i}{2} & 0 \\[2mm] \dfrac{1}{5} & \dfrac{1}{3} & 1 & 0 & 0 & \dfrac{\pi i}{2} \end{pmatrix}$$

If A is the operator corresponding to the lower left corner, then T_φ can be described by the 2×2 operator matrix

$$\frac{1}{\pi i}\begin{pmatrix} \dfrac{\pi i}{2} & AU \\[2mm] A & \dfrac{\pi i}{2} \end{pmatrix}.$$

Since $\psi = 1 - \varphi$, therefore $T_\psi = 1 - T_\varphi$, and therefore $T_\varphi T_\psi = T_\varphi - T_\varphi^2$. The rest is an easy computation: the operator matrix of $T_\varphi T_\psi$ is

$$\frac{1}{\pi i}\begin{pmatrix} \dfrac{\pi i}{2} & AU \\[2mm] A & \dfrac{\pi i}{2} \end{pmatrix} + \frac{1}{\pi^2}\begin{pmatrix} -\dfrac{\pi^2}{4} + AUA & \pi i AU \\[2mm] \pi i A & A^2U - \dfrac{\pi^2}{4} \end{pmatrix}.$$

Since $T_{\varphi\psi} = 0$, the relevant question is whether this is compact. The answer is obviously no: the lower left corner is a non-zero Toeplitz operator.

245 **Solution 245.** It is helpful to begin with some qualitative reflections. Consider a Laurent matrix written, as usual, so that the row index increases (from $-\infty$ to $+\infty$) as the rows go down, and the column index increases (from $-\infty$ to $+\infty$) as the columns go to the right. Fix attention on any particular entry on the main diagonal, and look at the unilaterally infinite matrix that starts there and goes down and to the right. All the matrices obtained in this way from one fixed Laurent matrix look the same; they all look like the associated Toeplitz matrix. Intuition suggests that as the selected diagonal entry moves up and left, the resulting Toeplitz matrices swell and tend to the original Laurent matrix.

An efficient non-matrix way of describing the situation might go like this. If P_n is the projection onto the span of $\{e_{-n}, \cdots, e_{-1}, e_0, e_1, e_2, \cdots\}$, $n = 1, 2, 3, \cdots$, then each Laurent operator L is the strong limit of the Toeplitz-like operators $P_n L P_n$. Since $P_n = W^{*n} P W^n$, and since W commutes with L (so that $WLW^* = L$), it follows that $W^{*n} PLP W^n \to L$ (strongly) as $n \to \infty$. This implies that if T is the Toeplitz operator corresponding to L, then $W^{*n} TP W^n \to L$ (strongly). It is instructive to compare this result with Solution 242 where weak convergence was enough.

The ground is now prepared for the proof of the spectral inclusion theorem for Toeplitz operators. It is sufficient to prove that if 0 is an approximate eigenvalue of L, then it is an approximate eigenvalue of T also; the non-zero values are recaptured by an obvious translation argument. Suppose therefore that to each positive number ε there corresponds a unit vector f_ε such that $\|Lf_\varepsilon\| < \varepsilon$. The preceding paragraph implies that $W^{*n} PW^n f_\varepsilon \to f_\varepsilon$ and $W^{*n} TP W^n f_\varepsilon \to Lf_\varepsilon$ (strongly). It follows that $\|PW^n f_\varepsilon\| \to 1$ and $\|TP W^n f_\varepsilon\| \to \|Lf_\varepsilon\|$. The first of these assertions says that $PW^n f_\varepsilon$ is, for large n, nearly a unit vector; the second one says that T nearly annihilates it. It follows, as promised, that 0 is an approximate eigenvalue of T.

Corollary 1 is now straightforward. Since L is normal, $\|L\| = r(L)$, and, by the result just proved, $r(L) \leqq r(T)$. It follows that $\|L\| \leqq \|T\|$. The reverse inequality was proved before, and the corollary follows from the known facts about the norm of a multiplication.

For Corollary 2: if the spectrum of T_φ consists of 0 alone, then the same is true of L_φ, and it follows that $\varphi = 0$.

The proof of Corollary 3 is similar to that of Corollary 2: if the spectrum of T_φ is real, then the same is true of L_φ, and it follows that φ is real.

The proof of Corollary 4 is the same as Solution 217: $\overline{W(L)} = \text{conv spec } L \subset \text{conv spec } T \subset \overline{W(T)} \subset \overline{W(L)}$.

Solution 246. *If one of φ and ψ is continuous (and the other is arbitrary in $\mathbf{L}^\infty$), then $T_\varphi T_\psi - T_{\varphi\psi}$ is compact.*

246

PROOF. Suppose first that $\varphi = e_j$ for some j. If $j < 0$, then T_φ is co-analytic, and, therefore, the difference $T_\varphi T_\psi - T_{\varphi\psi}$ is 0. If $j \geqq 0$, then

$$T_\varphi T_\psi = U^j T_\psi = U^j U^{*j} T_\psi U^j \quad \text{(because T_ψ is a Toeplitz operator)}$$
$$= T_\psi U^j - (1 - U^j U^{*j}) T_\psi U^j$$
$$= T_\psi T_\varphi + K \quad \text{(with K compact)}$$
$$= T_{\psi\varphi} + K \quad \text{(because T_φ is analytic)}.$$

343

Consequence: $T_\varphi T_\psi - T_{\varphi\psi}$ is compact whenever φ is a trigonometric polynomial. The general case can be inferred from the Weierstrass approximation theorem, as follows. Since $\|T_\varphi\| = \|\varphi\|_\infty$, it follows that if $\varphi_n \to \varphi$ in $\mathbf{L}^\infty$, then $T_{\varphi_n} \to T_\varphi$ in norm. Since at the same time $\varphi_n \psi \to \varphi\psi$ in $\mathbf{L}^\infty$, and, therefore, $T_{\varphi_n\psi} \to T_{\varphi\psi}$ in norm, it follows that

$$T_{\varphi_n} T_\psi - T_{\varphi_n\psi} \to T_\varphi T_\psi - T_{\varphi\psi}$$

in norm. If the terms of the approximating sequence are compact, then the limit is compact.

That settles the matter when φ is continuous. The result for ψ follows from the consideration of adjoints.

Problem 246 is a special case of a general question (when is $T_\varphi T_\psi - T_{\varphi\psi}$ compact?), and that took a long time to answer. The general answer had to take into account and to unify two special answers: the difference is compact if either of the given functions is continuous (a symmetric condition), and it is compact (in fact 0) if either the first function is co-analytic or the second one is analytic (an unsymmetric condition). For any θ in $\mathbf{L}^\infty$, let $\mathbf{H}^\infty[\theta]$ be the smallest closed subalgebra of $\mathbf{L}^\infty$ that contains θ and contains every function in $\mathbf{H}^\infty$. The ingenious unification goes as follows: $T_\varphi T_\psi - T_{\varphi\psi}$ is compact if and only if

$$\mathbf{H}^\infty[\varphi^*] \cap \mathbf{H}^\infty[\psi] \subset \mathbf{H}^\infty + \mathbf{C}$$

(where $\mathbf{C}$, in this context, is the set of all continuous functions, and $\mathbf{H}^\infty + \mathbf{C}$ is the set of all functions of the form $\theta + \gamma$ with θ in $\mathbf{H}^\infty$ and γ in $\mathbf{C}$). If either φ or ψ is in $\mathbf{C}$, the condition is clearly satisfied; if φ is co-analytic, then $\mathbf{H}^\infty[\varphi^*] = \mathbf{H}^\infty$, and if ψ is analytic, then $\mathbf{H}^\infty[\psi] = \mathbf{H}^\infty$, and, therefore, in either case, the condition is satisfied again.

References: [8, 147].

247   **Solution 247.** It is useful to remember that $\tilde{\mathbf{H}}^2$ is a functional Hilbert space, and, as such, it has a kernel function (Problem 37); it is not, however, important to know what that kernel function is. Let $\tilde{T}_\varphi$ be what T_φ becomes when it is transferred from $\mathbf{H}^2$ to $\tilde{\mathbf{H}}^2$; it follows from Solution 42 that $\tilde{T}_\varphi \tilde{f} = \tilde{\varphi} \cdot \tilde{f}$ for each $\tilde{f}$ in $\tilde{\mathbf{H}}^2$. If y is a complex (!) number, with $|y| < 1$, then $\tilde{f}(y) = (\tilde{f}, K_y)$; this implies that $\tilde{f}(y) = 0$ if and only if $\tilde{f} \perp K_y$. Fix y, put $\lambda = \tilde{\varphi}(y)$, temporarily fix an element $\tilde{f}$ in $\tilde{\mathbf{H}}^2$, and let $\tilde{g}$ be the function defined by $\tilde{g}(z) = (\tilde{\varphi}(z) - \lambda)\tilde{f}(z)$. Since $\tilde{g}(y) = (\tilde{\varphi}(y) - \lambda)\tilde{f}(y) = 0$, it follows that $\tilde{g} \perp K_y$. This implies that $(\tilde{T}_\varphi - \lambda)\tilde{\mathbf{H}}^2$ is included in the orthogonal complement of K_y, so that it is a proper subspace of $\tilde{\mathbf{H}}^2$, and hence that λ belongs to the (compression) spectrum of T_φ. Conclusion: $\tilde{\varphi}(D) \subset \operatorname{spec} T_\varphi$, and therefore $\overline{\tilde{\varphi}(D)} \subset \operatorname{spec} T_\varphi$.

The converse is even easier. If $|\tilde{\varphi}(z) - \lambda| \geqq \delta > 0$ whenever $|z| < 1$, then $1/(\tilde{\varphi} - \lambda)$ is a bounded analytic function in the open unit disc. It follows that its product with a function analytic in the disc and having a square-summable set of Taylor coefficients is another function with the same properties, i.e., that it induces a bounded multiplication operator on $\tilde{\mathbf{H}}^2$. Conclusion: $T_\varphi - \lambda$ is invertible, i.e., λ is not in spec T_φ.

Solution 248. *A Hermitian Toeplitz operator that is not a scalar has no eigenvalues.*   **248**

PROOF. It is sufficient to show that if φ is a real-valued bounded measurable function, and if $T_\varphi \cdot f = 0$ for some f in $\mathbf{H}^2$, then either $f = 0$ or $\varphi = 0$. Since $\varphi \cdot f^* = \varphi^* \cdot f^* \in \mathbf{H}^2$ (because $P(\varphi \cdot f) = 0$), and since $f \in \mathbf{H}^2$, it follows that $\varphi \cdot f^* \cdot f \in \mathbf{H}^1$ (Problem 34). Since, however, $\varphi \cdot f^* \cdot f$ is real, it follows that $\varphi \cdot f^* \cdot f$ is a constant (Solution 33). Since $\int \varphi \cdot f^* \cdot f \, d\mu = (\varphi \cdot f, f) = (T_\varphi f, f) = 0$ (because $T_\varphi f = 0$), the constant must be 0. The F. and M. Riesz theorem (Problem 138) implies that either $f = 0$ or $\varphi \cdot f^* = 0$. If $f \neq 0$, then f^* can vanish on a set of measure 0 only, and therefore $\varphi = 0$.

Solution 249. Yes, there are Toeplitz zero-divisors modulo the compact   **249** operators, and they can even be made positive.

Let φ be a smooth version of the characteristic function of the upper semicircle. More precisely: put $\varphi = 1$ on a subarc of the upper semicircle that reaches nearly down to the end points -1 and $+1$, put $\varphi = 0$ on an arc that is slightly longer than the lower semicircle at both ends, and join the upper and the lower parts continuously (through positive values if so desired). If ψ is the reflected version of φ (that is: $\psi(z) = \varphi(z^*)$), then both φ and ψ are continuous, and therefore $T_\varphi T_\psi - T_{\varphi\psi}$ is compact (Solution 246); since, however, $\varphi\psi = 0$, it follows that $T_\varphi T_\psi$ must be compact.

Solution 250. *If φ is a real-valued bounded measurable function, and if its*   **250** *essential lower and upper bounds are α and β, then spec T_φ is the closed interval $[\alpha, \beta]$.*

PROOF. If $\alpha = \beta$, then φ is constant, and everything is trivial. If $\alpha < \lambda < \beta$, it is to be proved that $T_\varphi - \lambda$ is not invertible. Assume the contrary, i.e., assume that $T_\varphi - \lambda$ is invertible, and, by an inessential change of notation, assume $\lambda = 0$. It follows, as an apparently very small consequence of invertibility, that e_0 belongs to the range of T_φ, and hence that there exists a (non-zero) function f in $\mathbf{H}^2$ such that $T_\varphi f = e_0$. This means that $\varphi \cdot f - e_0 \perp \mathbf{H}^2$. Equivalently (recall that $e_0(z) = 1$ for all z) the complex conjugate of $\varphi \cdot f$ is in $\mathbf{H}^2$; the next step is to deduce from this that sgn φ is constant (so that either $\varphi > 0$ almost everywhere or $\varphi < 0$ almost everywhere).

Since φ is real, it follows that $(\varphi \cdot f)^* = \varphi \cdot f^*$. Since both $\varphi \cdot f^*$ and f are in $\mathbf{H}^2$, Problem 34 implies that $\varphi \cdot f^* \cdot f \in \mathbf{H}^1$. Solution 33 becomes applicable and yields the information that $\varphi \cdot f^* \cdot f$ is a constant almost everywhere. Since $f \neq 0$, it follows that f is different from 0 almost everywhere (Problem 158), and consequently φ has almost everywhere the same sign as the constant value of $\varphi \cdot f \cdot f^*$.

In the original notation the result just obtained is that sgn($\varphi - \lambda$) is constant, and, since $\alpha < \lambda < \beta$, that is exactly what it is not. This contradiction proves that $[\alpha, \beta] \subset \operatorname{spec} T_\varphi$.

The reverse inclusion is easier. Since $\alpha \leqq \varphi \leqq \beta$, it follows that $\alpha \leqq L_\varphi \leqq \beta$; since $T_\varphi f = PL_\varphi f$ whenever $f \in \mathbf{H}^2$, it follows that

$$(T_\varphi f, f) = (PL_\varphi f, f) = (L_\varphi f, f),$$

and hence that $\alpha \leqq T_\varphi \leqq \beta$. This of course implies that spec $T_\varphi \subset [\alpha, \beta]$.

References

Each entry is followed by one or more numbers in brackets. Positive numbers refer to the problems or solutions in which the entry is cited. The number 0 refers to the preface.

1. N. I. Akhiezer, I. M. Glazman, *Theory of linear operators in Hilbert space*, Ungar, New York, 1961. [27]
2. A. A. Albert, B. Muckenhoupt, *On matrices of trace zero*, Michigan Math. J. **4** (1957) 1–3. [230]
3. T. Ando, *On a pair of commutative contractions*, Acta Szeged **24** (1963) 88–90. [229]
4. T. Ando, *On hyponormal operators*, Proc. A.M.S. **14** (1963) 290–291. [205, 206]
5. N. Aronszajn, *Theory of reproducing kernels*, Trans. A.M.S. **68** (1950) 337–404. [36]
6. N. Aronszajn, K. T. Smith, *Invariant subspaces of completely continuous operators*, Ann. Math. **60** (1954) 345–350. [191]
7. F. V. Atkinson, *The normal solubility of linear equations in normed spaces*, Mat. Sbornik **28** (1951) 3–14 [181]
8. S. Axler, S.-Y. A. Chang, D. Sarason, *Products of Toeplitz operators*, Integral Equations and Operator Theory **1** (1978) 285–309. [246]
9. S. K. Berberian, *Note on a theorem of Fuglede and Putnam*, Proc. A.M.S. **10** (1959) 175–182. [192]
10. S. K. Berberian, *A note on hyponormal operators*, Pac. J. Math. **12** (1962) 1171–1175. [206]
11. S. Bergman, *Sur les fonctions orthogonales de plusieurs variables complexes avec les applications à la théorie des fonctions analytiques*, Gauthier-Villars, Paris, 1947. [31]

12. A. R. Bernstein, A. Robinson, *Solution of an invariant subspace problem of K. T. Smith and P. R. Halmos*, Pac. J. Math. **16** (1966) 421–431. [191]

13. A. Beurling, *On two problems concerning linear transformations in Hilbert space*, Acta Math. **81** (1949) 239–255. [157, 191]

14. G. Birkhoff, G.-C. Rota, *On the completeness of Sturm-Liouville expansions*, Amer. Math. Monthly **67** (1960) 835–841. [12]

15. E. Bishop, *Spectral theory for operators on a Banach space*, Trans. A.M.S. **86** (1957) 414–445. [203]

16. J. Bram, *Subnormal operators*, Duke Math. J. **22** (1955) 75–94. [201, 203]

17. M. S. Brodskii, *On a problem of I. M. Gelfand*, Uspekhi Mat. Nauk **12** (1957) 129–132. [191]

18. A. Brown, *On a class of operators*, Proc. A.M.S. **4** (1953) 723–728. [137]

19. A. Brown, P. R. Halmos, *Algebraic properties of Toeplitz operators*, J. reine angew. Math. **231** (1963) 89–102. [243]

20. A. Brown, P. R. Halmos, C. Pearcy, *Commutators of operators on Hilbert space*, Can. J. Math. **17** (1965) 695–708. [231]

21. A. Brown, P. R. Halmos, A. L. Shields, *Cesàro operators*, Acta Szeged **26** (1965) 125–137. [173, 177]

22. A. Brown, C. Pearcy, *Structure of commutators of operators*, Ann. Math. **82** (1965) 112–127. [230, 235]

23. A. Brown, C. Pearcy, *Multiplicative commutators of operators*, Can. J. Math. **18** (1966) 737–749. [240]

24. S. W. Brown, *Some invariant subspaces for subnormal operators*, Integral Equations and Operator Theory **1** (1978) 310–333. [196]

25. J. R. Buddenhagen, *Subsets of a countable set*, Amer. Math. Monthly **78** (1971) 536–537. [7]

26. J. B. Conway, *Functions of one complex variable*, Springer, New York, 1978. [0]

27. J. B. Conway, B. B. Morrel, *Operators that are points of spectral continuity*, Integral Equations and Operator Theory **2** (1979) 174–198. [105]

28. M. J. Crabb, A. M. Davie, *Von Neumann's inequality for Hilbert space operators*, Bull. London Math. Soc. **7** (1975) 49–50. [229]

29. A. M. Davie, *The approximation problem for Banach spaces*, Bull. London Math. Soc. **5** (1973) 261–266. [175]

30. C. Davis, *The Toeplitz-Hausdorff theorem explained*, Can. Math Bull. **14** (1971) 245–246. [210]

31. G. De Barra, J. R. Giles, B. Sims, *On the numerical range of compact operators on Hilbert space*, J. London Math. Soc. **5** (1972) 704–706. [213]

32. L. de Branges, J. Rovnyak, *The existence of invariant subspaces*, Bull. A.M.S. **70** (1964) 718–721. [152]

33. L. de Branges, J. Rovnyak, *Correction to "The existence of invariant subspaces"*, Bull. A.M.S. **71** (1965) 396. [152]

34. D. Deckard, C. Pearcy, *Another class of invertible operators without square roots*, Proc. A.M.S. **14** (1963) 445–449. [145]

35. N. P. Dekker, *Joint numerical range and joint spectrum of Hilbert space operators*, Thesis, Free University of Amsterdam, 1969. [210]

36. W. F. Donoghue, *On the numerical range of a bounded operator*, Michigan Math. J. **4** (1957) 261–263 [210]

37. W. F. Donoghue, *The lattice of invariant subspaces of a completely continuous quasi-nilpotent transformation*, Pac. J. Math. **7** (1957) 1031–1035. [191]

38. R. G. Douglas, *On majorization, factorization, and range inclusion of operators on Hilbert space*, Proc. A.M.S. **17** (1966) 413–415. [59]

39. N. Dunford, J. T. Schwartz, *Linear operators, Part I: General theory*, Interscience, New York, 1958. [27, 58, 74, 86, 170]

40. N. Dunford, J. T. Schwartz, *Linear operators, Part II: Spectral theory*, Interscience, New York, 1963. [123, 191]

41. P. Enflo, *A counterexample to the approximation problem in Banach spaces*, Acta Math. **130** (1973) 309–317. [175]

42. P. A. Fillmore, J. G. Stampfli, J. P. Williams, *On the essential numerical range, the essential spectrum, and a problem of Halmos*, Acta Szeged **33** (1972) 179–192. [162]

43. C. Foiaş, *A remark on the universal model for contractions of G.-C. Rota*, Com. Acad. R. P. Romine **13** (1963) 349–352. [152]

44. J. M. Freeman, *Perturbations of the shift operator*, Trans. A.M.S. **114** (1965) 251–260. [184]

45. B. Fuglede, *A commutativity theorem for normal operators*, Proc. N.A.S. **36** (1950) 35–40. [192]

46. L. Gehér, *Cyclic vectors of a cyclic operator span the space*, Proc. A.M.S. **33** (1972) 109–110. [166]

47. C. Goffman, G. Pedrick, *First course in functional analysis*, Prentice-Hall, Englewood Cliffs, 1965. [11]

48. S. Gudder, *Inner product spaces*, Amer. Math. Monthly **81** (1974) 29–36. [54]

49. P. R. Halmos, *Normal dilations and extensions of operators*, Summa Brasil. Math. **2** (1950) 125–134. [203, 209, 222]

50. P. R. Halmos, *Introduction to Hilbert space and the theory of spectral multiplicity*, Chelsea, New York, 1951. [0, 1, 3, 5, 12, 13, 52, 53, 74, 86, 123, 146, 147, 160, 192, 218]

51. P. R. Halmos, *Spectra and spectral manifolds*, Ann. Soc. Pol. Math. **25** (1952) 43–49. [200]

52. P. R. Halmos, *Commutators of operators*, Amer. J. Math. **74** (1952) 237–240. [234]

53. P. R. Halmos, *Commutators of operators, II*, Amer. J. Math. **76** (1954) 191–198. [234]

54. P. R. Halmos, *Shifts on Hilbert spaces*, J. reine angew. Math. **208** (1961) 102–112. [155, 157]

55. P. R. Halmos, *A glimpse into Hilbert space*, pp. 1–22, Lectures on modern mathematics, vol. 1, Wiley, New York, 1963. [230]

56. P. R. Halmos, *What does the spectral theorem say?*, Amer. Math. Monthly **70** (1963) 241–247. [123, 192]

57. P. R. Halmos, *Numerical ranges and normal dilations*, Acta Szeged **25** (1964) 1–5. [211]

58. P. R. Halmos, *Invariant subspaces of polynomially compact operators*, Pac. J. Math. **16** (1966) 433–437. [191]

59. P. R. Halmos, *Ten problems in Hilbert space*, Bull. A.M.S. **76** (1970) 887–933. [154]

60. P. R. Halmos, *Finite-dimensional vector spaces*, Springer, New York, 1974. [228]

61. P. R. Halmos, *Measure theory*, Springer, New York, 1974. [0, 18]

62. P. R. Halmos, S. Kakutani, *Products of symmetries*, Bull. A.M.S. **64** (1958) 77–78. [143]

63. P. R. Halmos, G. Lumer, *Square roots of operators. II*, Proc. A.M.S. **5** (1954) 589–595. [56, 103]

64. P. R. Halmos, G. Lumer, J. J. Schäffer, *Square roots of operators*, Proc. A.M.S. **4** (1953) 142–149. [31, 145]

65. P. R. Halmos, J. E. McLaughlin, *Partial isometries*, Pac. J. Math. **13** (1962) 585–596. [130, 131, 133]

66. P. R. Halmos, V. S. Sunder, *Bounded integral operators on L^2 spaces*, Springer, Berlin, 1978. [173]

67. G. H. Hardy, J. E. Littlewood, G. Pólya, *Inequalities*, Cambridge University Press, Cambridge, 1934. [46]

68. P. Hartman, A. Wintner, *On the spectra of Toeplitz's matrices*, Amer. J. Math. **72** (1950) 359–366. [245]

69. F. Hausdorff, *Der Wertvorrat einer Bilinearform*, Math. Z. **3** (1919) 314–316. [210]

70. E. Heinz, *Ein v. Neumannscher Satz über beschränkte Operatoren im Hilbertschen Raum*, Göttingen Nachr. (1952) 5–6. [299]

71. H. Helson, *Lectures on invariant subspaces*, Academic Press, New York, 1964. [157, 191]

72. J. Hennefeld, *A nontopological proof of the uniform boundedness theorem*, Amer. Math. Monthly **87** (1980) 217. [27]

73. H. M. Hilden, L. J. Wallen, *Some cyclic and non-cyclic vectors of certain operators*, Indiana U. Math. J. **23** (1974) 557–565. [168]

74. E. Hille, R. S. Phillips, *Functional analysis and semi-groups*, A.M.S., Providence, 1957. [103]

75. K. Hoffman, *Banach spaces of analytic functions*, Prentice-Hall, Englewood Cliffs, 1962. [34, 158]

76. S. S. Holland, Jr., *A Hilbert space proof of the Banach-Steinhaus theorem*, Amer. Math. Monthly **76** (1969) 40–41. [27]

77. T. Ito, *On the commutative family of subnormal operators*, J. Fac. Sci. Hokkaido University **14** (1958) 1–15. [201]
78. T. Ito, T. K. Wong, *Subnormality and quasinormality of Toeplitz operators*, Proc. A.M.S. **34** (1972) 157–164. [209]
79. N. Jacobson, *Rational methods in the theory of Lie algebras*, Ann. Math. **36** (1935) 875–881. [232]
80. G. J. Johnson, *A crinkled arc*, Proc. A.M.S. **25** (1970) 375–376. [6]
81. G. J. Johnson, *Hilbert space problem four*, Amer. Math. Monthly **78** (1971) 525–527. [6]
82. R. V. Kadison, *Isometries of operator algebras*, Ann. Math. **54** (1951) 325–338. [136]
83. R. V. Kadison, *Strong continuity of operator functions*, Pac. J. Math. **26** (1968) 121–129. [116]
84. G. K. Kalisch, *On similarity, reducing manifolds, and unitary equivalence of certain Volterra operators*, Ann. Math. **66** (1957) 481–494. [191]
85. I. Kaplansky, *Products of normal operators*, Duke Math. J. **20** (1953) 257–260. [192]
86. T. Kato, *Some mapping theorems for the numerical range*, Proc. Japan Acad. **41** (1965) 652–655. [221]
87. J. L. Kelley, *General topology*, Springer, New York, 1975. [0]
88. D. C. Kleinecke, *On operator commutators*, Proc. A.M.S. **8** (1957) 535–536. [232]
89. P. D. Lax, *Translation invariant spaces*, Acta Math. **101** (1959) 163–178. [157]
90. A. Lebow, *On von Neumann's theory of spectral sets*, J. Math. Anal. Appl. **7** (1963) 64–90. [229]
91. V. I. Lomonosov, *Invariant subspaces for the family of operators which commute with a completely continuous operator*, Functional Anal. Appl. **7** (1973) 213–214. [191]
92. K. Löwner, *Grundzüge einer Inhaltslehre im Hilbertschen Raume*, Ann. Math. **40** (1939) 816–833. [18]
93. A. J. Michaels, *Hilden's simple proof of Lomonosov's invariant subspace theorem*, Adv. in Math. **25** (1977) 56–58. [191]
94. W. Mlak, *Unitary dilations of contraction operators*, Rozprawy Mat. **46** (1965). [227]
95. B. Sz.-Nagy, *Sur les contractions de l'espace de Hilbert*, Acta Szeged **15** (1953) 87–92. [227]
96. B. Sz.-Nagy, *Prolongements des transformations de l'espace de Hilbert qui sortent de cet espace*, Appendix to "Leçons d'analyse fonctionelle" by F. Riesz, B. Sz.-Nagy, Akadémiai Kiadó, Budapest, 1955. (*Extensions of linear transformations in Hilbert space which extend beyond this space*, Appendix to "Functional analysis" by F. Riesz, B. Sz.-Nagy, Ungar, New York, 1960.) [227, 228, 229]
97. B. Sz.-Nagy, *Sur les contractions de l'espace de Hilbert. II*, Acta Szeged **18** (1957) 1–14. [227]

98. B. Sz.-Nagy, *Suites faiblement convergentes de transformations normales de l'espace Hilbertien*, Acta Szeged **8** (1957) 295–302. [224]

99. B. Sz.-Nagy, C. Foiaş, *Sur les contractions de l'espace de Hilbert. V. Translations bilatérales*, Acta Szeged **23** (1962) 106–129. [155]

100. B. Sz.-Nagy, C. Foiaş, *Sur les contractions de l'espace de Hilbert. IX. Factorisations de la fonction caractéristique. Sous-espaces invariants*, Acta Szeged **25** (1964) 283–316. [191]

101. B. Sz.-Nagy, C. Foiaş, *On certain classes of power-bounded operators in Hilbert space*, Acta Szeged **27** (1966) 17–25. [221]

102. J. D. Newburgh, *The variation of spectra*, Duke Math. J. **18** (1951) 165–176. [105].

103. N. K. Nikolskii, *Invariant subspaces of certain completely continuous operators*, Vestnik Leningrad University Math. (1965) 68–77. [191]

104. R. F. Olin, *Functional relationships between a subnormal operator and its minimal normal extension*, Pac. J. Math. **63** (1976) 221–229. [198]

105. S. K. Parrott, *Unitary dilations for commuting contractions*, Pac. J. Math. **34** (1970) 481–490. [229]

106. C. Pearcy, *On commutators of operators on Hilbert space*, Proc. A.M.S. **16** (1965) 53–59. [234]

107. C. Pearcy, *An elementary proof of the power inequality for the numerical radius*, Michigan Math. J. **13** (1966) 289–291. [221]

108. C. R. Putnam, *On commutators of bounded matrices*, Amer. J. Math. **73** (1951) 127–131. [236]

109. C. R. Putnam, *On normal operators in Hilbert space*, Amer. J. Math. **73** (1951) 357–362. [192]

110. H. Radjavi, P. Rosenthal, *On invariant subspaces and reflexive algebras*, Amer. J. Math. **91** (1969) 683–692. [193]

111. H. Radjavi, P. Rosenthal, *Invariant subspaces*, Springer, Berlin, 1973. [191]

112. C. E. Rickart, *General theory of Banach algebras*, Krieger, Huntington, 1974. [103, 104]

113. F. Riesz, B. Sz.-Nagy, *Ueber Kontraktionen des Hilbertschen Raumes*, Acta Szeged **10** (1943) 202–205. [228]

114. F. Riesz, B. Sz.-Nagy, *Leçons d'analyse fonctionelle*, Akadémiai Kiadó, Budapest, 1952. (*Functional analysis*, Ungar, New York, 1955.) [12, 121, 130]

115. J. R. Ringrose, *On the triangular representation of integral operators*, Proc. London Math. Soc. **12** (1962) 385–399. [186]

116. J. B. Robertson, *On wandering subspaces for unitary operators*, Proc. A.M.S. **16** (1965) 233–236. [155]

117. S. Rolewicz, *On orbits of elements*, Studia Math. **32** (1969) 17–22. [168]

118. M. Rosenblum, *On a theorem of Fuglede and Putnam*, J. London Math. Soc. **33** (1958) 376–377. [192]

119. G.-C. Rota, *On models for linear operators*, Comm. Pure Appl. Math. **13** (1960) 469–472. [153]

120. H. L. Royden, *Real analysis*, Macmillan, New York, 1968. [0]
121. I. Z. Ruzsa, *A pseudonorm for unbounded transformations*, Acta Math. Acad. Sci. Hung. **34** (1979) 297–305. [50]
122. L. A. Sakhnovich, *On the reduction of Volterra operators to the simplest form and on inverse problems*, Izv. Akad. Nauk SSSR **21** (1957) 235–262. [191]
123. D. Sarason, *A remark on the Volterra operator*, J. Math. Anal. Appl. **12** (1965) 244–246. [191]
124. H. H. Schaefer, *Eine Bemerkung zur Existenz invarianter Teilräume linearer Abbildungen*, Math. Z. **82** (1963) 90. [191]
125. J. J. Schäffer, *On unitary dilations of contractions*, Proc. A.M.S. **6** (1955) 322. [227]
126. J. J. Schäffer, *More about invertible operators without roots*, Proc. A.M.S. **16** (1965) 213–219. [145]
127. R. Schatten, *Norm ideals of completely continuous operators*, Springer, Berlin, 1960 [173].
128. M. Schreiber, *Unitary dilations of operators*, Duke Math. J. **24** (1956) 579–594. [227]
129. J. Schur, *Bemerkungen zur Theorie der beschränkten Bilinearformen mit unendlich vielen Veränderlichen*, J. reine angew. Math. **140** (1911) 1–28. [45]
130. J. Schur, *Über Potenzreihen, die im Innern des Einheitskreises beschränkt sind*, J. reine angew. Math. **147** (1917) 205–232. [71]
131. I. E. Segal, *Equivalences of measure spaces*, Amer. J. Math. **73** (1951) 275–313. [123]
132. I. E. Segal, *Algebraic integration theory*, Bull. A.M.S. **71** (1965) 419–489. [18]
133. A. L. Shields, *Weighted shift operators and analytic function theory*, Mathematical Surveys **13** (1974) 49–128. (C. Pearcy, ed., *Topics in operator theory*, A.M.S., Providence, 1974.) [94]
134. F. V. Shirokov, *Proof of a conjecture of Kaplansky*, Uspekhi Mat. Nauk **11** (1956) 167–168. [232]
135. K. Shoda. *Einige Sätze über Matrizen*, Japanese J. Math. **13** (1936) 361–365. [230]
136. J. G. Stampfli, *Hyponormal operators*, Pac. J. Math. **12** (1962) 1453–1458. [205, 206]
137. J. G. Stampfli, *Perturbations of the shift*, J. London Math. Soc. **40** (1965) 345–347. [184]
138. M. H. Stone, *Linear transformations in Hilbert space and their applications to analysis*, A.M.S., New York, 1932. [27, 53]
139. O. Szász, *Über die Approximation stetiger Funktionen durch lineare Aggregate von Potenzen*, Math. Ann. **77** (1916) 482–496. [165]
140. R. C. Thompson, *On matrix commutators*, J. Wash. Acad. Sci. **48** (1958) 306–307. [236]
141. O. Toeplitz, *Zur Theorie der quadratischen Formen von unendlichvielen Veränderlichen*, Göttingen Nachr. (1910) 489–506. [44]

142. O. Toeplitz, *Das algebraische Analogon zu einem Satze von Fejér*, Math. Z. **2** (1918) 187–197. [210]

143. N. Tsao, *Approximate bases in a Hilbert space*, Amer. Math. Monthly **75** (1968) 750. [12]

144. F. A. Valentine, *Convex sets*, McGraw-Hill, New York, 1964. [216]

145. N. T. Varopoulos, *On an inequality of von Neumann and an application of the metric theory of tensor products to operator theory*, J. Funct. Anal. **16** (1974) 83–100. [229]

146. R. A. Vitale, *Representation of a crinkled arc*, Proc. A.M.S. **52** (1975) 303–304. [6]

147. A. L. Volberg, *Two remarks concerning the theorem of S. Axler, S.-Y. A. Chang, and D. Sarason*, J. Operator Theory **7** (1982) 209–219. [246]

148. J. von Neumann, *Zur Algebra der Funktionaloperationen und Theorie der normalen Operatoren*, Math. Ann. **102** (1929) 370–427. [27, 28]

149. J. von Neumann, *Proof of the quasi-ergodic hypothesis*, Proc. N.A.S. **18** (1932) 70–82. [228]

150. J. von Neumann, *Functional operators*, Princeton University Press, Princeton, 1950. [122]

151. J. von Neumann, *Eine Spektraltheorie für allgemeine Operatoren eines unitären Raumes*, Math. Nachr. **4** (1951) 258–281. [229]

152. P. Walters, *An introduction to ergodic theory*, Springer, New York, 1981. [228]

153. R. J. Whitley, *A note on hyponormal operators*, Proc. A.M.S. **49** (1975) 399–400. [207]

154. H. Widom, *Inversion of Toeplitz matrices, II*, Illinois J. Math. **4** (1960) 88–99. [250]

155. H. Widom, *On the spectrum of a Toeplitz operator*, Pac. J. Math. **14** (1964) 365–375. [250]

156. N. A. Wiegmann, *Normal products of matrices*, Duke Math. J. **15** (1948) 633–638. [192]

157. N. A. Wiegmann, *A note on infinite normal matrices*, Duke Math. J. **16** (1949) 535–538. [192]

158. H. Wielandt, *Ueber die Unbeschränktheit der Operatoren des Quantenmechanik*, Math. Ann. **121** (1949) 21. [230]

159. A. Wintner, *Zur Theorie der beschränkten Bilinearformen*, Math. Z. **30** (1929) 228–282. [219, 247]

160. A. Wintner, *The unboundedness of quantum-mechanical matrices*, Phys. Rev. **71** (1947) 738–739. [230]

List of Symbols

355

Index

Positive numbers refer to problems or solutions; the number 0 refers to the preface.

357

Graduate Texts in Mathematics

9 780387 906850